Lecture Notes in Computer Science   8994

Commenced Publication in 1973
Founding and Former Series Editors:
Gerhard Goos, Juris Hartmanis, and Jan van Leeuwen

More information about this series at http://www.springer.com/series/7407

Clarisse Dhaenens · Laetitia Jourdan
Marie-Eléonore Marmion (Eds.)

Learning and Intelligent Optimization

9th International Conference, LION 9
Lille, France, January 12–15, 2015
Revised Selected Papers

 Springer

Editors
Clarisse Dhaenens
Lille University
Villeneuve d'Ascq
France

Marie-Eléonore Marmion
Lille University
Villeneuve d'Ascq
France

Laetitia Jourdan
Lille University
Villeneuve d'Ascq
France

ISSN 0302-9743   ISSN 1611-3349 (electronic)
Lecture Notes in Computer Science
ISBN 978-3-319-19083-9   ISBN 978-3-319-19084-6   (eBook)
DOI 10.1007/978-3-319-19084-6

Library of Congress Control Number: 2015939426

LNCS Sublibrary: SL1 – Theoretical Computer Science and General Issues

Printed on acid-free paper

Springer International Publishing AG Switzerland is part of Springer Science+Business Media
(www.springer.com)

Preface

The large variety of heuristic algorithms for hard optimization problems raises numerous interesting and challenging issues. Practitioners are confronted with the burden of selecting the most appropriate method, in many cases through an expensive algorithm configuration and parameter tuning process, and subject to a steep learning curve. Scientists seek theoretical insights and demand a sound experimental methodology for evaluating algorithms and assessing strengths and weaknesses. A necessary prerequisite for this effort is a clear separation between the algorithm and the experimenter, who, in too many cases, is "in the loop" as a crucial intelligent learning component. Both issues are related to designing and engineering ways of "learning" about the performance of different techniques, and ways of using past experience about the algorithm behavior to improve performance in the future. This is the scope of the Learning and Intelligent OptimizatioN (LION) conference series.

This volume contains papers presented at the 9th LION (Learning and Intelligent OptimizatioN) conference held during January 12–15, 2015 in Lille, France.

This meeting, which continues the successful series of LION events (see LION 5 in Rome–Italy, LION 6 in Paris–France, LION 7 in Catania–Italy, and LION 8 in Gainesville–USA), is exploring the intersections and uncharted territories between machine learning, artificial intelligence, mathematical programming, and algorithms for hard optimization problems. The main purpose of the event is to bring together experts from these areas to discuss new ideas and methods, challenges and opportunities in various application areas, general trends, and specific developments. Optimization and machine learning researchers are now forming their own community and identity. The International Conference on Learning and Optimization is proud to be the premiere conference in the area.

A total of 58 papers were submitted to LION 9: 43 submissions of long papers and 15 submissions of short papers. Each manuscript was independently reviewed by at least three members of the Program Committee. 14 long papers and 17 short papers were accepted (some long submissions have been asked to be shortened). Hence, the selection rate for long papers is of 33 %.

During the conference, we were pleased to listen to four plenary speakers:

- **David Corne**, Heriot-Watt University, UK. *Psychic machines: mind-reading with machine learning*
- **Alex Freitas**, University of Kent, UK. *Automating the Design of Decision Tree Algorithms with Evolutionary Computation*
- **Daniel Le Berre**, Artois University, Lens, France. *From Boolean Satisfaction to Boolean Optimization: Application to Dependency Management*
- **Remi Munos**, Inria Lille Nord Europe, France. *The optimistic principle applied to function optimization and planning*

In addition, two tutorials were presented:

- **Thomas Stützle**, FNRS-IRIDIA, ULB, Belgium. *Automatic Algorithm Configuration: From Parameter Tuning to Automatic Design*
- **Sébastien Verel**, Littoral Côte d'Opale University, Calais, France. *Fitness landscape: the metaphor and beyond*

January 2015

Clarisse Dhaenens
Laetitia Jourdan
Marie-Eléonore Marmion

Organization

Program Committee

Hernán Aguirre	Shinshu University, Japan
Roberto Battiti	Università di Trento, Italy
Mauro Birattari	IRIDIA, Université Libre de Bruxelles, Belgium
Christian Blum	IKERBASQUE, Basque Foundation for Science, Spain
Juergen Branke	University of Warwick, UK
Dimo Brockhoff	Inria Lille - Nord Europe, France
Mauro Brunato	University of Trento, Italy
Philippe Codognet	JFLI - CNRS/UPMC/University of Tokyo, Japan
David Cornforth	University of Newcastle, Australia
Clarisse Dhaenens	CRIStAL/Inria/Lille University, France
Luca Di Gaspero	DIEGM - University of Udine, Italy
Karl Doerner	University of Vienna, Austria
Madalina Drugan	Vrije Universiteit Brussel, Belgium
Andries Engelbrecht	University of Pretoria, South Africa
Valerio Freschi	University of Urbino, Italy
Pablo Garcia	University of Granada, Spain
Deon Garrett	Icelandic Institute for Intelligent Machines, Iceland
Michel Gendreau	École Polytechnique de Montréal, Canada
Adrien Goëffon	LERIA - Université d'Angers, France
Walter Gutjahr	University of Vienna, Austria
Youssef Hamadi	Microsoft Research, UK
Jin-Kao Hao	University of Angers, France
Geir Hasle	SINTEF ICT, Norway
Alfredo Hernández-Díaz	Pablo de Olavide University (Seville), Spain
Francisco Herrera	University of Granada, Spain
Tomio Hirata	School of Information Science, Nagoya University, Japan
Holger Hoos	University of British Columbia, Canada
Frank Hutter	University of British Columbia, Canada
Helga Ingimundardóttir	University of Iceland, Iceland
Hisao Ishibuchi	Osaka Prefecture University, Japan
Yaochu Jin	University of Surrey, UK
Laetitia Jourdan	CRIStAL/Inria/Lille University, France
Zeynep Kiziltan	University of Bologna, Italy
Dario Landa-Silva	University of Nottingham, UK
Frederic Lardeux	LERIA - University of Angers, France
Julien Lepagnot	LMIA University of Haute Alsace, France

Xiaodong Li	Wharton School - University of Pennsylvania, USA
Arnaud Liefooghe	CRIStAL/Inria/Lille University, France
Manuel López-Ibáñez	IRIDIA/Brussels, Belgium
Marie-Eléonore Marmion	CRIStAL/Inria/Lille University, France
Franco Mascia	Université Libre de Bruxelles, Belgium
Francesco Masulli	University of Genoa, Italy
Basseur Matthieu	LERIA Angers, France
Bernd Meyer	Monash University, Australia
Marco A. Montes De Oca	University of Delaware, USA
Irene Moser	Swinburne University of Technology, Australia
Amir Nakib	Université Paris Est Creteil, France
Gabriela Ochoa	University of Stirling, UK
Luis Paquete	CISUC, University of Coimbra, Portugal
Panos Pardalos	University of Florida, USA
Andrew J. Parkes	University of Nottingham, UK
Vincenzo Piuri	University of Milan, Italy
Mike Preuss	TU Dortmund University, Germany
Günther Raidl	Vienna University of Technology, Austria
Steffen Rebennack	Colorado School of Mines, USA
Celso Ribeiro	Universidade Federal Fluminense, Brazil
Florian Richoux	LINA, University of Nantes, France
Eduardo Rodriguez-Tello	CINVESTAV-Tamaulipas, Mexico
Horst Samulowitz	IBM Research, USA
Frédéric Saubion	LERIA - University of Angers, France
Andrea Schaerf	University of Udine, Italy
Marc Schoenauer	Inria Saclay Île-de-France, France
Patrick Siarry	Université de Paris 12, France
Christine Solnon	INSA de Lyon, LIRIS, UMR 5205 CNRS, France
Thomas Stuetzle	IRIDIA, Université libre de Bruxelles, Belgium
El-Ghazali Talbi	CRIStAL/Inria/Lille University, France
Kiyoshi Tanaka	Shinshu University, Faculty of Engineering, Japan
Ke Tang	University of Science and Technology of China, China
Sébastien Verel	Université du Littoral Côte d'Opale, France
Stefan Voss	University of Hamburg, Germany
Markus Wagner	University of Adelaide, Australia
Toby Walsh	NICTA/University of New South Wales, Australia
David Woodruff	University of California, Davis, USA
Petros Xanthopoulos	University of Central Florida, USA
Ning Xiong	Mälardalen University, Sweden
Saba Yahyaa	Vrije Universiteit Brussel, Belgium
Saúl Zapotecas-Martinez	Cinvestav - IPN Mexico, Mexico
Rui Zhang	National University of Singapore, Singapore

Contents

From Sequential Algorithm Selection to Parallel Portfolio Selection

M. Lindauer[1]([✉]), Holger H. Hoos[2], and F. Hutter[1]

[1] University of Freiburg, Freiburg Im Breisgau, Germany
{lindauer,fh}@cs.uni-freiburg.de
[2] University of British Columbia, Vancouver, Canada
hoos@cs.ubc.ca

Abstract. In view of the increasing importance of hardware parallelism, a natural extension of per-instance algorithm selection is to select a set of algorithms to be run in parallel on a given problem instance, based on features of that instance. Here, we explore how existing algorithm selection techniques can be effectively parallelized. To this end, we leverage the machine learning models used by existing sequential algorithm selectors, such as *3S*, *ISAC*, *SATzilla* and *ME-ASP*, and modify their selection procedures to produce a ranking of the given candidate algorithms; we then select the top n algorithms under this ranking to be run in parallel on n processing units. Furthermore, we adapt the pre-solving schedules obtained by *aspeed* to be effective in a parallel setting with different time budgets for each processing unit. Our empirical results demonstrate that, using 4 processing units, the best of our methods achieves a 12-fold average speedup over the best single solver on a broad set of challenging scenarios from the algorithm selection library.

Keywords: Algorithm selection · Parallel portfolios · Constraint solving · Answer Set Programming

1 Introduction

For many challenging computational problems, such as SAT, ASP or QBF, there is no single dominant solver. Instead, the state of the art for these problems consists of a set of non-dominated solvers, each of which performs best on certain types of problem instances. In this situation, per-instance automated algorithm selection techniques can be used to leverage the strength of such complementary sets, or portfolios, of solvers (see, e.g., [16,27]). Fundamentally, for a new problem instance, these techniques map a set of cheaply computable instance features to a solver to be run. This mapping is typically learned, using machine learning techniques, from a representative set of training data. Unfortunately, even the best per-instance algorithm selection techniques do not always succeed in identifying the best solver for all problem instances, and their performance can suffer as a result of such incorrect selections.

© Springer International Publishing Switzerland 2015
C. Dhaenens et al. (Eds.): LION 9 2015, LNCS 8994, pp. 1–16, 2015.
DOI: 10.1007/978-3-319-19084-6_1

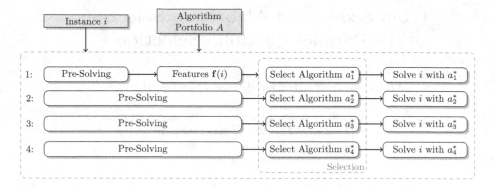

Fig. 1. Parallel portfolio selection with parallel pre-solving for four processing units.

Considering the fact that increases in computational power are nowadays primarily achieved through increased hardware parallelism, one approach for improving instance-based algorithm selection techniques is to select not one, but multiple solvers from a given portfolio, and to run these in parallel. The key idea behind this approach is to hedge against incorrect single-solver selections while exploiting readily available parallelism. There is some evidence in the literature that manually crafted per-instance parallel portfolio selectors can achieve impressive performance. For example, the portfolio SAT solver *CSHCpar* [21,22] won the open parallel track in the 2013 SAT Competition. The idea of *CSHCpar* is simple yet effective: It always runs, independently and in parallel, the parallel SAT solver *Plingeling* with 4 threads, the sequential SAT solver *CCASat*, and three per-instance selected solvers. These per-instance solvers are selected by three models that are trained on application, hard-combinatorial and random SAT instances, respectively. While *CSHCpar* is particularly designed for the SAT Competition with its 8 available cores and its three types of instances, in the following, we investigate a general, fully automated approach for selecting parallel portfolios without any of the special assumptions underlying *CSHCpar*.

Given the large variety of existing sequential algorithm selectors, we study the question how such existing selectors can be effectively parallelized. To this end, we use the learned models of sequential algorithm selectors and modify the selection procedure such that we rank algorithms for a given instance and select the top n algorithms for n processing units (e.g., processors or processor cores).

State-of-the-art algorithm selectors make extensive use of pre-solving schedules, i.e., they run a sequence of solvers prior to per-instance algorithm selection [14,31]. This makes it possible to solve easy instances quickly, without inducing the overhead of feature computation. To effectively use parallel resources and minimize sequential bottlenecks, our approach uses parallel rather than sequential pre-solving schedules, which can be obtained using parallel algorithm schedule systems, such as, *aspeed* [8] or *3S* [14].

Figure 1 shows the extension of sequential algorithm selection to parallel portfolio selection with pre-solving schedules. On the first processing unit, we execute the standard workflow of sequential algorithm selectors: to solve a given instance i, we run a pre-solving schedule for a limited amount of time (e.g., 10 % of the overall time budget [14]); if the pre-solving schedule fails to solve i, we compute instance features $\mathbf{f}(i)$ (i.e., numerical properties of the instance), and then, based on $\mathbf{f}(i)$, select the putatively best algorithm for the given instance. In the parallel workflow, we can spend the time used by feature computation for longer pre-solving schedules on all threads but the first.

To ensure the scalability of parallel portfolios selection to many algorithms and processing units, we aim to develop methods that satisfy the following requirement:

(i) the online selection of parallel portfolios has to be efficient, i.e., polynomial in the size of the parallel portfolio.

Our general methods for parallel portfolio selection are applicable in any scenario for which the following assumptions hold:

(ii) the algorithm portfolio consists of deterministic algorithms; and on each processing unit, we select a different algorithm;
(iii) algorithms running in parallel do not communicate (e.g., no clause sharing of SAT solvers); and
(iv) we do not have special structural knowledge about the problem domain (e.g., we do not know that SAT instances can be divided into three types).

Assumption (ii) simplifies the selection of a parallel portfolio because there is no noise in the training data, and repeated runs of algorithms are not increasing the chance of solving an instance such that we should select each algorithm at most once. Since communication between algorithms often results in stochastic behavior (e.g., nearly all parallel SAT solvers with clause sharing are stochastic), Assumption (iii) helps to satisfy Assumption (ii). Furthermore, if algorithms would communicate, the performance of an algorithm could not be estimated independently from the other algorithms in the portfolio. We note that Assumption (ii) does not allow the selection of parallel algorithms, e.g., the parallel SAT solver *Plingeling*. Last but not least, Assumption (iv) states that, in contrast to the *CSHCpar* solver, we have no structural knowledge about the problem domain, since such knowledge is only available for specific problems.

Given Assumption (ii), our approach cannot utilize more processing units than there are algorithms in our portfolio (4 to 31 algorithms in our experiments). Therefore, the approaches we consider focus on parallelization with a relatively modest number of processing units, as found in current off-the-shelf computing hardware. Other approaches exist for scaling to higher degrees of parallelism (see, e.g., [3]).

In the following, we will first discuss related work (Sect. 2). Next, in Sect. 3, we extend well-known algorithm selection approaches from *SATzilla* [31], *ME-ASP* [23], *ISAC* [15], and *3S* [14] to parallel portfolio selection respecting our requirements. Then, in Sect. 4, we adapt the algorithm (pre-solving) schedules of *aspeed* [8] to the setting of Fig. 1, with different time budgets for each processing unit. Finally, we present an evaluation of our per-instance parallel portfolios on scenarios of the algorithm selection library[1], which allows a fair and thorough evaluation on a set of 12 different constraint solving domains from SAT, MAXSAT, CSP, QBF, and ASP, and pre-marshalling.

2   Related Work

Our work draws on two lines of research reaching back to John Rice's seminal work on algorithm selection [27] and the work by Huberman et al. [10] on parallel algorithm portfolios. It addresses the *dynamic resource allocation* challenge, which has been identified as one of the seven challenges in parallel SAT solving [7].

Recently, the algorithm selection approach of *CSHC* [21], which is based on cost-sensitive hierarchical clustering, was extended to selection of parallel portfolios in *CSHCpar* [22][2]. This approach differs from ours in that it relies on explicitly identified, distinct classes of problem instances (as is the case with the different tracks of the SAT Competition) and provides no obvious way of adjusting the number of processing units.

The extension of *3S* [14] to parallel portfolio selection, dubbed *3Spar* [20] (see Footnote 2), selects a parallel portfolio using k-NN to find the k most similar instances in instance feature space. Using integer linear programming (ILP), *3Spar* constructs a static pre-solving schedule offline and a per-instance parallel algorithm schedule online, based on training data of the k most similar instances. The ILP problem that needs to be solved for every instance is *NP*-hard and its time complexity grows with the number of parallel processing units and number of available solvers. Unlike our approach, during the feature computation phase, *3Spar* runs in a purely sequential manner. Since feature computation can require a considerable amount of time (e.g., more than 100 s on industrial SAT instances), this can leave important performance potential untapped.

ISAC [15] combines algorithm configuration and algorithm selection by (i) clustering the training instances in the feature space and (ii) using an algorithm configuration procedure [2,12] to optimize a parametric solver on each cluster. For a new problem instance i to be solved, *ISAC* selects the configuration which was determined for the cluster closest to i. The most recent *ISAC* version, *ISAC* 2.0^3, performs only algorithm selection and uses the best of a fixed set of algorithms in Step (ii); it also provides a method for selecting parallel portfolios for each cluster of instances by searching over all $\binom{|A|}{n}$ combinations of $|A|$ algorithms and n processing units. As this approach quickly becomes infeasible for growing $|A|$ and n, Yuri Malitsky, author of *ISAC* 2.0, recommends to limit its use to at most 4 processing units (README file).

---

[1] aslib.net.

[2] Unfortunately, no implementation of *CSHCpar* and *3Spar* is publicly available.

[3] https://sites.google.com/site/yurimalitsky/downloads.

The *aspeed* system [8] solves a similar scheduling problem as *3Spar*, but generates a static algorithm schedule during an off-line training phase, thus avoiding overhead in the solving phase. Unlike *3Spar*, *aspeed* does not support including parallel solvers in the algorithm schedule, and the algorithm schedule is static and not per-instance selected. For this reason, *aspeed* is not directly applicable to per-instance selection of parallel portfolios; however, our approach uses it to effectively compute parallel pre-solving schedules.

RSR-WG [34] combines a case-based-reasoning approach from *CP-Hydra* [24] with greedy construction of parallel portfolio schedules via *GASS* [28] for CSPs. Since the schedules are constructed on a per-instance basis, *RSR-WG* relies on instance features. In the first step, a schedule is greedily constructed to maximize the number of instances solved within a given cutoff time, and in the second step, the components of the schedule are distributed over the available processing units. In contrast to our approach, *RSR-WG* optimizes the number of timeouts and is not directly applicable to arbitrary performance metrics. Since the schedules are optimized online on a per-instance base, *RSR-WG* has to solve an *NP*-hard problem for each instance, which is done heuristically.

Finally, there is some work on parallel portfolios with dynamically adjusted timeshares (see e.g., [5]). Such approaches could eventually be used to dynamically adjust a portfolio determined by any of the methods we study in the following.

3   Selection of Parallel Portfolios

In this section, we show how to extend existing sequential algorithm selection approaches to handle parallel portfolio selection. Formally, the selection of parallel portfolios is an extension of the per-instance algorithm selection problem, in which not only one algorithm is selected, but rather a set of algorithms to be run in parallel.

Definition 1. *A per-instance parallel portfolio selection problem can be defined by a 5-tuple $\langle I, \mathcal{D}, A, U, m \rangle$, where*

- *I is a set of instances of a problem,*
- *$\mathcal{D}$ is a probability distribution over I,*
- *A is a set of algorithms for instances in I,*
- *U is a set of parallel processing units available, and*
- *$m : I \times A \to \mathbb{R}$ is a performance metric measuring the performance of algorithm $a \in A$ on instance $i \in I$.*

A solution of this problem is a mapping $\phi_u : I \to A$ for each processing unit $u \in U$; we refer to such a mapping as a parallel selection portfolio. *The performance metric we aim to minimize across the possible parallel selection portfolios is $\mathbb{E}_{i \sim \mathcal{D}} \min_{u \in U} m(i, \phi_u(i))$.*

Since we assume that the algorithms do not communicate with each other (Assumption (iii)), the performance of a parallel selection portfolio is the performance of the best algorithm in the selected portfolio. Therefore, a perfect parallel selection portfolio would, for each instance, select a set of algorithms containing the best algorithm for that instance. Ultimately, we would therefore like to model the per-instance correlations between solvers to select complementary sets of solvers for each instance.

In this work, however, we pursue a different approach, namely that of generically extending the various existing sequential selection strategies to the parallel selection setting, with the goal of assessing the merit of this overall approach and of empirically studying which sequential selection strategies lend themselves well to this setting. Since these existing sequential selection strategies do not model per-instance correlation between the algorithms, we restrict ourselves to constructing the portfolio in a greedy fashion, choosing the n solvers individually predicted to be best for a parallel portfolio on n processing units. Such a ranking of algorithms is admitted by most algorithm selection approaches [17].

Our approach requires, for each sequential algorithm selection mechanism under consideration, a scoring function

$$s : I \times A \to \mathbb{R} \tag{1}$$

that ranks the candidate algorithms for a given instance to be solved, such that the putatively best algorithm receives the lowest score value, the second best the second lowest score, etc. Then we simply sort the algorithms in A based on their scorses (breaking ties arbitrarily), using time $\mathcal{O}(|A| \log |A|)$. Thus, if we can compute the scores efficiently, we obtain a computationally efficient approach to parallel algorithm selection, satisfying Requirement (i). In the following, we show that we can indeed efficiently compute such scores for five prominent algorithm selection approaches.

Performance-Based Nearest Neighbor (PNN). The algorithm selection approach in *3S* [21] in its simplest form uses a k-nearest neighbour approach. For a new instance i with features $\mathbf{f}(i)$, it finds the k nearest training instances $I_k(i)$ in the feature space F and selects the algorithm that has the best training performance on them. Formally, given a performance metric $m : I \times A \to \mathbb{R}$, we define $m_k(i, a) = \sum_{i' \in I_k(i)} m(i', a)$ and select algorithm $\arg\min_{a \in A} m_k(i, a)$.

To extend this approach to parallel portfolios, we determine the same k nearest training instances $I_k(i)$ and simply select the n algorithms with the best performance for I_k. Formally, our scoring function in this case is simply:

$$s_{PNN}(i, a) = m_k(i, a). \tag{2}$$

In terms of complexity, identifying the k nearest instances costs time $\mathcal{O}(\#f \cdot |I| \cdot \log |I|)$, with $\#f$ denoting the number of used instance features; and averaging the performance values over the k instances costs time $\mathcal{O}(k \cdot |A|)$.

Distance-Based Nearest Neighbor (DNN). *ME-ASP* [23] implements an interface for different machine learning approaches used in its selection framework, but its released version uses a simple nearest neighbour approach with neighbourhood size 1, which also worked best empirically [23]. At training time, this approach memorizes the best algorithm $a^*(i')$ on each training instance $i' \in I$. For a new instance i, it finds the nearest training instance i' in the feature space and selects the algorithm $a^*(i')$ associated with that instance.

To extend this approach to parallel portfolios, for a new test instance i, we score each algorithm a by the minimum of the distances between i and any training instance associated with a. Formally, letting $d(\mathbf{f}(i), \mathbf{f}(i'))$ denote the distance in feature space between instance i and i', we have the following scoring function:

$$s_{DNN}(i, a) = \min\{d(\mathbf{f}(i), \mathbf{f}(i')) \mid i' \in I \wedge a^*(i') = a\}. \tag{3}$$

If $\{i' \in I \mid a^*(i') = a\}$ is empty (because algorithm a was never the best algorithm on an instance) then $s_{DNN}(i, a) = \infty$ for all instances i. Since we memorize the best algorithm for each instance in the training phase, the time complexity of this method is dominated by the cost of computing the distance of each training instance to the test instance, $\mathcal{O}(|I| \cdot \#f)$, where $\#f$ is the number of features.

Clustering. The selection part of *ISAC* [15][4] uses a technique similar to *ME-ASP*'s distance-based NN approach, with the difference that it operates on clusters of training instances instead of on single instances. Specifically, *ISAC* clusters the training instances, memorizing the cluster centers Z (in the feature space) and the best algorithms $\hat{a}(z)$ for each cluster $z \in Z$. For a new instance, similar to *ME-ASP*, it finds the nearest cluster z in the feature space and selects the algorithm associated with z.

To extend this approach to parallel portfolios, for a new test instance i, we score each algorithm a by the minimum of the distances between i and any cluster associated with a. Formally, using $d(\mathbf{f}(i), z)$ to denote the distance in feature space between instance i and cluster center z, we have the following scoring function:

$$s_{Clu}(i, a) = \min\{d(\mathbf{f}(i), z) \mid z \in Z \wedge \hat{a}(z) = a\}. \tag{4}$$

As for DNN, if $\{z \in Z \mid \hat{a}(z) = a\}$ is empty (because algorithm a was not the best algorithm on any cluster) then $s_{Clu}(i, a) = \infty$ for all instances i. The time complexity is as for DNN, replacing the number of training instances $|I|$ with the number of clusters $|Z|$.

Regression. The first version of *SATzilla* [31] used a regression approach, which, for each $a \in A$, learns a regression model $r_a : F \to \mathbb{R}$ to predict performance on new instances. For a new instance i with features $\mathbf{f}(i)$, it selected the algorithm with the best predicted performance, i.e., $\arg\min_{a \in A} r_a(\mathbf{f}(i))$.

---

[4] In its original version, *ISAC* is a combination of algorithm configuration and selection, but only the selection approach was used in later publications.

This approach trivially extends to parallel portfolios; we simply use scoring function

$$s_{Reg}(i, a) = r_a(\mathbf{f}(i)) \tag{5}$$

to select the n algorithms predicted to perform best. The complexity of model evaluations differs across models, but it is a polynomial for all models in common use; we denote this polynomial by P_{reg}. Since we need to evaluate one model per algorithm, the time complexity to select a parallel portfolio is then $\mathcal{O}(P_{reg} \cdot |A|)$.

Pairwise Voting. The most recent *SATzilla* version [32] uses cost-sensitive random forest classification to learn for each pair of algorithms $a_1 \neq a_2 \in A$ which of them performs better for a given instance; each such classifier c_{a_1,a_2} : $F \rightarrow \{0,1\}$ votes for a_1 or a_2 to perform better, and *SATzilla* then selects the algorithms with the most votes from all pairwise comparisons. Formally, let $v(i, a) = \sum_{a' \in A \setminus \{a\}} c_{a,a'}(\mathbf{f}(i'))$ denote the sum of votes algorithm a receives for instance i; then, *SATzilla* selects $\arg\max_{a \in A} v(i, a)$.

To extend this approach to parallel portfolios, we simply select the n algorithms with the most votes by defining our scoring function to be minimized as:

$$s_{Vote}(i, a) = -v(i, a). \tag{6}$$

As for regression models, the time complexity for evaluating a learned classifier differs across classifier types, but it is polynomial for all commonly-used types; we denote this polynomial function by P_{class}. Since we need to evaluate pairwise classifiers for all algorithm pairs, the time complexity to select a parallel portfolio is in $\mathcal{O}(P_{class} \cdot |A|^2)$.

4   Parallel Pre-Solving Schedules

State-of-the-art algorithm selectors commonly make use of algorithm schedules for pre-solving, i.e., they run a sequence of solvers prior to per-instance algorithm selection [14,31]. If one of the pre-solvers already solves a given instance, we do not need to compute instance features for the algorithm selection phase and save the time induced by the feature computation.

Malitsky et al. [21] and Hoos et al. [8] have already presented how to find timeout-optimal parallel algorithm schedules. In their settings, the schedules on all processing units get the same amount of runtime. However, as shown in Fig. 1, the computation of instance features is limited to one processing unit, and we can run longer pre-solving schedules on all other units. The feature computation time differs from instance to instance, but since we compute our presolving schedule offline, we require a constant estimate of the feature computation runtime, $FeatT$. To err on the pessismistic side, in each algorithm selection scenario we estimate $FeatT$ as the maximal feature computation time observed across the scenario's training instances.

We added a constraint to the flexible and declarative Answer Set Programming (ASP [4]) encoding of *aspeed* [8][5] to ensure that the pre-solving schedule

---

[5] Since *3S* [21] is not publicly available, using it was not an option.

on the first core is limited by a maximal pre-solving time budget, $PreT$. All pre-solving schedules on the other processing units are given an additional budget of $FeatT$ to ensure we use them while the first core computes features. Please refer to Listing 1.1 for our ASP encoding.

```
:- not [ slice(1,A,T) = T ] PreT.
:- not [ slice(U,A,T) = T : U != 1 ] PreT+FeatT, unit(U).
```

Listing 1.1. ASP constraints in the language of the ASP grounder gringo [6]. `slice(U,A,T)` denotes that algorithm A has a runtime slice T on processing unit U. The first integrity contraint limits the sum of runtimes T of all algorithms A on processing unit 1 by the maximal pre-solving runtime `PreT` (an external constant). The second integrity constraint does the same for all other units, but extends the maximal pre-solving runtime by the feature computation time `FeatT` (external constant).

The problem of optimizing an algorithm schedule is NP-hard. However, the empirical results of Hoos et al. [8] indicated that the problem of optimizing parallel schedules gets easier with more processing units. In contrast, $ISAC$ has to solve a problem offline that gets more complex with more processing units and is not applicable with more than 4 units.

5 Empirical Evaluation

We now turn to an empirical assessment of our parallel portfolio selection approaches on twelve algorithm selection scenarios that make up the Algorithm Selection Library ($ASlib$).[6] These scenarios involve a wide range of hard combinatorial problems; each of them includes the performance data of a range of solvers for a set of instances, instance features[7] organized in feature groups (we use the default feature groups), and associated costs for these features (see Table 1). We refer to the $ASlib$ website (aslib.net) for the details on all scenarios; we chose $ASlib$ as the basis for our evaluation since this allows us to compare our approach in a fair and uniform way against other algorithm selection methods. Since all experiments are based on the data in the scenarios, we did not need to run any of the algorithms in the portfolio. This ensures repeatability of our experiments, but it also means that resource contention between algorithms running in parallel are not reflected in our results. Depending on the hardware used (e.g., multi-core vs. multi-processor systems), performance may be reduced when running too many algorithms in parallel.

---

[6] Since the *CSP-2010* scenario consists of only 2 algorithms, it already admits a perfect portfolio using two processing units. Therefore, we excluded it from our experiments.

[7] Instance features typically consist of cheap syntactic features, such as number of variables and number of clauses, and probing features, i.e., extracting runtime statistics by running an algorithm for a short time on a given instance.

Table 1. The *ASlib* algorithm selection scenarios – information on the number of instances $|I|$, number of unsolvable instances $|U|$ ($U \subset I$), number of algorithms $|A|$, number of features $\#f$, number of feature groups $\#f_g$, the average feature computation cost of the used default features $\varnothing t_f$, and runtime cutoff t_c

Scenario	$|I|$	$|U|$	$|A|$	$\#f$	$\#f_g$	$\varnothing t_f$	t_c	Ref
ASP-POTASSCO	1294	82	11	138	4	1.3	600	[9]
MAXSAT12-PMS	876	129	6	37	1	0.1	2100	[1,13]
PREMARSHALLING	527	0	4	16	1	0	3600	[29]
PROTEUS-2014	4021	428	22	198	4	6.4	3600	[11]
QBF-2011	1368	314	5	46	1	0	3600	[18,26]
SAT11-HAND	296	77	15	115	10	41.2	5000	[13,32]
SAT11-INDU	300	47	18	115	10	135.3	5000	[13,32]
SAT11-RAND	600	108	9	115	10	22.0	5000	[13,32]
SAT12-ALL	1614	20	31	115	10	40.5	1200	[13,33]
SAT12-HAND	767	229	31	115	10	39.0	1200	[13,33]
SAT12-INDU	1167	209	31	115	10	80.9	1200	[13,33]
SAT12-RAND	1362	322	31	115	10	9.0	1200	[13,33]

Setup. We implemented our parallel selection approach in the open-source and flexible algorithm selection framework of *claspfolio 2* (2.1.0; using *scikit-learn* 0.14.1 [25]).[8] For the choice of machine learning models, *claspfolio 2* follows the implementations of well-known algorithm selectors; we used random forests for pairwise voting (*SATzilla11* [32]), ridge regression for regression (*SATzilla'09* [31]) and *k*-means for clustering (*ISAC* [15]).[9]

Within *claspfolio 2*, we use *aspeed* [8] with the ASP tools *gringo* (3.0.5) and *clasp* (2.2) [6] with a time budget of at most 300 CPU seconds to effectively compute pre-solving schedules. Our pre-solving schedules are limited to at most 256 s on the first processing unit and an additional 10 % of the runtime cutoff on the other processing units (10 % of the runtime cutoff is the maximal feature computation runtime - parameter of *claspfolio 2*; the runtime cutoff differs across the *ASlib* scenarios).

Since speedup is a commonly used performance metric in parallelization and PAR10 (penalized average runtime, counting each timeout as 10 times the runtime cutoff) is a commonly used performance metric in algorithm selection, we assessed our approaches based on PAR10-speedups over the (sequential) single best algorithm (*SB*) in the given algorithm portfolios. We note that the possible speedup is bounded from above by the performance of a perfect algorithm selector (the vir-

---

[8] www.cs.uni-potsdam.de/claspfolio/.

[9] The original *ISAC* [15] uses *g*-means, which automatically determines the number of clusters. In preliminary experiments, we observed that the square root of the number of instances gives a good upper bound for the number of clusters; therefore, we did not used *g*-means but *k*-means with this cluster bound.

tual best solver VBS) that always selects the best algorithm for a given instance without inducing feature computation costs. We used 10-fold cross validation (i.e., 10 different training and test splits) to obtain an unbiased performance estimate for *claspfolio 2*, as given in *ASlib*. To avoid artificially inflating PAR10 scores, we removed from the test sets all instances that could be solved neither by any of the candidate algorithms nor during feature computation. Furthermore, to verify which approaches performed statistically indistinguishable from the best approach, we used permutation tests with 100 000 permutations at significance level $\alpha = 0.05$.

Table 2. Speedup on PAR10 (wallclock) in comparison to *SB* with one processing unit (U). Entries for which the number of processing units exceed the number of candidate algorithms are marked 'NA'. Entries shown in bold-face are statistically indistinguishable from the best speedups obtained for the respective scenario and number of processing units (according to a permutation test with 100 000 permutations and $\alpha = 0.05$). If more processing are available than algorithms, we run all algorithms and achieve a perfect VBS score (number in parentheses)

$|U|$	1	2	4	8	$|U|$	1	2	4	8	$|U|$	1	2	4	8
ASP-POTASSCO (VBS: 25.1)					**QBF-2011 (VBS: 95.6)**					**SAT12-ALL (VBS: 31.6)**				
DNN	2.0	3.8	6.3	18.7	DNN	6.7	14.7	33.6	(95.6)	DNN	2.4	4.2	6.3	10.7
clustering	2.7	3.7	5.8	15.2	clustering	4.5	10.5	22.4	(95.6)	clustering	1.5	1.9	2.6	3.5
PNN	4.1	5.0	8.9	8.6	PNN	6.3	23.0	40.6	(95.6)	PNN	2.2	2.9	3.9	7.7
pairwise-voting	3.2	4.7	7.3	10.8	pairwise-voting	8.9	23.1	93.7	(95.6)	pairwise-voting	2.8	4.1	6.1	9.0
regression	2.3	3.9	8.6	18.2	regression	4.7	11.9	60.2	(95.6)	regression	2.1	2.9	4.3	7.1
SB	1.0	3.0	7.4	8.8	SB	1.0	2.7	13.6	(95.6)	SB	1.0	1.0	1.4	1.7
MAXSAT12-PMS (VBS: 51.8)					**SAT11-HAND (VBS: 37.2)**					**SAT12-HAND (VBS: 34.7)**				
DNN	8.4	21.4	51.5	(51.8)	DNN	3.2	5.2	9.6	23.9	DNN	3.7	6.2	11.4	14.3
clustering	4.8	10.4	21.2	(51.8)	clustering	1.6	2.9	4.2	7.0	clustering	1.8	2.3	3.3	4.6
PNN	4.7	7.2	11.6	(51.8)	PNN	2.3	2.8	8.4	10.8	PNN	2.0	2.8	4.9	7.5
pairwise-voting	7.7	15.8	31.1	(51.8)	pairwise-voting	3.4	4.8	8.6	10.9	pairwise-voting	4.2	5.4	9.0	12.4
regression	4.7	6.4	21.6	(51.8)	regression	2.9	4.5	8.4	12.5	regression	2.9	4.2	7.0	9.8
SB	1.0	1.3	1.8	(51.8)	SB	1.0	1.2	1.9	6.2	SB	1.0	1.0	1.4	1.9
PREMARSHALLING (VBS: 30.8)					**SAT11-INDU (VBS: 21.4)**					**SAT12-INDU (VBS: 15.4)**				
DNN	2.7	5.8	(30.8)	(30.8)	DNN	1.4	1.9	2.6	7.8	DNN	2.0	2.4	3.4	5.0
clustering	2.5	7.0	(30.8)	(30.8)	clustering	1.3	1.9	3.3	5.3	clustering	1.3	2.1	2.8	4.6
PNN	2.4	4.4	(30.8)	(30.8)	PNN	1.1	1.5	2.6	5.2	kNN	1.6	2.3	3.9	5.7
pairwise-voting	2.8	7.6	(30.8)	(30.8)	pairwise-voting	2.0	2.4	3.6	4.7	pairwise-voting	2.4	3.0	3.8	5.4
regression	2.6	4.7	(30.8)	(30.8)	regression	1.3	2.0	3.6	7.8	regression	1.9	2.5	3.5	6.3
SB	1.0	1.4	(30.8)	(30.8)	SB	1.0	1.7	2.9	7.2	SB	1.0	1.5	2.5	4.8
PROTEUS-2014 (VBS: 408.9)					**SAT11-RAND (VBS: 65.7)**					**SAT12-RAND (VBS: 12.1)**				
DNN	5.2	0.2	19.0	46.0	DNN	3.8	11.0	42.2	60.5	DNN	0.8	1.5	4.7	8.6
clustering	7.4	7.7	14.2	30.2	clustering	6.1	9.5	32.3	42.7	clustering	1.3	1.7	2.7	4.9
PNN	5.9	9.5	23.3	50.9	PNN	6.5	9.3	10.7	60.2	PNN	1.2	2.1	4.8	7.8
pairwise-voting	5.8	11.0	20.8	42.0	pairwise-voting	4.4	8.3	11.4	60.4	pairwise-voting	1.1	1.7	2.8	6.4
regression	6.4	9.8	20.4	53.8	regression	5.9	7.8	8.3	60.3	regression	1.3	1.8	5.2	8.3
SB	1.0	2.7	5.9	10.4	SB	1.0	5.9	6.8	64.8	SB	1.0	1.5	4.0	6.8

Comparison of Approaches Within Claspfolio 2. In Table 2, we report performance results for the approaches presented in Sect. 3 as implemented in *claspfolio 2*, for 1 to 8 processing units (U). For 4 processing units, the speedup over the single best algorithm is between 2.6 (*SAT11-INDU* with PNN) and 93.7 (*QBF-2011* with pairwise voting). The best parallelization approach differed between the scenarios, which is not too surprising since it is known that there is no single dominant approach for sequential per-instance algorithm selection. We note that the recent work on *autofolio* [19] proposes using algorithm configuration to determine a well-performing algorithm selection approach and its parameters for a given scenario. On average, DNN and pairwise voting performed best across our scenarios; for 4 processing units, both approaches achieved performance levels that were was not significantly worse than the best approach on 10 out of 12 scenarios. The geometric average speedup was 11.89 for DNN and 10.90 for pairwise voting, respectively.[10] In contrast, on one processing unit, DNN performed best on only 3 scenarios and pairwise voting on 9 scenarios. We note that performance differences between the approaches decreased as the number of processing units increased: all approaches got closer to the optimal speedup achieved when running all candidate algorithms in parallel.

Overall, our approaches and also the *VBS* do not scale as well on some of the SAT scenarios as they do on the other scenarios (e.g., the maximal *VBS* speedup is 95.6 on *QBF-2011* but only 12.1 on *SAT12-RAND*). We speculate that this is due to (i) the relatively large number of SAT solvers (which makes it harder to perform as well as the *VBS*) and (ii) the relatively low performance correlation between some of those solvers.

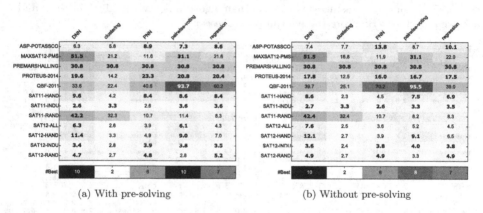

(a) With pre-solving   (b) Without pre-solving

Fig. 2. Heatmap for PAR10 speedups (wallclock) against sequential *SB* on 4 processing units. A value is printed in bold-face if a statistical test (i.e., permutation test with 100 000 and $\alpha = 0.05$) cannot find evidence that it is significantly lower than that of the best approach for a given number of threads. The last row shows how often an approach was en par with the best.

As can be seen in Fig. 2, using pre-solving schedules improved the performance on 4 out of our 5 approaches on 4 processing units. Surprisingly, the distance-based nearest neighbor approach (DNN) performed slightly better without pre-solving schedules, which we believe may be caused by over-fitting to the training data.

Comparison with Other Systems. While the previous experiment fixed all design decisions except the selection strategy, we now compare the results

---

[10] We note that we have to use a geometric average instead of an arithmetic average, because we aggregate over speedup *factors*. This can be seen when considering a case with speedups of 2 and 0.5, where the arithmetic average gives a misleading 1.25.

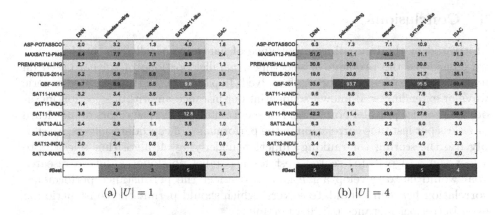

(a) $|U| = 1$ (b) $|U| = 4$

Fig. 3. Comparison of parallelization approaches of different algorithm selection mechanisms on 1 and 4 processing units; we can't assess statistical difference since *ISAC* only outputs a single performance value.

for our two best results (DNN and pairwise-voting) with three other strategies: SATzilla'11-like, a variant of our pairwise-voting approach in which we restrict the number of presolvers and their time limit to resemble more closely the strategy used in *SATzilla* 2011 [32]; the *aspeed* system, which does not perform per-instance selection, but produces static parallel schedules and is used for pre-solver scheduling in *claspfolio 2*; and the *ISAC* system, for which we have written a converter from the *ASlib* format into its native input format. Since *ISAC* determines its cross-validation folds internally and only outputs a single performance number, we cannot perform statistical tests for this experiment and only report the number of times each method performed best, as well as the methods' (geometric) mean speedups.

Figure 3 shows the performance of these systems on 1 and 4 processing units. In the sequential case, SATzilla'11-like performed best overall (best on 5 of our 12 benchmarks; average speedup 3.80), followed by pairwise voting (best on 3 benchmarks; average speedup of 3.49), and *aspeed* (best on 4 benchmarks; average speedup 2.34). Using 4 processing units, while SATzilla'11-like still performed best (best on 5 benchmarks; average speedup 12.27), it was now closely followed by the other two approaches: DNN (also best on 5 benchmarks, average speedup 11.89) and ISAC (best on 4 benchmarks, average speedup 10.72).

We conclude that, while SATzilla'11-like yields stable performance, the performance of different methods scales differently as the number of processing units grows. We also note that, going up to 4 processing units, the best average speedups obtained were roughly linear in the number of units. While *ISAC* should not be used with more than 4 processing units (due to its exponential time requirements in the number of units), Table 2 shows that our methods (especially DNN) even improved further based on 8 processing units, without increasing the effort to train or use them.

6 Conclusions

Motivated by the increasing importance of hardware parallelism, in this work, we considered the problem of selecting a parallel portfolio of solvers based on features of a problem instance to be solved. In particular, we investigated generic ways of extending well-known sequential per-instance algorithm selection methods to produce parallel portfolios. Since current algorithm selectors do not learn or assess per-instance correlation in the performance of candidate solvers, we simply use the scoring (or ranking) function underlying all algorithm selectors to select the n algorithms scored best for a parallel portfolio. A future research goal would be to develop a method to consider the per-instance performance correlation between candidate solvers, which should permit the construction of even better per-instance parallel portfolios.

Our extensive empirical study demonstrated that all methods we considered performed quite well on the large range of scenarios from the algorithm selection library, with speedups from 2.6 to 93.7 on 4 processing units in comparison to running only the single best available algorithm sequentially. Overall, we found our distance-based nearest neighbor (DNN) and pairwise-voting approaches to perform better than other approaches.

However, as for any algorithm selection approach, the performance of our parallel portfolio selectors is bounded by that of an oracle selector, i.e., a perfect algorithm selector that always selects the single best algorithm for a given instance. We see two ways to overcome this obstacle, (i) use of per-instance algorithm configuration [15,30] to improve the performance of the candidate set of algorithms and hence of the oracle; and (ii) to permit communication between the algorithms in the parallel portfolio (e.g., clause sharing between SAT solvers). Both avenues can potentially be pursued by extending the techniques investigated here.

References

1. Ansótegui, C., Malitsky, Y., Sellmann, M.: Maxsat by improved instance-specific algorithm configuration. In: Proceedings of AAAI 2014, pp. 2594–2600 (2014)
2. Ansótegui, C., Sellmann, M., Tierney, K.: A gender-based genetic algorithm for the automatic configuration of algorithms. In: Gent, I.P. (ed.) CP 2009. LNCS, vol. 5732, pp. 142–157. Springer, Heidelberg (2009)
3. Arbelaez, A., Codognet, P.: From sequential to parallel local search for SAT. In: Middendorf, M., Blum, C. (eds.) EvoCOP 2013. LNCS, vol. 7832, pp. 157–168. Springer, Heidelberg (2013)
4. Baral, C.: Knowledge Representation, Reasoning and Declarative Problem Solving. Cambridge University Press, New York (2003)
5. Gagliolo, M., Schmidhuber, J.: Towards distributed algorithm portfolios. In: Corchado, J.M., Rodríguez, S., Llinas, J., Molina, J.M. (eds.) DCAI 2008. Advances in Soft Computing, vol. 50, pp. 634–643. Springer, Heidelberg (2008)
6. Gebser, M., Kaminski, R., Kaufmann, B., Ostrowski, M., Schaub, T., Schneider, M.: Potassco: the Potsdam answer set solving collection. AI Commun. **24**(2), 107–124 (2011)

7. Hamadi, Y., Wintersteiger, C.: Seven challenges in parallel SAT solving. AI Mag. **34**(2), 99–106 (2013)
8. Hoos, H., Kaminski, R., Lindauer, M., Schaub, T.: aspeed: solver scheduling via answer set programming. TPLP **15**, 117–142 (2015)
9. Hoos, H., Lindauer, M., Schaub, T.: claspfolio 2: advances in algorithm selection for answer set programming. TPLP **14**, 569–585 (2014)
10. Huberman, B., Lukose, R., Hogg, T.: An economic approach to hard computational problems. Science **275**, 51–54 (1997)
11. Hurley, B., Kotthoff, L., Malitsky, Y., O'Sullivan, B.: Proteus: a hierarchical portfolio of solvers and transformations. In: Simonis, H. (ed.) CPAIOR 2014. LNCS, vol. 8451, pp. 301–317. Springer, Heidelberg (2014)
12. Hutter, F., Hoos, H., Leyton-Brown, K., Stützle, T.: ParamILS: an automatic algorithm configuration framework. JAIR **36**, 267–306 (2009)
13. Hutter, F., Xu, L., Hoos, H., Leyton-Brown, K.: Algorithm runtime prediction: methods and evaluation. J. Artif. Intell. **206**, 79–111 (2014)
14. Kadioglu, S., Malitsky, Y., Sabharwal, A., Samulowitz, H., Sellmann, M.: Algorithm selection and scheduling. In: Lee, J. (ed.) CP 2011. LNCS, vol. 6876, pp. 454–469. Springer, Heidelberg (2011)
15. Kadioglu, S., Malitsky, Y., Sellmann, M., Tierney, K.: ISAC - instance-specific algorithm configuration. In: Proceedings of ECAI 2010, pp. 751–756 (2010)
16. Kotthoff, L.: Algorithm selection for combinatorial search problems: a survey. AI Mag. **35**(3), 48–60 (2014)
17. Kotthoff, L.: Ranking algorithms by performance. In: Pardalos, P.M., Resende, M.G.C., Vogiatzis, C., Walteros, J.L. (eds.) LION 2014. LNCS, vol. 8426, pp. 16–20. Springer, Switzerland (2014)
18. Kotthoff, L., Gent, I., Miguel, I.: An evaluation of machine learning in algorithm selection for search problems. AI Commun. **25**(3), 257–270 (2012)
19. Lindauer, M., Hoos, H., Hutter, F., Schaub, T.: Autofolio: algorithm configuration for algorithm selection. In: Proceedings of Workshops at AAAI 2015 (2015)
20. Malitsky, Y., Sabharwal, A., Samulowitz, H., Sellmann, M.: Parallel SAT solver selection and scheduling. In: Milano, M. (ed.) Principles and Practice of Constraint Programming. LNCS, pp. 512–526. Springer, Heidelberg (2012)
21. Malitsky, Y., Sabharwal, A., Samulowitz, H., Sellmann, M.: Algorithm portfolios based on cost-sensitive hierarchical clustering. In: Rossi, F. (ed.) Proceedings of IJCAI 2013, pp. 608–614 (2013)
22. Malitsky, Y., Sabharwal, A., Samulowitz, H., Sellmann, M.: Parallel lingeling, ccasat, and csch-based portfolio. In: Proceedings of SAT Competition 2013, pp. 26–27 (2013)
23. Maratea, M., Pulina, L., Ricca, F.: A multi-engine approach to answer-set programming. TPLP **14**, 841–868 (2014)
24. O'Mahony, E., Hebrard, E., Holland, A., Nugent, C., O'Sullivan, B.: Using case-based reasoning in an algorithm portfolio for constraint solving. In: Proceedings of AICS 2008 (2008)
25. Pedregosa, F., Varoquaux, G., Gramfort, A., Michel, V., Thirion, B., Grisel, O., Blondel, M., Prettenhofer, P., Weiss, R., Dubourg, V., Vanderplas, J., Passos, A., Cournapeau, D., Brucher, M., Perrot, M., Duchesnay, E.: Scikit-learn: machine learning in Python. JMLR **12**, 2825–2830 (2011)
26. Pulina, L., Tacchella, A.: A self-adaptive multi-engine solver for quantified boolean formulas. Constraints **14**(1), 80–116 (2009)
27. Rice, J.: The algorithm selection problem. Adv. Comput. **15**, 65–118 (1976)

28. Streeter, M., Golovin, D., Smith, S.: Combining multiple heuristics online. In: Proceedings of AAAI 2007, pp. 1197–1203 (2007)
29. Tierney, K.: An algorithm selection benchmark of the container pre-marshalling problem. Tech. Rep. DS&OR Working Paper 1402, DS&OR Lab, University of Paderborn (2014)
30. Xu, L., Hoos, H., Leyton-Brown, K.: Hydra: automatically configuring algorithms for portfolio-based selection. In: Proceedings of AAAI 2010, pp. 210–216 (2010)
31. Xu, L., Hutter, F., Hoos, H., Leyton-Brown, K.: SATzilla: portfolio-based algorithm selection for SAT. JAIR **32**, 565–606 (2008)
32. Xu, L., Hutter, F., Hoos, H., Leyton-Brown, K.: Evaluating component solver contributions to portfolio-based algorithm selectors. In: Cimatti, A., Sebastiani, R. (eds.) SAT 2012. LNCS, vol. 7317, pp. 228–241. Springer, Heidelberg (2012)
33. Xu, L., Hutter, F., Shen, J., Hoos, H., Leyton-Brown, K.: SATzilla2012: improved algorithm selection based on cost-sensitive classification models. In: Proceedings of SAT Challenge 2012, pp. 57–58 (2012)
34. Yun, X., Epstein, S.L.: Learning algorithm portfolios for parallel execution. In: Hamadi, Y., Schoenauer, M. (eds.) LION 2012. LNCS, vol. 7219, pp. 323–338. Springer, Heidelberg (2012)

An Algorithm Selection Benchmark
of the Container Pre-marshalling Problem

Kevin Tierney[1]([⊠]) and Yuri Malitsky[2]

[1] Decision Support and Operations Research Lab,
University of Paderborn, Paderborn, Germany
tierney@dsor.de
[2] IBM T.J Watson Research Center, Yorktown Heights, USA
ymalits@us.ibm.com

Abstract. We present an algorithm selection benchmark based on optimal search algorithms for solving the container pre-marshalling problem (CPMP), an NP-hard problem from the field of container terminal optimization. Novel features are introduced and then systematically expanded through the recently proposed approach of latent feature analysis. The CPMP benchmark is interesting, as it involves a homogeneous set of parameterized algorithms that nonetheless result in a diverse range of performances. We present computational results using a state-of-the-art portfolio technique, thus providing a baseline for the benchmark.

1 Introduction

The container pre-marshalling problem (CPMP) is a well-known NP-hard problem in the container terminals and stacking literature [2, 7, 10], first introduced in [6]. The CPMP deals with the sorting of containers in a set of stacks (called a *bay*) of intermodal containers based on their exit times from the stacks, such that containers that must leave the stacks first are placed on top of containers that must leave later. This prevents *mis-overlaid* containers from blocking the timely exit of other containers. The goal of the CPMP is to find the minimal number of container movements necessary to ensure that all of the stacks are sorted by the exit time of each container without exceeding the maximum height of each stack. Solving the CPMP assists container terminals in reducing delays and increasing the efficiency of their operations.

A recent approach for solving the CPMP to optimality [11] presents two state-of-the-art approaches, based on A* and IDA*. We use parameterized versions of these approaches to form a benchmark for algorithm selection. We introduce 22 novel features to describe CPMP instances and show how the approach of latent feature analysis (LFA) [8] can assist domain experts in developing useful features for algorithm selection approaches. Finally, we augment the existing CPMP instances with extra instances from a new instance generator[1].

---

[1] An extended version of this paper is available at https://bitbucket.org/eusorpb/cpmp-as/downloads/asl_pm_extended.pdf.

© Springer International Publishing Switzerland 2015
C. Dhaenens et al. (Eds.): LION 9 2015, LNCS 8994, pp. 17–22, 2015.
DOI: 10.1007/978-3-319-19084-6_2

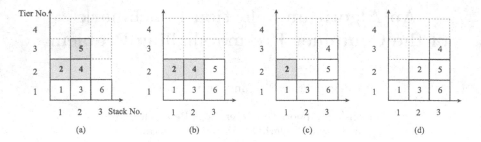

Fig. 1. An example solution to the CPMP with mis-overlays highlighted (Reproduced from [11]).

2   The Container Pre-marshalling Problem

Given an initial layout of a bay with a fixed number of stacks and tiers (stack height), the goal of the CPMP is to find the minimal number of container movements (or *rehandles*) necessary to eliminate all mis-overlays in the bay. Every container is assigned a group that indicates when it must leave the bay. A mis-overlaid container is defined as a container with a group that is higher than the group of any container underneath it, or a container above a mis-overlaid container.

Consider the simple example of Fig. 1, which shows a bay composed of three stacks of containers in which containers can be stacked at most four tiers high. Each container is represented by a box with its corresponding group[2]. This is not an ideal layout as the containers with groups 2, 4 and 5 will need to be relocated in order to retrieve the containers with higher groups (1 and 3). That is, containers with groups 2, 4 and 5 are mis-overlaid. Consider a container movement (f, t) defining the relocation of the container on top of the stack f to the top position of the stack t. The containers in the initial layout of Fig. 1 can reach the final layout (d) with three relocation moves: $(2, 3)$ reaching layout (b), $(2, 3)$ reaching layout (c) and $(1, 2)$ reaching layout (d) where no mis-overlays occur.

Pre-marshalling is important both in terms of operational and tactical goals at a container terminal. In particular, effective pre-marshalling of containers can help reduce delays moving containers from the terminal yard onto vessels, as well as from the yard onto trucks or trains. We refer to [11] for more information and a discussion of related work.

3   Latent Feature Analysis (LFA)

Given a set of solvers for a problem and a set of instance, algorithm selection is the study of finding the best performing solver for each instance. There are a variety of approaches that can be used to make this decision, including machine learning techniques as well as scheduling algorithms. For an overview of this area,

---

[2] We note that multiple containers may have the same group, but in order to make containers easily identifiable, in this example we have assigned a different group to each container.

we refer the reader to a recent survey [5]. Although there are many algorithm selection approaches, they are not the only important component in a selection approach. The quality of features in differentiating instances is critical to the success or failure of any algorithm selection strategy.

Features are normally created based on the knowledge of domain experts. In [8], the authors theorize how latent features gathered from a matrix decomposition can systematically help domain experts augment a set of features with more effective ones. Specifically, [8] shows that latent features can be determined using an existing set of structural features. Features that assist algorithm selection techniques in making correct predictions can then be identified, thus guiding a domain expert towards the features that work best on his or her problem.

The idea proposed by [8] uses singular value decomposition (SVD) to find the latent features that best describe the changes in the actual performance of solvers on instances. SVD is a method for identifying and ordering the dimensions along which data points exhibit the most variation, which is mathematically represented by the following equation: $M = U \Sigma V^T$, where M is the $m \times n$ matrix of solver performance. In our case, we consider an M where there are m instances each described by the performance of n solvers. This means that the $m \times n$ orthonormal columns of U can be interpreted as a latent feature that describes that instance. The columns of the V^T matrix refer to each solver, with each row presenting how active, or important a particular feature is for that solver.

If for a given instance it were possible to predict the latent features, using this decomposition we could multiply the feature vectors by the existing Σ and V^T matrices to get back the performance of each solver. While this is of course impossible in practice, we can use an existing set of structural features to predict these latent features. By then studying these predictions, we can identify exactly which latent features are currently difficult to predict accurately and even identify which latent feature we should focus on getting right to maximize the quality of the resulting prediction.

It is assumed that if we are unable to accurately predict a latent feature using our existing features, then our feature set is missing something critical about the underlying structure of an instance. By computing the correct value for this latent feature and sorting all training instances based on it, we assume that there must be something different for the instances where the latent feature value is large and those instances where the value is small. It is then up to a domain expert to try to analyze this difference and propose a new, expanded set of features for the algorithm selection approach to take advantage of.

4   Algorithm Selection Benchmark

We now describe our benchmark in detail[3]. Four optimal parameterizations of the A* and IDA* approaches in [11] form the basis of the benchmark. Due to space limits, we refer interested readers to [11] for the algorithm and heuristic details.

---

[3] This benchmark is available in the algorithm selection library (www.aslib.net) under the name "PREMARSHALLING-ASTAR-2013".

1. Number of stacks
2. Number of tiers
3. Tiers/stacks ratio
4. Container density
5. Empty stack percentage
6,7. Percent of all {slots, stacks} that are mis-overlaid
8. Bortfeldt & Forster lower bound
9–12. Min/max/mean/stdev container group counts
13–16. Min/max/mean/stdev group of top non-mis-overlaid
      container in each stack

17. Container density in stacks 1 through $\#Stacks/3$
18. Tier-weighted groups
19. Largest group L1 distance from top left (average)
20. Pct. contiguous empty space including one empty stack

21. Mis-overlaid stack (≥ 2 containers) percentage
22. Low-group containers near stack tops (percentage)

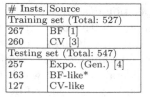

# Insts.	Source
Training set (Total: 527)	
267	BF [1]
260	CV [3]
Testing set (Total: 547)	
257	Expo. (Gen.) [4]
163	BF-like*
127	CV-like

Fig. 2. Features for the CPMP.   **Fig. 3.** Instances used.

An interesting aspect of the pre-marshalling benchmark in relation to other benchmarks, such as those based on SAT, CSP, QBF, etc. is that the portfolio of algorithms is not particularly diverse (very similar algorithm parameterizations), but performance variations are nonetheless significant.

We create a set of training and test instances out of existing pre-marshalling instances from [1] (BF) and from [3] (CV) as well as instances we generated. We filter out instances where all algorithms timeout/memout or are too easy. An overview is provided in Fig. 3. Our BF-generated instances are not exactly the same as in [1] because their instance generation is not completely described.

The features used in our dataset are given in Fig. 2, split into three categories. Features 1 through 16 were designed before performing latent feature analysis. Features 17 through 20 were created based on our first iteration of latent feature analysis, and features 21 and 22 using our second iteration. All of the features can be computed quickly. Our feature generation code (and instance generator) is available at https://bitbucket.org/eusorpb/cpmp-as. We note that other features are certainly possible, such as probing features.

Original features are created in the standard way for algorithm selection benchmarks, based on domain knowledge. The first 5 features address the problem size and density of containers. Feature 6 counts the number of mis-overlaid containers, a naive lower bound to the problem, whereas Feature 7 counts how many stacks contain mis-overlaid containers. Feature 8 provides the lower bound from [1], analyzing indirect container movements in addition to the mis-overlays present in feature 7. Features 9 through 12 offer information on how many containers belong to each group. Features 12 through 15 attempt to uncover the structure of the groups of the top non-mis-overlaid container on each stack.

LFA features are constructed based on the suggestions of the latent features. Feature 17 is the density of containers on the "left" side of the instance. We note that this feature is likely "overtuned" to the algorithms in our benchmark. Feature 18 measures whether containers with high group values are on high or low tiers by multiplying the tier of a container by its group, summing these values together and dividing by the maximum this value could take (namely if the highest group container was in each slot). Feature 19 measures the L1 (manhattan) distance from the top left of a problem to each container in the latest exit time, averaging these distances if there are multiple containers in the latest exit group. The final feature from iteration 1 computes the percentage of empty space in the instance in which an area of contiguous empty space includes at least one empty stack. Features 21 and 22 come from LFA iteration 2. Feature 21 counts how many stacks with more than two containers are mis-overlaid, and Feature 22 counts "low" ($\leq max\text{-}group/4$) valued containers on the top of stacks.

5   Computational Results

We evaluate our features using the cost-sensitive hierarchical clustering (CSHC) approach from [9]. Table 1 provides the performances[4] of a CSHC based portfolio when trained on the three datasets versus the best single solver (BSS) and the virtual best solver (VBS), which is a portfolio that always picks the correct solver. CSHC using just the initial arbitrary features already performs significantly better than the BSS, indicating even the original features have descriptive value.

When a CSHC portfolio is trained on the first iteration of features, the performance improves not only in the number of instances solved, but also on the average time taken to solve each instance. This shows that by utilizing the latent feature analysis, a researcher is able to develop a richer set of features to describe the instances. Furthermore, the process can be repeated, as is evidenced by the performance of CSHC on the second iteration of features. Note that the overall performance is again improved not only in the number of instances solved, but the time taken to solve them on average. Thus, multiple iterations of the latent feature analysis process can lead to even better features, although there are clearly diminishing returns.

Table 1. Performance of CSHC trained on the three feature sets.

Solver	Avg.	PAR-10	Solved
BSS	78.6	5,923	458
Original Features	51.6	3,469	495
LFA Iteration 1 Features	46.6	2,741	506
LFA Iteration 2 Features	45.4	2,543	509
VBS	12.8	12.8	547

---

[4] All runtime data was generated on an AMD Opteron 2425 HE processor running at 2.1 GHz with a 1 h timeout.

6   Conclusion

We presented an algorithm selection benchmark for the container pre-marshalling problem, a well-known problem from the container terminals literature. Our benchmark includes novel features and instances. We further showed that latent feature analysis can help in augmenting problem features. We hope that this benchmark will help further algorithm selection research on real-world problems. For future work, the latent feature analysis process could be more formalized. A number of open questions remain, such as what criteria to use to gauge the performance of a new feature during a single iteration of the latent feature analysis process. Further challenges are to determine the number of iterations to perform and what kind of performance/man-hour trade-off exists for each iteration past the first.

References

1. Bortfeldt, A., Forster, F.: A tree search procedure for the container pre-marshalling problem. Eur. J. Oper. Res. **217**(3), 531–540 (2012)
2. Carlo, H., Vis, I., Roodbergen, K.: Storage yard operations in container terminals: literature overview, trends, and research directions. Eur. J. Oper. Res. **235**(2), 412–430 (2014)
3. Caserta, M., Voß, S.: A corridor method-based algorithm for the pre-marshalling problem. In: Giacobini, M., Brabazon, A., Cagnoni, S., Di Caro, G.A., Ekárt, A., Esparcia-Alcázar, A.I., Farooq, M., Fink, A., Machado, P. (eds.) EvoWorkshops 2009. LNCS, vol. 5484, pp. 788–797. Springer, Heidelberg (2009)
4. Expósito-Izquierdo, C., Melián-Batista, B., Moreno-Vega, M.: Pre-marshalling problem: heuristic solution method and instances generator. Expert Syst. Appl. **39**(9), 8337–8349 (2012)
5. Kotthoff, L.: Algorithm selection for combinatorial search problems: a survey. AI Mag. **35**(3), 48–60 (2014)
6. Lee, Y., Hsu, N.: An optimization model for the container pre-marshalling problem. Comput. Oper. Res. **34**(11), 3295–3313 (2007)
7. Lehnfeld, J., Knust, S.: Loading, unloading and premarshalling of stacks in storage areas: survey and classification. Eur. J. Oper. Res. **239**(2), 297–312 (2014)
8. Malitsky, Y., O'Sullivan, B.: Latent features for algorithm selection. In: Symposium on Combinatorial Search (2014)
9. Malitsky, Y., Sabharwal, A., Samulowitz, H., Sellmann, M.: Algorithm portfolios based on cost-sensitive hierarchical clustering. In: IJCAI (2013)
10. Stahlbock, R., Voß, S.: Operations research at container terminals: a literature update. OR Spectrum **30**(1), 1–52 (2008)
11. Tierney, K., Pacino, D., Voß, S.: Solving the pre-marshalling problem to optimality with A* and IDA*. Technical report WP#1401, DS&OR Lab, University of Paderborn (2014)

ADVISER: A Web-Based Algorithm Portfolio Deviser

Mustafa Mısır[✉], Stephanus Daniel Handoko, and Hoong Chuin Lau

School of Information Systems, Singapore Management University,
Singapore, Singapore
{mustafamisir,dhandoko,hclau}@smu.edu.sg

1 Introduction

The basic idea of algorithm portfolio [1] is to create a mixture of diverse algorithms that complement each other's strength so as to solve a diverse set of problem instances. Algorithm portfolios have taken on a new and practical meaning today with the wide availability of multi-core processors: from an enterprise perspective, the interest is to make best use of parallel machines within the organization by running different algorithms simultaneously on different cores to solve a given problem instance. Parallel execution of a portfolio of algorithms as suggested by [2,3] a number of years ago has thus become a practical computing paradigm.

However, algorithm portfolios to date has remained largely a research pursuit among algorithm designers. For algorithm portfolios to become truly usable by enterprises, we need to enable an end-user to easily obtain an algorithm portfolio when he/she provides a raw set of algorithms and has at his/her disposal a K-core machine. This raises an interesting research challenge: given n target algorithms—some parameter-less and some parameterized—as well as a reference set of problem instances (hereinafter will be referred to as the training instances), how do we automatically construct an algorithm portfolio with a maximum size of K such that together the algorithms in the portfolio are capable of solving the problem instances in the reference set effectively when executed in parallel? Our goal is to generate a portfolio of $k \leq K$ algorithms that are sufficiently diverse from each other and altogether solve the reference instances effectively.

Several software libraries or frameworks have been already introduced in the literature. Hydra [4] is a tuning-based portfolio building strategy that allows incorporating existing parameter tuning and algorithm portfolio techniques. ISAC [5] constructs parameter tuning-based portfolios via instance clustering. SufTra [6] employ problem-independent features to perform instance-specific tuning. LLAMA[1] [7] is an algorithm portfolio selection toolkit implemented in R. HyFlex[2] [8] is a hyper-heuristic framework with iterative heuristic selection methods to solve optimisation problems in a problem-independent manner. All of these frameworks to our knowledge are targeted for use by algorithm developers and not for an end-user in mind.

---

[1] https://bitbucket.org/lkotthoff/llama.
[2] http://www.hyflex.org/.

© Springer International Publishing Switzerland 2015
C. Dhaenens et al. (Eds.): LION 9 2015, LNCS 8994, pp. 23–28, 2015.
DOI: 10.1007/978-3-319-19084-6_3

We present in this paper the ADVISER, an automated Algorithm portfolio DeVISER service that combines ideas from algorithm configuration [9], algorithm selection [10], and portfolio generation within a single framework. To maximize usability by an end-user, ADVISER is a web interface system. Providing such a system over the web is inspired from a another web-based platform dedicated to algorithm configuration, called AutoParTune[3] [6].

The remaining of this paper is organized as follows. Section 2 describes the proposed ADVISER in greater detail. Section 3 presents the success of parallel portfolio recommended by ADVISER through a use-case. Section 4 briefly describes our web-based system. Finally, Sect. 5 concludes this work and presents directions for future works.

2   ADVISER

Figure 1 summarizes the workflow of ADVISER through block diagram. Given a mixture of n parameter-less and parameterized target algorithms as well as a set of training instances as the input, ADVISER first performs *algorithm configuration* and *algorithm selection* to generate a portfolio of $k \leq K$ (configured) algorithms as the output. Parameter-less algorithms directly gets included in the initial portfolio, whereas ADVISER performs algorithm configuration (such as applying ParamILS [11], F-Race [12] and Post-Selection [13]) for each parameterized algorithm to determine the best configuration to be included in the initial portfolio. Performance data is then obtained by executing all algorithms in the initial portfolio on the training instances. Performance data of an algorithm when it runs on an instance refers to a number representing solution quality. The algorithms in the initial portfolio are then clustered based on their performance data and the time taken to achieve such performance. A simple k-means clustering is used for this purpose. Finally, a representative algorithm is chosen from each cluster via algorithm selection. In this work, we consider choosing the single best algorithm in each cluster for simplicity, where "single best" refers to the algorithm which performs best among the other algorithms in the cluster on most training instances.

3   Case Study

In the following, we present results with $K = 4$ for two parametric algorithms on the Quadratic Assignment Problem (QAP). The first is a population-based a memetic algorithm (MA), and the second is a single-point simulated annealing-tabu search (SA-TS) [14] hybrid metaheuristic.

Table 1 shows the algorithms and their configuration spaces. Both algorithms have three parameters to be set. MA has two categorial parameters and one integer parameter. These categorical parameters are used to represent which crossover and local search operators are to be used while the integer parameter indicates

---

[3] http://research.larc.smu.edu.sg/autopartune/.

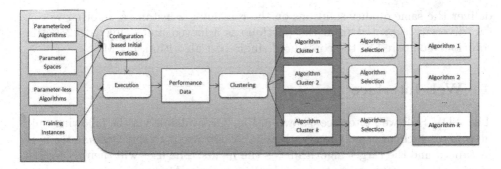

Fig. 1. Workflow of ADVISER

Table 1. Configuration spaces of MA and SA-TS

Method	Type	Name	Range
MA	Categorical	Crossover (C)	[0, 1, 2, 3, 4]
	Continuous	Mutation Rate (M)	[0, 1]
	Categorical	Local Search (L)	[0, 1, 2, 3]
SA-TS	Integer	Initial Temperature (T)	[4000, 6500]
	Continuous	Cooling Factor (C)	[0.85, 0.95]
	Integer	Tabu List Length (L)	[5, 10]

the mutation level. The upper bound values of these categorical parameters refer to the cases where no operator of that type is applied. The two parameters of SA-TS including initial temperature and cooling factor, are for simulated annealing. For the tabu search part, only an integer parameter specifying the tabu list length needs to be set.

Table 2. Portfolio suggested by ADVISER for the QAP using MA and SA-TS

Method	Configuration
MA	-C 4 -M 0.4 -L 2
SA-TS	-T 6500 -C 0.9 -L 5

Table 2 shows the resulting portfolio constructed by using 20 QAP instances. The portfolio is composed of MA and SA-TS with one configuration each $k = 2$) instead of four ($K = 4$) since ADVISER detected that there is no need to run that many configurations in parallel. Since the single best algorithm-configuration pair is selected from each cluster, the overall single best which is the MA configuration, automatically is a part of the portfolio.

The portfolio of MA and SA-TS is then tested on 42 QAP instances. The results revealed that MA with the given configuration finds superior results on 28 instances while SA-TS outperforms MA on 12 instances. Both algorithms deliver the same quality solutions on the remaining 2 instances. In other words, the diversity expected from the portfolio is achieved and delivered 12 better solutions compared to the configured single best algorithm, i.e. MA.

4   Web Interface

The ADVISER web interface, shown in Fig. 2, is available via the following link: http://research.larc.smu.edu.sg/adviser/. A user needs to specify some training instances and the target algorithms as the inputs. The user will then receive an email with the instructions to verify his/her request. After verification, a process involving *algorithm configuration* and *algorithm selection* described in Sect. 2 is started to build the portfolio. Once the process is completed, the user will receive a notification email along with the portfolio of $k \leq K$ (configured) algorithms as the output.

Each target algorithm should be provided in .exe which can be run as follows. After calling an algorithm, it should return a value representing the quality of the resulting solution.

$$\texttt{algorithm.exe} - \texttt{I instance\_file} - \texttt{S seed} \dots \texttt{OtherParameters}$$

Alongside with each parametric algorithm, a parameter space file should be given in the following form. In a parameter space file, for each parameter, there should be a parameter name (e.g. `INITIAL_TEMPERATURE`), a parameter

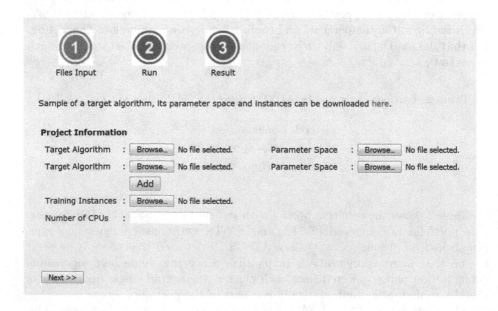

Fig. 2. ADVISER web interface

argument (e.g. "-T"), parameter type information (i: integer, r: continuous, c: categorical) and the range of values (lower and upper bounds for integer and continuous parameters) to be set.

INITIAL_TEMPERATURE" − T" i [4000, 6500]

ADVISER has been developed in Java. In addition to the presented system, a number of existing parameter tuning related components are integrated. Among those components, a Design of Experiments (DOE) [15] implementation is used to reduce the initial parameter configuration space of each parametric algorithm. SufTra [6] is incorporated for determining similar instances in order to fasten a training process by using a small yet representative instance set. Post-Selection [13] is embedded as a parameter tuner.

5 Conclusion

We believe ADVISER is the first step towards unifying the concepts of algorithm configuration, selection, and portfolio generation with an end-user in mind. The workflow of ADVISER shows how the three components play different yet inter-related roles. Moving forward, we hope to incorporate various techniques of algorithm configuration and selection and allow some degrees of customizations. Options to use instance or algorithmic features, whenever available, will also be explored.

References

1. Huberman, B., Lukose, R., Hogg, T.: An economics approach to hard computational problems. Science **275**(3), 51–54 (1997)
2. Gomes, C., Selman, B.: Algorithm portfolios. Artif. Intell. **126**, 43–62 (2001)
3. Petrik, M., Zilberstein, S.: Learning parallel portfolios of algorithms. Ann. Math. Artif. Intell. **48**, 85–106 (2006)
4. Xu, L., Hoos, H., Leyton-Brown, K.: Hydra: Automatically configuring algorithms for portfolio-based selection. In: Proceedings of the Twenty-Fourth AAAI Conference on Artificial Intelligence (AAAI 2010), pp. 210–216 (2010)
5. Kadioglu, S., Malitsky, Y., Sellmann, M., Tierney, K.: ISAC-instance-specific algorithm configuration. In: Proceedings of the 19th European Conference on Artificial Intelligence (ECAI 2010), pp. 751–756 (2010)
6. Lindawati, Yuan, Z., Lau, H.C., Zhu, F.: Automated parameter tuning framework for heterogeneous and large instances: case study in quadratic assignment problem. In: Nicosia, G., Pardalos, P. (eds.) LION 7. LNCS, vol. 7997, pp. 423–437. Springer, Heidelberg (2013)
7. Kotthoff, L.: LLAMA: leveraging learning to automatically manage algorithms. Technical Report (2013). arXiv:1306.1031
8. Ochoa, G., et al.: HyFlex: a benchmark framework for cross-domain heuristic search. In: Hao, J.-K., Middendorf, M. (eds.) EvoCOP 2012. LNCS, vol. 7245, pp. 136–147. Springer, Heidelberg (2012)

9. Hutter, F., Hoos, H., Stutzle, T.: Automatic algorithm configuration based on local search. In: Proceedings of the National Conference on Artificial Intelligence. vol. 22, pp. 1152. AAAI Press, Menlo Park, CA. MIT Press, Cambridge, MA; London (2007)
10. Rice, J.: The algorithm selection problem. Adv. Comput. **15**, 65–118 (1976)
11. Hutter, F., Hoos, H., Leyton-Brown, K., Stützle, T.: ParamILS: an automatic algorithm configuration framework. J. Artif. Intell. Res. **36**, 267–306 (2009)
12. Birattari, M., Yuan, Z., Balaprakash, P., Stützle, T.: F-race and iterated f-race: An overview. Exp. methods Anal. Optim. Algorithms **153**, 311–336 (2010)
13. Yuan, Z., Stützle, T., Montes de Oca, M.A., Lau, H.C., Birattari, M.: An analysis of post-selection in automatic configuration. In: Proceeding of the 15th Annual Conference on Genetic and Evolutionary Computation Conference (GECCO 2013), pp. 1557–1564. ACM (2013)
14. Ng, K., Gunawan, A., Poh, K.: A hybrid algorithm for the quadratic assignment problem. In: Proceedings of International Conference on Scientific Computing, Nevada, USA (2008)
15. Gunawan, A., Lau, H.C., Lindawati.: Fine-Tuning algorithm parameters using the design of experiments approach. In: Coello, C.A.C. (ed.) LION 2011. LNCS, vol. 6683, pp. 278–292. Springer, Heidelberg (2011)

Identifying Best Hyperparameters for Deep Architectures Using Random Forests

Zhen-Zhen Li[✉], Zhuo-Yao Zhong, and Lian-Wen Jin

School of Electronic and Information Engineering,
South China University of Technology,
Wushan Road 381, Guangzhou 510641, China
betty@scut.edu.cn
http://www.hcii-lab.net

Abstract. A major problem in deep learning is identifying appropriate hyperparameter configurations for deep architectures. This issue is important because: (1) inappropriate hyperparameter configurations will lead to mediocre performance; (2) little expert experience is available to make an informed decision. Random search is a straightforward choice for this problem; however, expensive time cost for each test has made numerous trails impractical. The main strategy of our solution has been based on data modeling via random forest, which is used as a tool to analyze data characteristics of performance of deep architectures with respect to hyperparameter variants and to explore underlying interactions of hyperparameters. This is a general method suitable for all types of deep architecture. Our approach is tested by using deep belief network: the error rate reduced from 1.2 % to 0.89 % by merely replacing three hyperparameter values.

1 Introduction

In 2006, G.E. Hinton et al. published two important papers and created a new era of deep learning [1,2]. It has been considered as taking a big step towards true artificial intelligence [3]. In deep learning, the elements are artificial neural cells, which are modeled loosely on the simple properties of biological neurons [4]. Before 2006, neural networks were generally wide and shallow, with fewer than three hidden layers, and error backpropagation methods were the dominant learning algorithms. However, these methods suffer from a severe problem known as error diffusion, which is one of the main factors that prevent networks from going deeper. The problems of these dominant methods were overcome by Hinton's team. Inspired by the deep structure of human brain, they developed a novel unsupervised layerwised learning method to train deep networks. Ever since then, *Deep Architectures* have dominated the fields of artificial intelligence, such as computer vision and natural language comprehension [3].

With networks growing from shallow to deep, architectures become increasingly more complex, such as deep belief networks [1], convolutional neural networks [5], and stacked autoencoders [6]. Variants of these as well as new architectures are turning up one after another [7–9].

© Springer International Publishing Switzerland 2015
C. Dhaenens et al. (Eds.): LION 9 2015, LNCS 8994, pp. 29–42, 2015.
DOI: 10.1007/978-3-319-19084-6_4

All deep architectures are affected by an unavoidable problem of configuring architecture, i.e., identifying an appropriate hyperparameter configuration for a specific architecture. A hyperparameter is an inner architecture parameter, such as depth, the number of neural units in a particular hidden layer, or learning rate. Figure 1 shows a selection of hyperparameters for a deep belief network architecture.

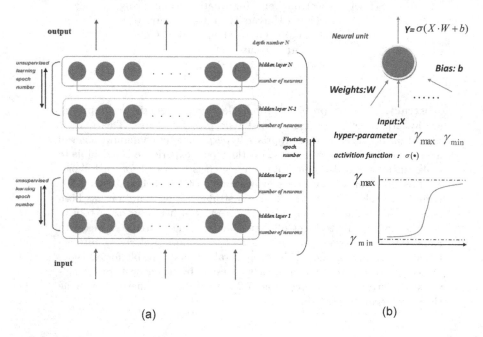

(a)   (b)

Fig. 1. Examples of hyperparameters in a deep belief network: (a) a schematic diagram of a deep belief network, where red words indicate various hyperparameters; (b) a neuron unit, where the activity function has two hyperparameters. Note that only a subset of hyperparameters is shown.

Architecture configuration is an important issue because different configurations significantly affect overall performance, where performance improvement can be modest and sometimes smaller than the performance differences due to architecture (hyper) parameters [10]. Moreover, for some specific tasks, it is important to identify the most suitable architecture for that task.

Furthermore, no expert opinion or other hints are available to facilitate a selection of an appropriate configuration for a new specific task. For example, a researcher who addresses the task of Chinese handwritten character recognition using a successful deep architecture (e.g., GPU-MCCNN (Ref. [11])) may not know how to assign specific values to hyperparameters for that deep architecture.

In this paper, we propose a method using random forests to address the problem of hyperparameter configuration. The remainder of this paper is organized as follows. We pose the issue of hyperparameter search and related work in

Sect. 2. In Sect. 3, we provide a brief introduction to random forests and propose a method for hyperparameters searching based on random forests. In Sect. 4, we discuss the application of our method to deep belief networks and convolutional networks as well. Our concluding remarks are given in Sect. 5.

2 Problem Statement and Related Works

Given a recognition task, a *deep learning architecture* A can be trained using a data set χ by adapting parameters with weights W. After that a set of test data can be sent into A to obtain a generalization performance. However, a deep learning architecture itself needs to be configured by a set of parameters, called hyperparameters Λ.

Suppose that Λ has N elements or dimensions, i.e., $\Lambda = [\lambda_1, \lambda_2, \ldots, \lambda_N]$, $\lambda_i \subseteq s_i, s_i \subset \mathbb{R}, \mathbb{S}^N = s_1 \times s_2 \times \cdots \times s_N$, and thus $\Lambda \subseteq \mathbb{S}^N$ and $\mathbb{S}^N \subset \mathbb{R}^N$. Generalization performance, usually evaluated by generalization error η, is defined as a proportion of falsely recognized test samples in all test data. The performance η is achieved by a configured deep learning architecture : $\eta = A_\Lambda(\chi)$.

Our goal is to find a way to choose Λ to minimize generalization error η, as written in Eq. (1):

$$\Lambda^* = \arg\min_{\Lambda \in \mathbb{S}^N} \mathrm{E}[\eta] = \arg\min_{\Lambda \in \mathbb{S}^N} \mathrm{E}[A_\Lambda(\chi)] \tag{1}$$

Brute Random Search Method is the baseline method for hyperparameter search [13]. Random search can be used to configure a deep architecture by Λ_o, obtained by a random selection function $\Re$ defined on a distribution p : $\Lambda_o = \Re_p(\Lambda)$. With respect to Λ_o, actually compute performance η_{true} using the deep architecture A_{Λ_o}, to get $\eta_{true} = A_{\Lambda_o}(\chi)$. Best configuration is picked out by ranking η_{true}. This search method was applied by Pinto et al. with 2500 trials for each particular task [12]. However, this type of method is very costly, a single trial usually requires hours or even days.

Bayesian Optimization Methods have been carried out to automatically tune configuration hyperparameters. Hutter investigated the problem of automated configuration of algorithms in 2009 in his PhD thesis. He is the first to research this problem within a framework of Bayesian optimization, and proposed sequential model-based optimization (SMBO) approaches for general algorithm configuration [16]. Coincidentally, Snoek et al. stated a Bayesian optimization method, which involved two aspects: an assumed performance distribution and an acquisition function $a(\Lambda)$. The acquisition function $a(\Lambda)$ was maximized under the assumed performance distribution to select the next configuration for testing $\Lambda^{(next)} = \mathrm{argmax}(a(\Lambda))$ [20]. Bergstra went further to use Gaussian Process (GP) and Tree-structured Parzen Estimator (TPE) approaches for hyperparameter optimization of deep architectures, such as deep belief networks (DBNs) [17]. Hutter et al. are putting their effect to construct online projects to automatically search best hyperparameter configurations for various kinds of applications [14,15,18,24]. Ghahramani's team employed a Bayesian nonparametric method to optimize the weights $\boldsymbol{W}$ and hyperparameters Λ simultaneously. They assumed that better configurations of a deep belief network would conform with a stochastic process called the Indian Buffet Process [21]. For a more detailed review of hyper-parameter optimization methods in machine learning, see Ref. [22]. These optimization methods share one common feature: the performance is computed by $\eta = A_{\Lambda_o}(\chi)$, which provides insights into performance situations along the route rather than an overview of the entire hyperparameter space, as shown in Fig. 2(b).

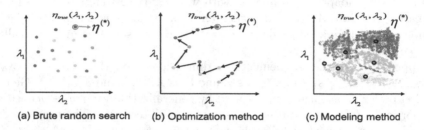

| (a) Brute random search | (b) Optimization method | (c) Modeling method |

Fig. 2. A comparison of three hyper-parameter search methods. From left to right: Brute random search method; optimization method and modeling method.

Our method was inspired by Hutter et al.'s work, in which they used random forests models and a functional ANOVA framework to assess the importance of each hyperparameter [22]. We introduce a *prediction modeling* approach to search hyperparameters within a novel framework. Illustrations of the three types of hyperparameter searching methods are shown in Fig. 2.

Using **a prediction modeling method** to search hyperparameters replaces the actual computation of the deep architecture with a prediction modeling function $\boldsymbol{M}$. Suppose that the difference between the performance predicted by model $\boldsymbol{M}$ and the true performance is sufficiently small, i.e., smaller than a given positive number ζ, then we can define the performance predicted by model $\boldsymbol{M}$ as $\eta_{predicted} \equiv \boldsymbol{M}(\Lambda, \chi) \approx \boldsymbol{A}_\Lambda(\chi)$, Therefor, the best configuration Λ^* can be obtained using Eqs. (2) and (3):

$$\Lambda^* = \arg\min_{\Lambda \in S^N} \mathrm{E}[\eta_{predicted}] \equiv \arg\min_{\Lambda \in S^N}(M(\Lambda, (\chi)) \tag{2}$$

where,

$$\varepsilon = \mathop{\mathrm{E}}_{\Lambda \in S^N}[|M(\Lambda, \chi) - A_\Lambda(\chi)|] < \zeta \tag{3}$$

Random forests is a powerful prediction model that was proposed by Breiman in 2001 [23]. We hypothesize that it is effective to use random forests as a model to predict the overall performance with respect to all possible hyperparameter configurations based on some (very few) performance results obtained using random configurations of an actual deep architecture program. Thus, we need to select these configurations based on the best performance estimates for further validation. We applied this method to deep belief networks and convolutional networks to validate the effectiveness of our method.

3   Random Forest Model

3.1   Random Forests

The ingenious idea of *Random Forests* is derived from decision trees (forests) with randomness injected [23]. A *random forest* F is defined as a predictor that comprises a collection of tree structured predictors $\{f(x, \Theta_k), k = 1, \cdots, l\}$, where x is an input vector and $\{\Theta_k, k = 1, \ldots, l\}$ are independent identically distributed random vectors. Parameter l is the number of trees in the forest. Each tree contributes to a final prediction result mapped by a voting function $V(\cdot)$, as follows:

$$F_l(x) = V(f(x, \Theta)) \tag{4}$$

Random forests are a kind of ensembled trees. It is generally considered to be an accurate and robust method for modeling real world data and it appears to have the ability to capture underlying and tangled structures of data. Breiman has proved this in a mathematical manner, where it is the randomness that helps improve accuracy. See Ref. [23] for a comprehensive introduction to random forests. We selected Random Forests to model M based on three reasons:

1. There can be no prior assumptions about the performance distributions. Assumptions of ideal distributions do not sit well with highly varied and nonlinear real-world tasks
2. They are fast. In our experiments, only 1 min is required to predict over 10,000 performance data points using random forests compared to approximately 1000 min for one performance data point when using the deep architecture program.
3. They have few parameters. We prefer not to introduce further parameters into the model M. As showed Eq. (3), the random forest used by us has only one parameter l: which is the number of trees. Furthermore, random forests converge, [23] i.e., a large number of trees will definitely not lead to overfitting, therefore, we can safely set a relative high number for this parameter.

3.2   Model

Our goal is to identify an appropriate configuration Λ^* with the best true performance $\eta^*_{true} \leq \eta_{true}(\Lambda)$, for all $\Lambda \subseteq \mathbb{S}^N$. In this subsection, we describe the process to achieve this configuration using random forests. Given a deep architecture, a random forest model F is generated with a training set $\gamma^{(train)}$, and the model moves through the entire hyper-parameter space $\mathbb{S}^N$ to obtain an overview of the predicted performance. Select the configuration with the best predicted performance. We can confirm if this is a good estimate by computing the performance using the deep architecture. The overall process comprises three steps. The flow of the three steps in our method is illustrated in Fig. 3.

Step 1. Random Search: This step establishes a training set $\gamma^{(train)} = \{\Lambda_i^{(train)}, \eta_{true,i}^{(train)}\}_{i=1}^K$ to construct a random forest model $\boldsymbol{F}$. The inputs of $\gamma^{(train)}$ are some randomly selected configurations $\Lambda^{(train)}$, which are obtained by a random selection function $\Re_p$ in the same manner as brute random search, but with very few samples. The outputs of $\Lambda^{(train)}$ are the performance levels that are computed by the deep architecture $\eta_{true}^{(train)}$, i.e.,

$$\eta_{true}^{(train)} = \boldsymbol{A}_{\Lambda^{(train)}}(\chi) \tag{5}$$

where

$$\Lambda_i^{(train)} = \Re_{\dot{p}}(\Lambda), i = 1, 2, \cdots, K \tag{6}$$

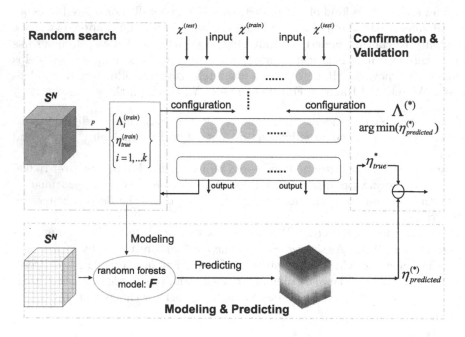

Fig. 3. Diagram flow of the proposed hyperparameters searching method. The proposed method comprises three stages: random search (top left), modeling and predicting (bottom), and confirmation and validation (top right).

The function $\Re_p$ is defined most frequently on a uniform distribution. However, for some special cases where prior knowledge about the configuration is available, Poisson distribution is a good alternative for highlighting specific ranges in the hyperparameter space.

Step 2. Modeling and Predicting: The random forests model $\boldsymbol{F}$ is grown with the training data $\Lambda^{(train)}$. Each tree in $\boldsymbol{F}$ is an unpruned classic classification and regression tree (CART). The simplest random forest is formed using our method by selecting each node at random where only one dimension of the input is split.

After the model F is obtained, performance predictions $\eta_{predicted}(\Lambda)$ are performed on grid points in the entire hyperparameter space. This provides us with a panorama of the performance for all possible configurations. A global illustration of performance with various hyperparameters can easily identify where the best performance level is located. This corresponds to the expected configuration, but it also provides insights into the interaction between performance and the hyperparameters. This is a unique advantage of our modeling method compared with the other two methods. With adequate information for $\eta_{predicted}(\Lambda)$, the best configuration will be achieved according to Eq. (7):

$$\Lambda^{(*)} = \arg\min_{\Lambda \in \mathbb{S}^N}(F(\Lambda_{grid}, l)) \tag{7}$$

where l is the number of trees grown in the random forests model, which is set sufficiently high.

Step 3. Confirmation and Validation: In this step, the deep architecture is configured according to the best estimated configuration Λ^* to produce the true performance result η^*_{true} thereby checking whether Eq. (8) holds.

$$\eta^*_{true} = A_{\Lambda^*}(\chi) \leq \eta_{true}(\Lambda) \tag{8}$$

To validate the predication accuracy of the random forests model, we need to compute the prediction error using Eq. (3).

Theoretically, both Eqs. (3) and (8) need to be held for all $\Lambda \subseteq \mathbb{S}^N$, which is apparently intractable. Fortunately, the ultimate goal is to achieve a better performance. Thus, η^*_{true} is considered to be the best if the performance of a deep architecture $A_{\Lambda^*}(\chi)$ is lower than all others. In the same manner, a validation is performed with a limited number of sample test data, where we use grid selection for samples in the space of the hyperparameters.

4   Experiments

To validate this method, we performed experiments based on an open-source code provided by Hinton, where the classification task was implemented for a standard MNIST database using a three hidden layer deep belief network.[1] The original configuration neural units in each layer was: $\Lambda_o = [500\ \ 500\ \ 2000]$, and the error rate $\eta_{true} = 1.2\%$, We were working on one PC with a $Core^{TM}$ I7-4770 CPU@3.4 GHz, and with a version of Matlab R2013 64 Bit.

4.1   Modeling Results

First, we needed to acquire a training set to construct our random forests. In this case $K = 27$, $\mathbb{S}^N = [100, 3000] \times [100, 3000] \times [100, 3000]$, $\Lambda = [\lambda_1, \lambda_2, \lambda_3]$ and points in $\mathbb{S}^N$ were distributed uniformly. By applying Eqs. (5) and (6), a training set $\{\gamma_i^{(train)}\}_{i=1}^K$ has been obtained, as showed in Fig. 4(a).

---

[1] http://www.cs.toronto.edu/hinton/MatlabForSciencePaper.html.

M. files for constructing random forests are available at googlecode online.[2]
Constructing random forests model F is fairly fast (less than a second).
Grid searching of Λ was performed during this stage: $\eta_{predicted} = F(\Lambda_{grid})$.
An overview of $\eta_{predicted}$ was showed in Fig. 4(b), where the colors present the
performance levels: the darker the blue the better the performance. The performance varied slowly but not monotonously as the hyperparameter configuration
changed and the best performance was achieved in a specific area. Clearly, the
original configuration $\Lambda_o = [500 \quad 500 \quad 2000]$ was not located in the darkest
blue area.

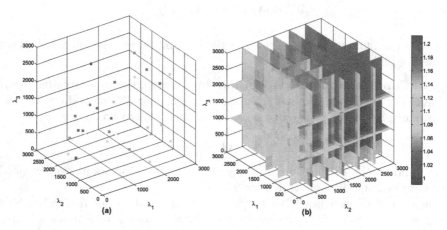

Fig. 4. Illustration of modeling results. Left: training set $\{\gamma_i^{(train)}\}_{i=1}^K$ for constructing
random forests model. Right: Overview of performance $\eta_{predicted}$ with respect to all Λ.
Dots in the three-dimension axis with their position: $\Lambda = [\lambda_1, \lambda_2, \lambda_3]$ indicates values
of hyperparameters and their color stands for value of generalization error η (Color
figure online).

We tested the top configurations Λ^* by running the deep architecture programs. These configurations obtained better performance, as indicated in Table
reftab:1. We tested each Λ^* twice, as denoted by *test 1*, *test 2* in the table
below, the values of different trials varied. Performance results of different
tests vibrated with a maximum range of $\pm 0.05\%$. Clearly, the hyperparameters
$\Lambda^* = [950 \quad 2500 \quad 800]$, obtained the best performance for the MNIST database,
with an error rate of 0.89% (Table 1).

4.2   Modeling Accuracy

We identified a better hyperparameter configuration and better performance
with MNIST, but one question still remained: Did the random forests actually

---

[2] https://randomforest-matlab.googlecode.com/files/Windows-Precompiled-RF_Mex
standalone-v0.02-.zip.

Table 1. Experiments results showing the actual performance of Λ^*.

$\eta_{predicted}(\%)$	Λ^*			$\eta_{true}(test1)(\%)$	$\eta_{true}(test2)(\%)$
0.99	[700	2600	800]	0.97	0.98
0.99	[700	2500	850]	0.98	0.95
0.99	[700	2300	800]	0.93	1.02
0.99	[700	2300	800]	0.97	0.96
0.99	[950	2500	800]	0.89	0.95
0.99	[950	2300	800]	0.96	0.99
0.99	[650	2300	800]	0.97	0.98

model the sample data well? We tested some grid configurations of Hinton's code and compared $\eta_{predicted}$ and η_{true} to check the prediction accuracy of the random forests models. The prediction accuracy levels obtained using our method based on the experimental results are illustrated in Fig. 5. We tested each grid configuration twice. Most of the values of the actual performance levels agreed with the predicted values. Note that we used $K = 27$ samples which was less than adequate for obtaining accurate predictions using random forests. Better predictions shall be obtained with more samples. Figure 5(b) shows obtaining one data of generalization error by deep architecture takes hours which is hundreds of hundred times slower than modeling prediction.

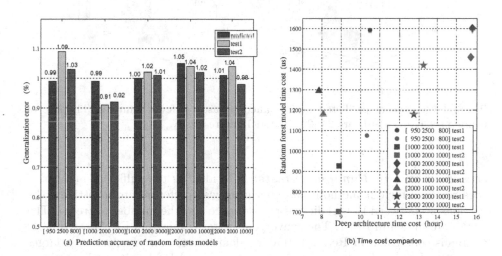

(a) Prediction accuracy of random forests models   (b) Time cost comparion

Fig. 5. Experiments results of prediction accuracy of our method. Left: accuracy comparison between predicted value and true values obtained. Right: time cost comparison between prediction and actually computation throughout the deep architecture. Note that abscissa values are in *Hour* while ordinate values are in *Millisecond*.

4.3   Modeling Parameters

Size of Training Set: K. More than one week was required to collect the $K = 27$ pairs of samples used to construct the random forests models, which is unsatisfactory. Thus, we tested whether this method could still perform well with fewer samples. We decreased the sample size from 24 to 12 with an interval of four to obtain an overview for $\eta_{predicted}(\Lambda)$, as shown in Fig. 6. The experiment results reveal that random forests can still provide some searching clues even with very few samples. The clues progressed from blurred to clear with increasing numbers of training data K. This is important if we are not concerned about the modeling accuracy, because it allows us to model data using random forests and to test some promising candidate configurations at the same time. We can probably achieve an acceptable performance rapidly. However, the use of significantly few samples would be misleading, such as the case shown in Fig. 6(d) where the position of the best performance has moved.

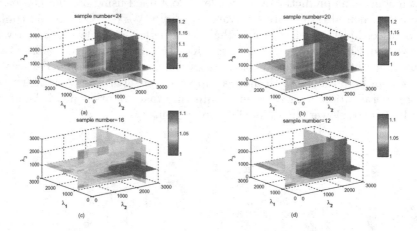

Fig. 6. Overviews of predicted performance in cases of fewer samples. The size of training set K in (a)-(d) are 24, 20, 16 and 12 decreasingly.

Tree Numbers in Random Forests: l. To identify the sufficient number of trees required for our method, an experiment was performed to investigate the relationship between the prediction performance and the number of trees l. The results are shown in Fig. 7. The convergent patterns shown in Fig. 7(a) supports Breimans conclusion in Ref. [23] that random forests converge, where a larger number of l did not cause overfitting with our model. Moreover, the increase in computational time was reasonable with respect to the increasing number of trees. In the case $l = 5000$, this model predicts each data within less than 0.2 s. Thus, a large number of trees maintained the efficiency.

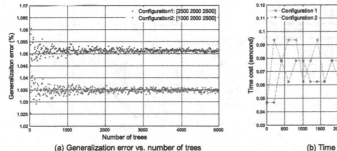

(a) Generalization error vs. number of trees   (b) Time cost vs. number of trees

Fig. 7. Experiment results of investigation on relationship between $\eta_{predicted}(\Lambda)$ and parameter l using two configurations. Left: performance vs tree numbers. Right: time cost of increased tree numbers.

4.4   Importance of Hyperparameters

An additional benefit of random forests is the ability to evaluate the importance of variables, which helps us to understand the interactions among variables. This benefit can be included in our method. Although a space with a small dimension of $N = 3$ was tested using our method, it is applicable to higher dimension hyperparameters and has the potential to uncover the underlying interactions among hyperparameters.

The method used to evaluate the importance of hyperparameters comprises several steps: given a constructed model $\boldsymbol{F}$, superpose some random perturbation on the ith element of the hyperparameters $\lambda_i : \lambda_i \rightarrow \lambda_i(1 + \tau)$, where τ is a perturbation amplitude parameter, and $0 < \tau < 1$. Next, compute the increase in the error compared with the case with no perturbation. A greater increase indicates higher importance. Figure 8 shows the increases in the errors depending on the numbers in the three layers of the deep belief network. In this case, the generalization errors for the variables in layer 2 increased greatly compared with all others, thus the number of neural units in the second layer was more important than that in the other two layers.

4.5   An Extended Experiment for Convolutional Networks

We applied this method to convolutional networks based on an open-source code named Caffe, available at [25], where the classification task was implemented for the same MNIST database using a plain convolutional neural network. Five hyperparameters, $N = 5$, has been chosen, showed in Table 2. The original configuration in Caffe was $\Lambda_o = [20, 5, 50, 5, 500]$, with a baseline generalization accuracy of 1.03 %. We were working on one GeForce GTX TITAN Black 6 GB GPU and a PC with a $Core^{TM}$ I7-4770CPU@3.4 GHz a version of Matlab R2013 64 Bit.

In this case: $\mathbb{S}^N = [100, 500] \times [3, 11] \times [100, 500] \times [3, 11] \times [100, 2000]$. K=127. The best configuratio we obtained was $\Lambda^* = [250, 5, 100, 7, 1600]$ with a generalization error of 0.78 %.

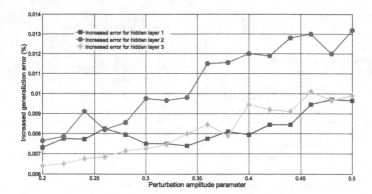

Fig. 8. Experimental evaluation of the importance of hyperparameters. It shows generalization error increased with respect to increased amplitude parameter, and generalization error in layers 2 (indicated by tiny circles) are all higher than other errors.

Table 2. Five hyperparameters chosen for experiments.

Λ	Meanings
λ_1	Number of feature maps in convolutional layer 1
λ_2	Kernel size in convolutional layer 1
λ_3	Number of feature maps in convolutional layer 2
λ_4	Kernel size in convolutional layer 2
λ_5	Number of neurons in fully connected layer

To investigate the minimum size of the training set needed to achieve an acceptable generalization accuracy, we randomly chose several samples from the training set to form a new smaller training set. Then we carried out this prediction and confirmation based on each small training set. Results are showed in Table 3. Runtime per training sample was around 20 to 30 min, and the entire experiment was finished within 80 h. The results of this extended experiment indicate acceptable generalization error could be obtained with few samples, in line with that of former experiments.

Table 3. Generalization error based on different sizes of training sets.

K	$(Test1)(\%)$ Λ^*	$\eta_{predicted}$	η_{true}	$(Test2)(\%)$ Λ^*	$\eta_{predicted}$	η_{true}
40	$[300, 7, 100, 3, 400]$	0.76	1.02	$[300, 5, 150, 3, 1500]$	0.77	0.82
80	$[300, , 5100, 7, 1600]$	0.57	0.88	$[250, 5, 100, 7, 1600]$	0.54	0.87
120	$[250, 5, 100, 7, 1600]$	0.56	0.84	$[250, 5, 100, 7, 1600]$	0.54	0.84

5 Conclusions and Remarks

Far less than enough work has been done on the issue of hyperparameter configuration. To address the problem, we propose a solution based on random forests. This study shows that our method is effective to identify better hyperparameter configurations for deep architectures. The advantages of this method are as follows: (1) the idea is straightforward; (2) the implementation is simple; and (3) it is suitable for all cases.

Although we only tested small dimensions of hyperparameters, this concept is applicable to larger dimensions since random forests have no restrictions on data dimensions. Thus, we will apply this method to search hyperparameter configurations with high-dimensions in our future work.

During our research, we identified one disadvantage of this method where the predictions obtained by the random forests models were highly reliant on the samples used for training, i.e., the range of outputs for the samples defined the range for the predictions. However, random forests models are good at predicting the trends in output data. Thus these models are useful for indicating the values of hyperparameters with the best performance output. This explains why the actual performance was generally better than the predicted results.

Acknowledgments. This research is supported in part by NSFC (Grant No.: 61201348, 61472144), National Science and Technology Support plan (Grant No.:2013B AH65F01 -2013BAH65F04), GDNSF (Grant No.: S2011020000541, S201204 0008016), GDSTP (Grant No.: 2012A010701001), Research Fund for the Doctoral Program of Higher Education of China (Grant No.: 20120172110023).

References

1. Hinton, G.E., Salakhutdinov, R.R.: Reducing the dimensionality of data with neural networks. Science **313**, 504–507 (2006)
2. Hinton, G.E., Osindero, S., Teh, Y.W.: A fast learning algorithm for deep belief nets. Neural comput. **18**, 1527–1554 (2006)
3. Jones, N.: Computer science: the learning machines. Nature **505**, 146–148 (2014)
4. Arbib, M.A.: The elements of brain theory and neural networks, part I: background. In: Arbib, M.A. (ed.) The Handbook of Brain Theory and Neural Networks, pp. 3–7. MIT press, Cambridge (1995)
5. LeCun, Y., Bengio, Y.: Convolutional networks for images, speech, and time-series. In: Arbib, M.A. (ed.) The Handbook of Brain Theory and Neural Networks, pp. 255–257. MIT press, Cambridge (1995)
6. Hinton, G.E., Zemel, R.S.: Autoencoders, minimum description length, and helmholtz free energy. Adv. Neural Inf. Process. Syst. **6**, 3–10 (1994)
7. Lopes, N., Riberio, B.: Towards adaptive learning with improved convergence of deep belief networks on graphic processing units. Pattern Recogn. **47**, 114–127 (2014)
8. Hinton, G.E., Deng, L., Yu, D., et al.: Deep neural networks for acoustic modeling in speech recognition. IEEE signal Process. Mag. **11**, 82–97 (2012)

9. Graves, A., Mohamed, A.R., Hinton, G.E.: Speech recognition with deep recurrent neural networks. In: Proceedings of 2013 IEEE International Conference on Acoustics, Speech and Signal Processing, pp. 6645–6649. IEEE Press, New York (2013)

10. Saxe, A.M., Koh, P.W., Chen, Z.: On random weights and unsupervised feature learning. In: 2011 International Conference on Machine Learning, pp. 1089–1096. IEEE Press, New York (2011)

11. Ciresan, D., Meier, U., Schmidhuber, J.: Multi-column deep neural networks for image classifcation. In: 2012 IEEE Conference on Computer Vision and Pattern Recognition, pp. 3642–3649. IEEE Press, New York (2012)

12. Pinto, N., Cox, D., DiCarlo, J.: A high-throughput screening approach to discovering good forms of biologically inspired visual representation. PLoS Comput. Biol. **5**, 1–12 (2009)

13. Bergstra, J., Bengio, Y.: Random search for hyper-parameter optimization. J. Mach. Learn. **13**, 281–305 (2012)

14. Hutter, F., Lopez-Ibanez, M., Fawcett, C., Lindauer, M., Hoos, H., Leyton-Brown, K., Stutzle, T.: AClib: a benchmark library for algorithm configuration. In: Pardalos, P.M., Resende, M.G.C., Vogiatzis, C., Walteros, J.L. (eds.) Lion8. LNCS, vol. 8426, pp. 36–40. Springer, Heidelberg (2014)

15. Hutter, F., Hoos, H.H., Leyton-Brown, K.: Sequential model-based optimization for general algorithm configuration. In: Coello, C.A. (ed.) LION 5. LNCS, vol. 6683, pp. 507–523. Springer, Heidelberg (2011)

16. Hutter, F.: Automated configuration of algorithm for solving hard computational problems. Ph.D. thesis, Department of computer science, University of British Columbia (2009)

17. Bergstra, J., Bardenet, R., Bengio, Y., Kegl, B.: Algorithms for hyperparameter optimization. Adv. Neural Inf. Process. Syst. **24**, 2546–2554 (2011)

18. Bergstra, J., Yamins, D., Cox, D.D.: Making a science of model search: hyperparameter optimization in hundreds of dimensions for vision architectures. In: 30th International Conference on the Machine Learning (2013)

19. Thornton, C., Hutter, F., Hoos, H., Leyton-Brown, K.: Auto-WEKA: combined selection and hyperparameter optimization of clasification algorithms. In: 19th ACM SIGKDD International Conference on Knowledge Discovery and Data Mining (2013)

20. Snoek, J., Larochelle, H., Adams, R.P.: Practical bayesian optimization of machine learning algorithms. Advances in neural information processing system 4 (2012)

21. Adams, R.P., Wallach, H.M., Ghahramani, Z.: Learning the structure of deepsparse graphical models. J. Mach. Learn. **9**, 1–8 (2010)

22. Hutter, F., Hoos, H., Leyton-Brown, K.: An efficient approach for assessing hyperparameter importance. In: Proceedings of the 2014 International Conference on Machine Learning (2014)

23. Breiman, L.: Random forests. Mach. Learn. **45**, 5–32 (2001)

24. Automated algorithm configuration project. http://www.cs.ubc.ca/labs/betaProjects/AAC/index.html

25. Caffe. https://github.com/BVLV/caffe

Programming by Optimisation Meets Parameterised Algorithmics: A Case Study for Cluster Editing

Sepp Hartung[1](✉) and Holger H. Hoos[2]

[1] Institut Für Softwaretechnik und Theoretische Informatik,
TU Berlin, Berlin, Germany
sepp.hartung@tu-berlin.de
[2] Department of Computer Science, University of British Columbia,
Vancouver, Canada
hoos@cs.ubc.ca

Abstract. Inspired by methods and theoretical results from parameterised algorithmics, we improve the state of the art in solving CLUSTER EDITING, a prominent NP-hard clustering problem with applications in computational biology and beyond. In particular, we demonstrate that an extension of a certain preprocessing algorithm, called the $(k+1)$-data reduction rule in parameterised algorithmics, embedded in a sophisticated branch-&-bound algorithm, improves over the performance of existing algorithms based on Integer Linear Programming (ILP) and branch-&-bound. Furthermore, our version of the $(k+1)$-rule outperforms the theoretically most effective preprocessing algorithm, which yields a $2k$-vertex kernel. Notably, this $2k$-vertex kernel is analysed empirically for the first time here. Our new algorithm was developed by integrating Programming by Optimisation into the classical algorithm engineering cycle – an approach which we expect to be successful in many other contexts.

1 Introduction

CLUSTER EDITING is a prominent NP-hard combinatorial problem with important applications in computational biology, e.g. to cluster proteins or genes (see the recent survey by Böcker and Baumbach [6]). In machine learning and data mining, weighted variants of CLUSTER EDITING are known as CORRELATION CLUSTERING [4] and have been the subject of several recent studies (see, e.g., [8,12]). Here, we study the unweighted variant of the problem, with the goal of improving the state of the art in empirically solving it. Formally, as a decision problem it reads as follows:

Sepp Hartung—Major parts of this work were done during a research visit of SH at the University of British Columbia in Vancouver (Canada), supported by a "DFG Forschungsstipendium" (HA 7296/1-1).

C. Dhaenens et al. (Eds.): LION 9 2015, LNCS 8994, pp. 43–58, 2015.
DOI: 10.1007/978-3-319-19084-6_5

CLUSTER EDITING
Input: An undirected graph $G = (V, E)$ and a positive integer $k \in \mathbb{N}$.
Question: Is there a set of at most k edge insertions and deletions that transform G into a cluster graph, that is, a graph in which each connected component is a complete graph?

CLUSTER EDITING corresponds to the basic clustering setting in which pairwise similarities between the entities represented by the vertices in G are expressed by unweighted edges, and the objective is to find a pure clustering, in the form of a cluster graph, by modifying as few pairwise similarities as possible, i.e., by removing or adding a minimal number of edges. Notably, this clustering task requires neither the number of clusters to be specified, nor their sizes to be bounded.

Related Work. The CLUSTER EDITING problem is known to be APX-hard [10] but can be approximated in polynomial time within a factor of 2.5 [25]. Furthermore, various efficient implementations of exact and heuristic solvers have been proposed and experimentally evaluated (see the references in [6]). These methods can be divided into exact algorithms, which are guaranteed to find optimal solutions to any instance of CLUSTER EDITING, given sufficient time, and inexact algorithms, which provide no such guarantees, but can be very efficient in practice. State-of-the-art exact CLUSTER EDITING algorithms are based on integer linear programming (ILP) or specialised branch-&-bound methods (i.e., search tree) [6,7]. Theoretically, the currently best *fixed-parameter* algorithm runs in $\mathcal{O}(1.62^k + |G|)$ time and it is based on a sophisticated search tree method [5].

Our work on practical exact algorithms for CLUSTER EDITING makes use of so-called *data reduction rules* [11,16,17,19] – preprocessing techniques from parameterised algorithmics that are applied to a given instance with the goal of shrinking it before attempting to solve it. Furthermore, when solving the problem by a branch-&-bound search, these data reduction rules can be "interleaved" [23], meaning that they can be again applied within each recursive step. If after the exhaustive application of data reduction rules the size of the remaining instance can be guaranteed to respect certain upper bounds, those instances are called *problem kernels* [14,23]. Starting with an $\mathcal{O}(k^2)$-vertex problem kernel [17], the best state-of-the-art kernel for CLUSTER EDITING contains at most $2k$-vertices [11].

Our Contribution. Starting from a search tree procedure originally developed for a more general problem called M-HIERARCHICAL TREE CLUSTERING (M-Tree Clustering) [20], and making heavy use of data reduction rules, we developed a competitive state-of-the-art exact solver for (unweighted) CLUSTER EDITING[1].

To achieve this goal, and to study the practical utility of data reduction rules for CLUSTER EDITING, we employed multiple rounds of an algorithm engineering cycle [24] that made use of the *Programming by Optimisation (PbO)*

---

[1] Notably, our implementation is still able to solve M-Tree Clustering. However, here our focus is on improving over state-of-the-art exact solvers for CLUSTER EDITING.

paradigm [21]. In a nutshell, PbO is based on the idea to consider and expose design choices during algorithm development and implementation, and to use automated methods to make those choices in a way that optimises empirical performance for given use contexts, characterised by representative sets of input data.

We show that, using a clever implementation of a well-known (from a theoretical point of view, out-dated) reduction rule, called $(k+1)$-Rule, we can achieve improvements over existing state-of-the-art exact solvers for CLUSTER EDITING on challenging real-world and synthetic instances. For example, for the synthetic data with a timeout of 300 s our so-called Hier solver times out only on 8 % of the 1476 instances, while the best previously known solver has a rate of 22 %. Furthermore, we demonstrate that on the hardest instances the $(k+1)$-Rule dominates on aggregate all other data reduction rules we considered, and that using the best known data reduction rules [9,11] (yielding the best known kernel of size $2k$) does not yield further significant improvements.

Achieving these results involved multiple rounds of optimizing the implementation of the $(k+1)$-Rule as well as the use of automated algorithm configuration tools in conjunction with a new method for selecting the sets of training instances used in this context. It is based on the coefficient of variation of the running time observed in preliminary runs, which we developed in the context of this work, but believe to be more broadly useful.

Overall, our work demonstrates that the adoption of the Programming by Optimisation paradigm, and in particular, the use of automated algorithm configuration methods can substantially enhance the "classical" algorithm engineering cycle and aid substantially in developing state-of-the-art solvers for hard combinatorial problems, such as CLUSTER EDITING. We note that a similar approach has been taken by de Oca et al. [13] to optimise a particle swarm optimization algorithm.

2   Preliminaries

We use standard graph-theoretic notations. All studied graphs are undirected and simple without self-loops and multi-edges. For a given graph $G = (V, E)$ with vertex set V and edge set E, a set consisting of edge deletions and additions over V is called an *edge modification set*. For a given CLUSTER EDITING-instance (G, k) an edge modification set S over V is called a *solution*, if it is of size at most k and transforms G into a cluster graph, which we denote by $G \otimes S$. For convenience, if two vertices $\{u, v\}$ are not adjacent, we call $\{u, v\}$ a *non-edge*.

It is well-known that a graph $G = (V, E)$ is a cluster graph if, and only if, it is *conflict-free*, where three vertices $\{u, v, w\} \subseteq V$ form a *conflict* if $\{u, v\}, \{v, w\} \in E$, but $\{u, w\} \notin E$ – in other words, a conflict consists of three vertices with two edges and one non-edge. We denote by $\mathcal{C}(G)$ the set of all conflicts of G. Branching into either deleting one of the two edges in a conflict or adding the missing edge is a straight-forward search tree-strategy that results in a $\mathcal{O}(3^k + |V|^3)$ algorithm to decide an instance $((V, E), k)$ [17]. This algorithm can be generalised to M-Tree Clustering [20] and is the basic algorithm implemented in our Hier solver.

Parametrised Algorithmics. Since our algorithm makes use of data reduction rules known from parametrised algorithmics, and CLUSTER EDITING has been intensely studied in this context, we briefly review some concepts from this research area (see [14,23]). A problem is *fixed-parameter tractable* (FPT) with respect to a parameter k if there is a computable function f such that any instance (I, k), consisting of the "classical" problem instance I and parameter k, can be exactly solved in $f(k) \cdot |I|^{O(1)}$ time. In this work k always refers to the "standard" parameter solution size.

The term *problem kernel* formalizes the notion of effective and (provably) efficient preprocessing. A *kernelization algorithm* reduces any given instance (I, k) in polynomial time to an equivalent instance (I', k') with $|I'| \leq g(k)$ and $k' \leq g(k)$ for some computable function g. Here, equivalent means that (I, k) is a yes-instance if, and only if, (I', k') is a yes-instance. The instance (I', k') is called *problem kernel* of size g. For example, the smallest problem kernel for CLUSTER EDITING consists of at most $2k$ vertices [11]. A common way to derive a problem kernel is by the exhaustive application of *data reduction rules*. A data reduction rule is a polynomial-time algorithm which computes for each instance (I, k) an equivalent *reduced* instance (I', k') and it has been applied *exhaustively* if applying it once more would not change the instance.

PbO and Automated Algorithm Configuration. Programming by Optimisation (PbO) is a software design approach that emphasises and exploits choices encountered at all levels of design, ranging from high-level algorithmic choices to implementation details [21]. PbO makes use of powerful machine learning and optimisation techniques to find instantiations of these choices that achieve high performance in a given application situation, where application situations are characterised by representative sets of input data, here: instances of the CLUSTER EDITING problem. In the simplest case, all design choices are exposed as algorithm parameters and then optimised for a given set of training instances using an automated algorithm configurator. In this work, we use SMAC [22] (in version 2.08.00), one of the best-performing general-purpose algorithm configurators currently available. SMAC is based on sequential model-based optimisation, a technique that iteratively builds a model relating parameter settings to empirical performance of a given (implementation of a) target algorithm $\mathcal{A}$, here: our CLUSTER EDITING solver Hier, and uses this model to select promising algorithm parameter configurations to be evaluated by running $\mathcal{A}$ on training instances.

By following a PbO-based approach, using algorithm configurators such as SMAC, algorithm designers and implementers no longer have to make *ad-hoc* decisions about heuristic mechanisms or settings of certain parameters. Furthermore, to adapt a target algorithm to a different application context, it is sufficient to re-run the algorithm configurator, using a set of training instances from the new context.

We note that the algorithm parameters considered in the context of automated configuration are different from the problem instance features considered in parameterised algorithmics, where these features are also called parameters.

3 Our Algorithm

Basic Algorithm Design. The algorithm framework underlying our Hier solver is outlined in Algorithm 1; the actual implementation has several refinements of this three-step approach, and many of them are exposed as algorithm parameters (in total: 49) to be automatically configured using SMAC.

Given a graph G as input for the optimization variant of CLUSTER EDITING, we maintain a lower and upper bound, called k_{LB} and k_{UB}, on the size of an optimal solution for G. As long as lower and upper bound are not equal, we call our branch-&-bound search procedure (Line 8) to decide whether (G, k_{LB}) is a yes-instance. At the heart of our solver lies the following recursive procedure for solving the (decision variant) CLUSTER EDITING-instance (G, k_{LB}). First, a set of data reduction rules is applied to the given instance (see Line 2 in decisionSolver). Next, a lower bound is computed on the size of a minimum solution using our LP-based lower bound algorithm. If this lower bound is larger than k, then we abort this branch, otherwise we proceed with the search. Afterwards, if there are still conflicts in the resulting graph, one of these is chosen,

---

ALGORITHM 1. Pseudo code of our Hier solver.

```
1  Algorithm Hier ()
   │  Input: Graph G.
   │  Output: The size k_OPT of a minimum edge modification set S such
   │          that G ⊗ S is a cluster graph.
3  │  Compute a lower bound k_LB ≤ k_OPT
5  │  Compute an upper bound k_OPT ≤ k_UB
7  │  while k_LB < k_UB do
8  │    │  if decisionSolver(G, k_LB) = YES then
9  │    │    │  return k_LB
10 │    │  else
11 │    │    │  increase k_LB  //details are subject to two algorithm parameters
12 │    │  end
13 │  end

1  Procedure decisionSolver (G, k)
   │  Input: Graph G and integer k.
   │  Output: YES/NO whether there is a size-at-most-k edge modification set
   │          for G.
2  │  (G, k) ← Apply data reduction rules to (G, k)
3  │  if LP-based lower bound on modification cost for G > k then return NO
4  │  {u, v, w} ← a conflict in G
5  │  if {u, v} is unmarked ∧ decisionSolver (G − uv, k − 1) = YES then
   │     return YES
6  │  else Mark edge {u, v} unmodifiable
7  │  if {v, w} is unmarked ∧ decisionSolver (G − vw, k − 1) = YES then
   │     return YES
8  │  else Mark edge {v, w} unmodifiable
9  │  if decisionSolver(G + uw, k − 1) = YES then return YES
10 │  else return NO
```

---

say $\{u, v, w\}$. Then the algorithm branches into the three possibilities to resolve the conflict: Delete the edge $\{u, v\}$, delete $\{v, w\}$, or add the edge $\{u, w\}$.

On top of this, if the branch of deleting edge $\{u, v\}$ has been completely explored without having found any solution, then in all other branches this edge can be marked as unmodifiable (the branch for deleting $\{v, w\}$ is handled analogously). Moreover, in all three recursive steps, the (non-)edge that was introduced to solve the conflict $\{u, v, w\}$ gets marked as unmodifiable. Furthermore, the choice of the conflict to resolve prefers conflicts involving unmodifiable (non-) edges, since this reduces the number of recursive calls by one or, in the best case, completely determines how to resolve the conflict. Combining this with solving "isolated" conflicts is known to reduce the (theoretical) time complexity from $\mathcal{O}(3^k + |V|^3)$ to $\mathcal{O}(2.27^k + |V|^3)$ [17]. Our empirical investigation revealed that this improvement is also effective in practice.

Data Reduction Rules. In total, we considered seven data reduction rules and implemented them such that each of them can be individually enabled or disabled via an algorithm parameter. We first describe three rather simple data reduction rules. First, there is a rule (Rule 2 in Hier) that deletes all vertices not involved in any conflict (see [20] for the correctness). A second simple rule (Rule 4 in Hier) checks all sets of three vertices forming a triangle, and in case two of the edges between them are already marked as unmodifiable it also marks the third one (deleting this edge would result in a unresolvable conflict). The last simple rule (Rule 6 in Hier) checks each conflict and resolves it in case of there is only one way to do this as a result of already marked (non-)edges.

We describe the remaining "sophisticated" data reduction rules in chronological order of their invention. Each of it either directly yields or is the main data reduction rule of a problem kernel.

$(k+ 1)$**-Rule:** Gramm et al. [17] provide a problem kernel of size $\mathcal{O}(k^3)$ that can be computed in $\mathcal{O}(n^3)$ time. More specifically, the kernel consists of at most $2k^2 + k$ vertices and at most $2k^3 + k^2$ edges. At the heart of this kernel lies the following so-called $(k + 1)$-Rule (Rule 1 in [17]):

> Given a CLUSTER EDITING-instance (G, k), if there are two vertices $\{u, v\}$ in G that are contained in at least $k + 1$ conflicts in $\mathcal{C}(G)$, then in case of $\{u, v\} \notin E$ add the edge $\{u, v\}$ and otherwise delete the edge $\{u, v\}$.

The $(k + 1)$-Rule is correct, since a solution that is not changing the (non-)edge $\{u, v\}$ has to resolve all the $\geq k+1$ conflicts containing $\{u, v\}$ by pairwise disjoint edge modifications; however, this cannot be afforded with a "budget" of k.

We heuristically improved the effectiveness of the $(k + 1)$-Rule by the following considerations: For a graph G denote by $\mathcal{C}(\{u, v\}) \subseteq \mathcal{C}(G)$ all conflicts containing $\{u, v\}$. If $|\mathcal{C}(\{u, v\})| \geq k + 1$, then the $(k + 1)$-Rule is applicable. Otherwise, let $\mathcal{C}_{\overline{u,v}}(G) \subseteq \mathcal{C}(G) \setminus \mathcal{C}(\{u, v\})$ be all conflicts that are (non-)edge-disjoint with $\mathcal{C}(\{u, v\})$, meaning that any pair of vertices occurring in a conflict in $\mathcal{C}_{\overline{u,v}}(G)$ does not occur in a conflict in $\mathcal{C}(\{u, v\})$. By the same argument as for the correctness of the $(k + 1)$-Rule, it follows that if any lower bound on the number of edge modifications needed to solve all conflicts in $\mathcal{C}_{\overline{u,v}}(G)$ plus $|\mathcal{C}(\{u,v\})|$ exceeds k, then the (non-)edge $\{u,v\}$ needs to be changed (all these conflicts require pairwise disjoint edge modifications). We use our heuristic algorithm described below to compute a (heuristic) lower bound on the modification cost of $\mathcal{C}_{\overline{u,v}}(G)$.

As our experimental analysis reveals, the heuristically improved version of the $(k+1)$-Rule is the most successful one in Hier. Its operational details are configurable by three algorithm parameters (not counting the parameters to enable/disable it), and we implemented two different versions of it (Rule 0 & 1 in Hier). These versions differ in their "laziness": Often it is too time consuming to exhaustively apply the $(k+1)$-Rule, as any edge modification requires an update on the lower bound for $\mathcal{C}_{\overline{u,v}}(G)$. In addition to various heuristic techniques, we implemented a priority queue that (heuristically) delivers the (non-)edges that are most likely reducible by the $(k+1)$-Rule.

$\mathcal{O}(M \cdot k)$-vertex Kernel: There is a generalisation of CLUSTER EDITING called M-Tree Clustering, in which the input data is clustered on M levels [2]. The parametrised complexity of M-Tree Clusteringhas been first examined by Guo et al. [20], who introduced a $(2k \cdot (M+2))$-vertex kernel which is computable in $\mathcal{O}(M \cdot n^3)$ time. This kernel basically corresponds to a careful and level-wise application of the $4k$-vertex kernel by Guo [19] for CLUSTER EDITING. The underlying technique is based on so-called *critical cliques* – complete subgraphs that have the same neighbourhood outside and never get split in an optimal CLUSTER EDITING-solution. We refer to Guo et al. [20] for a detailed description of the implemented $\mathcal{O}(M \cdot k)$ kernel (Rule 3 in Hier).

$2k$-vertex Kernel: The state-of-the-art problem kernel for CLUSTER EDITING has at most $2k$-vertices and is based on so called *edge-cuts* [11]. In a nutshell, for the closed neighbourhood N_v of each vertex v, the cost of completing it to a complete graph (adding all missing edges into N_v) and cutting it out of the graph (removing all edges between a vertex in N_v and a vertex not in N_v) is accumulated. If this cost is less than the size of N_v, then N_v is completed and cut out. This kernel has been generalised to M-Tree Clusteringwithout any increase in the worst-case asymptotic size bound [9]. We implemented this kernel in its generalized form for M-Tree Clustering(Rule 7), but omitted a rule that basically merges N_v after it has been completed and cut out of the graph; although this rule is necessary for the bound on the kernel size, as it removes vertices from the graph, Hier will not deal with these vertices again and thus simply ignores them.

Lower- and Upper-Bound Computation. We implemented two lower-bound algorithms (LP-based and heuristic) and one upper-bound heuristic. Our preliminary experiments revealed that high-quality lower- and upper-bound algorithms are a key ingredient for obtaining strong performance in our CLUSTER EDITING solver. In total, these algorithms expose twenty-two algorithm parameters that influence their application and behaviour.

LP-based Lower Bound Computation: We implemented the ILP-formulation for M-Tree Clusteringproposed by Ailon and Charikar [3], which corresponds to the "classical ILP-formulation" for CLUSTER EDITING in case of $M = 1$ [6]. The formulation involves a 0/1-variable for each vertex of the graph and a cubic number of constraints. Our LP-based lower bound algorithm simply solves the relaxed LP-formulation where all variables take real values from the interval [0, 1], which provides a lower bound on any ILP-solution. If after having solved the relaxed LP-formulation the time limit (set via an algorithm parameter) has not been exceeded, then we require a small fraction of the variables to be 0/1-integers and try to solve the resulting mixed-integer-linear-program (MIP) again. Surprisingly, to obtain optimal integer solutions, in many cases, one only needs to require a small fraction of the variables ($\approx$10%) to be 0/1-integers. Using this mechanism, we are frequently able to provide optimal bounds on the solution size, especially for small instances where the LP-formulation can be solved quickly.

Heuristic Lower Bound Computation: Given a set of conflicts C (not necessarily all, as in the application of the $(k + 1)$-Rule), our second lower bound algorithm heuristically determines a maximum-size set of independent conflicts based on the following observation. Consider the conflict graph for C, which contains a vertex for each conflict in C and an edge between two conflicts if they have a (non-)edge in common. A subset of vertices is an *independent set* if there is no edge between any two vertices in it. Similarly to the correctness argument for the $(k + 1)$-Rule, it follows that the size of an independent set in the conflict graph of C is a lower bound on the number of edge modifications needed to resolve all conflicts in C. Computing a maximum-size independent set in a graph is a classical NP-hard problem, and we thus implemented the commonly known "small-degree heuristic" to solve it: As long as the graph is not empty, choose one of the vertices with smallest degree, put it into the independent set and delete it and all its neighbours. We apply this small-degree heuristic multiple times with small (random) perturbations on the order in which the vertices get chosen (not necessarily a smallest degree vertex is chosen, but only one with small degree). In total, there are four algorithm parameters which determine the precise way in which the order is perturbed and how often the heuristic is applied.

Heuristic Upper Bound Computation: Given a graph G and the set of conflicts $C(G)$ in G, we use the following heuristic algorithm to compute an upper bound on the minimum modification cost for G. The *score* of an (non-)edge is the number of its occurrences in $C(G)$, and the score of a conflict is simply the maximum over the scores of all its modifiable (non-)edges. The algorithm proceeds as follows: While there are still conflicts in $C(G)$, choose a conflict with highest *score* in $C(G)$ and among the modifiable (non-)edges change (delete if it is an edge otherwise add) one of those with highest score. Furthermore, mark the corresponding (non-)edge as unmodifiable. Before solving the next conflict, we exhaustively apply Rule 6, which solves all conflicts for which two of its (non)-edges have been marked as unmodifiable.

In our implementation, the score of an edge is randomly perturbed, and thus we run the algorithm described above multiple times and return the minimum over all these runs. The time limit for this computation as well as the maximum number of rounds are exposed as algorithm parameters.

4 Experimental Results

Algorithms and Datasets. We compare our solver, Hier, with two other exact solvers for (weighted) CLUSTER EDITING: The Peace solver by Böcker et al. [7] applies a sophisticated branching strategy based on merging edges, which yields a search tree of size at most $\mathcal{O}(1.82^k)$. This search tree algorithm is further enhanced by a set of data reduction rules that are applied in advance and during branching. Böcker et al. [7] compared the empirical performance of Peace against that obtained by solving an ILP-formulation (due to Grötschel and Wakabayashi [18]) using the commercial CPLEX solver 9.03. In August 2013, a new version 2.0 of this ILP-based approach has become available, which now directly combines data reduction rules with an ILP-formulation. We refer to this solver as Yoshiko[2] (developed by G. Klau and E. Laude, VU University Amsterdam).

We compare our algorithm to Peace and Yoshiko (version 2.0) on the synthetic and biological datasets provided by Böcker et al. [7]. The (unweighted) synthetic dataset consists of 1475 instances that are generated from randomly disturbed cluster graphs with 30–1040 vertices (median: 540) and densities of 11–99 %. These instances have been observed to be substantially harder than the biological dataset, which consists of 3964 instances that have been obtained from a protein similarity network.[3] The number of vertices in the biological dataset range from 3 to 3387, but the median is only 10, and thus, most instances are rather easy. Since the biological instances are weighted CLUSTER EDITING-instances and Hier is restricted to unweighted CLUSTER EDITING (as a result of its ability to solve the general M-Tree Clusteringproblem), we transformed them into unweighted instances by setting edges only for the $c\%$ of the pairs with highest weight (corresponds to highest similarity). Using three different values of $c = 33, 50$, and 66, we obtained 11 889 biological instances in total.

Implementation and Execution Environment. All our experiments were run on an Intel Xeon E5-1620 3.6 Ghz machine (4 Cores + Hyper-Threading) with 64 GB memory under the Debian GNU/Linux 6.0 operating system, with a time limit of 300 s per problem instance. Our Hier solver was implemented in Java and is run under the OpenJDK runtime environment in version 1.7.0_25 with 8 GB heap space. We use the commercial Gurobi MIP solver in version 5.62 to compute our LP-based lower bound [1]. The source code along with the scenario file used for configuration with SMAC is freely available.[4] For Yoshiko, we used the binary provided by the authors, and we compiled Peace using the provided Make file with gcc, version 4.7.2. Our Hier solver sets up parallel threads for computing

---

[2] http://www.mi.fu-berlin.de/w/LiSA/YoshikoCharles.

[3] We removed the largest instance with 8836 vertices from the dataset. It is more than two times larger than the second largest instance and could not be solved.

[4] http://fpt.akt.tu-berlin.de/cluEdit/.

the lower and upper bounds, but otherwise runs in only one thread. Peace uses a single thread, while Yoshiko makes extensive use of the parallel processing capabilities of the CPU (according to its output, Yoshiko sets up 8 threads). All running times were measured in wall-clock seconds.

Results for Synthetic Dataset. Table 1 and the scatter plots in Fig. 1 provide an overview of our experimental findings on the synthetic dataset. Hier-Opt$_S$ refers to Hier with the best configuration found by SMAC. Before discussing how we obtained this configuration we first discuss the performance of Hier's default configuration (always referred to simply as Hier) to that of Yoshiko and Peace.

As can be seen from these results, Hier clearly outperforms both Yoshiko and Peace (see columns 4–6 in Table 1). Furthermore, it seems that search-tree based algorithms, such as Peace and Hier, generally perform better than the ILP-based Yoshiko-solver. We suspect that this is mainly due to the instance sizes which are considerably larger than for the biological dataset. As can be seen in the top left scatter plot in Fig. 1, Peace is on average faster than Hier for instances solvable within ≤25 s by both solvers. However, the higher the time required by both solvers, the more Hier starts to dominate on average, and, of course, its overall success is heavily due to the smaller timeout-rate of 7.8 % (Peace: 21 %).

The bottom two scatter plots in Fig. 1 show that Hier-Opt$_S$ clearly dominates Yoshiko and Peace on most instances (also on instances solvable in a couple of seconds). We obtained Hier-Opt$_S$ by using SMAC; however, not by a single "shot", but rather by using SMAC repeatedly within an algorithm engineering cycle. This means that we performed multiple rounds of tweaking the implementation, testing it, and analysing it on our experimental data. Therein, in each round we performed multiple SMAC runs in order to analyse not only the default configuration of our current solver but also its optimized variant. We then used an ablation analysis [15] to further pinpoint the crucial parameter adjustments made by SMAC. This was important, because it revealed which algorithm parameters – and thus, which parts of the algorithm – are particularly relevant for the overall performance of our solver. For example, we learned that by allowing more time for the application of our original implementation of the $(k+1)$-Rule, we can reduce the number of timeouts. We thus spent serious effort on tweaking the implementation of the $(k+1)$-Rule and making more of its details accessible

Table 1. Running time (wall time in s) comparison of four different solvers on the synthetic dataset (performance on disjoint training and test instances).

| | Training (#=196) | | Test (#=953) | | | | |
	Hier	Hier-Opt$_S$	Peace	Yoshiko	Hier	Hier-Opt$_S$	Hier-Opt$_S$-Rule7
Par-10	187.4	127.2	662.2	904.1	255.2	252.7	265.8
Mean	49.7	30.7	92.8	142.0	45.6	40.2	42.0
Median	28.4	7.5	26.4	96.6	18.6	10.1	9.6
% Timeouts	5.1 %	3.6 %	21.1 %	28.2 %	7.8 %	7.9 %	8.3 %

to get optimized by SMAC. Of course, if one parameter setting clearly had been identified by SMAC to be beneficial, then we adjusted the default values of this parameter for the next round. This is the main reason why the final default configuration of Hier is already quite competitive (for example, we started with a version of Hier that had more than 30 % timeouts on the synthetic data).

In each round of the algorithm engineering cycle, we performed at least five independent SMAC runs, each with a wall-clock time limit of 36 hours and a cut-off time of 300 s per run of Hier. In each SMAC run about 160–200 configurations were evaluated and about 1200–1500 runs of Hier have been performed. We not only started SMAC from the default configuration, but also with the best configuration that had been obtained in previous runs (we obtained our final best configuration from one of these runs). We chose a validation set of 368 instances uniformly at random from the entire synthetic dataset, and we selected the best configurations from multiple SMAC runs based on their performance on this set. Our training set was initially also chosen uniformly at random from the entire synthetic data set. However, we found that SMAC found better configurations when selecting the training set as follows: We had, from multiple rounds of the algorithm engineering cycle, multiple performance evaluations for default and optimised configurations, and we observed that on many instances, these running times did not vary. More specifically, there were many instances whose solving times only seemed to improve due to some general improvements (e. g. parallelizing the lower and upper bound computation) but appeared to be almost entirely uncorrelated with algorithm parameters. Surprisingly, this was true not only for rather quickly solvable instances, where one would expect only minor differences, but also for harder instances. For example, we found instances that were almost completely unaffected by the data reduction rules and that were solved by exploring a (more less constant) number of search-tree nodes. In light of this observation, we computed for each instance the coefficient of variation (standard deviation divided by the mean) of the running times measured for different configurations we had run on it. We then selected only the instances with the highest coefficient of variation into a training set of size 196.

As can be seen in the top right scatter plot in Fig. 1, the configuration Hier-Opt$_S$ clearly dominates Hier on average. Furthermore, according to columns 6 and 7 in Table 1, although Hier-Opt$_S$ improves the timeout-rate only slightly from 5.1 % to 3.6 % (on training data), the mean and PAR-10 running times are considerable smaller and the median is less than half.[5] Notably, Hier-Opt$_S$ enables the $(k + 1)$-Rule but disables all other data reduction rules. While this was already observed for Rule 3 (computing the $\mathcal{O}(M \cdot k)$ kernel) in previous studies [20], this was surprising for Rule 7, which computes the $2k$-vertex kernel [9]. The last column in Table 1 provides the results for Hier-Opt$_S$ with Rule 7 enabled. Interestingly, while it slightly decreases the running time (mean and PAR-10) due to slightly more timeouts, the median is even lower than for Hier-Opt$_S$. This shows that Rule 7, in principle, reduces the running

---

[5] PAR-10 is the average with timeouts counted as 10 times the cut-off time.

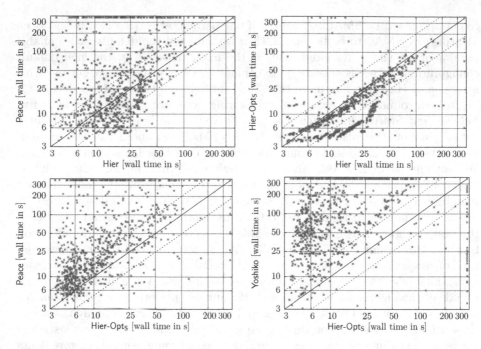

Fig. 1. Scatter plots of the running time of all solvers on the test instances of the synthetic dataset (full synthetic set minus training and validation instances). Timeouts (>300 s) are plotted at 360 s.

time on many instances, but the cost of applying it is overall not amortised by its benefits.

Results for Biological Dataset. Our experimental findings for the biological dataset are summarized in Table 2 and in the scatter plots in Fig. 2.

Unlike for the synthetic dataset, the ILP-based solver Yoshiko clearly outperforms Peace and Hier. However, comparing results for the latter two revealed that Hier is still better than Peace (see the upper-right plot in Fig. 2), especially, on harder instances. In general, since the median of the running times is pretty small (for Hier ≤ 0.18 s and for Peace and Yoshiko even ≤ 0.01 s), we suspect

Table 2. Running time (wall time in s) comparison of five solvers on the biological dataset with different "density" parameters c. The median of all solvers is less than 0.2 s.

c	Peace			Hier			Hier-Opt$_B$			Yoshiko			Yoshiko & Hier-Opt$_B$		
	33	50	66	33	50	66	33	50	66	33	50	66	33	50	66
Par-10	109	124	126	101	94.9	84.4	78.8	78.1	65.8	72.8	82.1	66.4	68.7	68.8	53
Mean	11.9	13.9	14	11.2	11.1	10.1	9.3	9.3	8.6	8.8	9.9	8.5	8.1	8.2	6.7
Timeouts	142	161	164	132	123	109	102	101	84	94	106	85	89	89	68

that our Hier solver suffers from the fact that on extremely easy instances the
initialization cost of the Java VM dominates the running time.

While the default configuration of Hier is not competitive with Yoshiko, our
SMAC-optimized configuration, called Hier-Opt$_B$, considerably closes this gap.
Although, being greatly slower for density value $c = 33$, Hier-Opt$_B$ clearly beats
Yoshiko for $c = 50$ and even slightly for $c = 66$. The bottom right plot in Fig. 2
stresses this point by clearly demonstrating that starting from instances that
require at least 10 s on both solvers, Hier-Opt$_B$ begins to dominate on average. This behaviour goes together with the observations that can be made from
directly comparing Hier-Opt$_B$ with Hier (see the bottom-left plot in Fig. 2): For
instances up to 1 s, Hier and Hier-Opt$_B$ roughly exhibit the same performance, but
the higher the running times get, the clearer Hier-Opt$_B$ is dominating on average.
We suspect that this is mainly caused by an algorithm parameter adjustments
made in Hier-Opt$_B$ that heavily increases the time fraction spend to compute the
initial lower bound. While easy instances do not largely benefit from computing
a slightly better lower bound, on large instances this might save expensive calls
of the search-tree solver for the decision variant. Even better performance can
be obtained by running Hier-Opt$_B$ and Yoshiko in parallel on the same instances,
as evident from the bottom right plot of Fig. 2. To demonstrate the potential of

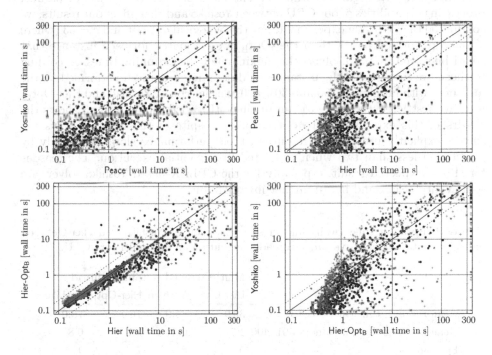

Fig. 2. Scatter plots of the running time of all solvers on the biological dataset (point
colour/value for c: black/33, blue/50, red/66). Timeouts (>300 s) are plotted at 360 s
(Color figure online).

this approach, the last column in Table 2 shows the running times of a virtual solver that takes the minimum of Yoshiko and Hier-Opt$_B$ for each instance.

To obtain Hier-Opt$_B$, SMAC was used in the same way as for the synthetic data, but could typically perform about 7500 algorithm runs and evaluate 3500 different configurations, because the instances tend to be easier. Due to the small median running time, we once again selected the training set based on the coefficient of variation but only among those instances, where at least one previous run needed at least 0.5 s. On the 327 training instances, the PAR-10 running time value of Hier is 850 s and could be improved to 149 s for Hier-Opt$_B$. This improvement was mainly due to a reduction in the number of timeouts from 90 down to 12.

We note that Hier-Opt$_B$ enables all data reduction rules, except the two simple Rules 4 & 6. However, Rule 7 (computing the $2k$-vertex kernel) is also almost disabled, since it is applied only in every 88th recursive step (adjusted by an algorithm parameter) of the search tree. For all other enabled rules, this "interleaving constant" is at most 13. Overall, having a more heterogeneous set of data reductions seems to be important on the biological dataset, but not for synthetic data, where only the $(k+1)$-Rule was enabled. Our default Hier enables all rules except Rules 4 and 7.

Finally, to investigate to which extent the difference in the use of parallel processing capabilities of our CPU between Yoshiko and Hier affect our results, we conducted the following experiment: For the biological dataset and $c = 33$ (where Yoshiko performed better than Hier-Opt$_B$) we computed for each instance that could be solved by both solvers the maximum of their running times. According to these, we then sorted the instances in descending order and performed on the instances with number 1–100 and 301–400 another run of Yoshiko and Hier-Opt$_B$, were we restricted the CPU to run in single-threaded mode. Table 3 shows the results of this experiment. To our surprise, despite of the different ways the solvers explicitly use parallel resources, their performance slows down only by a factor of less than two when restricted to sequential execution. The reasons for this unexpected result, especially for the CPLEX-based Yoshiko solver, are somewhat unclear and invite further investigation.

Table 3. Running time (wall time in s) comparison of Yoshiko and Hier-Opt$_B$ on biological data for multi- vs single-threaded execution on our multi-core CPU.

	Multi-threaded		Single thread	
	Hier-Opt$_B$	Yoshiko	Hier-Opt$_B$	Yoshiko
Mean running time, instances 1–100	39.9	44.9	76.1	70.0
Mean running time, instances 301–400	1.7	0.4	3.7	0.8
Timeouts	0	0	7	4

5   Conclusions and Future Work

We have shown how, by combining data reduction rules known from parameterised algorithmics with a heuristically enhanced branch-&-bound procedure, we can solve the NP-hard (unweighted) CLUSTER EDITING problem more efficiently in practice than the best known approaches known from the literature. This success was enabled by integrating Programming by Optimisation into the classical algorithm engineering cycle and, as a side effect, lead to a new method for assembling training sets for effective automated algorithm configuration.

It would be interesting to see to which extent further improvements could be obtained by automatically configuring the LP solver used in our algorithm, or the MIP solver used by Yoshiko. Furthermore, we see potential for leveraging the complementary strengths of the three algorithms studied here, either by means of per-instance algorithm selection techniques, or by deeper integration of mechanisms gleaned from each solver. We also suggest to study more sophisticated methods, such as multi-armed bandit algorithms, to more fine-grainely determine in which depths of the search tree a data reduction rule should be applied. Finally, we see considerable value in extending our solver to weighted CLUSTER EDITING, and in optimising it for the general M-HIERARCHICAL TREE CLUSTERING problem.

Acknowledgement. We thank Tomasz Przedmojski who provided, as part of his bachelor thesis, an accelerated implementation of the $\mathcal{O}(M \cdot k)$ kernel [20].

References

1. Gurobi 5.62. Software (2014)
2. Agarwala, R., Bafna, V., Farach, M., Narayanan, B., Paterson, M., Thorup, M.: On the approximability of numerical taxonomy (fitting distances by tree matrices). SIAM J. Comput. **28**(3), 1073–1085 (1999)
3. Ailon, N., Charikar, M.: Fitting tree metrics: hierarchical clustering and phylogeny. In: Proceedings of the 46th FOCS, pp. 73–82 (2005)
4. Bansal, N., Blum, A., Chawla, S.: Correlation clustering. Mach. Learn. **56**(1–3), 89–113 (2004)
5. Böcker, S.: A golden ratio parameterized algorithm for cluster editing. J. Discrete Algorithms **16**, 79–89 (2012)
6. Böcker, S., Baumbach, J.: Cluster editing. In: Bonizzoni, P., Brattka, V., Löwe, B. (eds.) CiE 2013. LNCS, vol. 7921, pp. 33–44. Springer, Heidelberg (2013)
7. Böcker, S., Briesemeister, S., Klau, G.W.: Exact algorithms for cluster editing: evaluation and experiments. Algorithmica **60**(2), 316–334 (2011)
8. Bonchi, F., Gionis, A., Gullo, F., Ukkonen, A.: Chromatic correlation clustering. In: Proceedings of 18th ACM SIGKDD (KDD 2012), pp. 1321–1329. ACM Press (2012)
9. Cao, Y., Chen, J.: On parameterized and kernelization algorithms for the hierarchical clustering problem. In: Chan, T.-H., Lau, L., Trevisan, L. (eds.) TAMC 2013. LNCS, vol. 7876, pp. 319–330. Springer, Heidelberg (2013)

10. Charikar, M., Guruswami, V., Wirth, A.: Clustering with qualitative information. J. Comput. Syst. Sci. **71**(3), 360–383 (2005)
11. Chen, J., Meng, J.: A $2k$ kernel for the cluster editing problem. J. Comput. Syst. Sci. **78**(1), 211–220 (2012)
12. Chierichetti, F., Dalvi, N., Kumar, R.: Correlation clustering in MapReduce. In: Proceedings of 20th ACM SIGKDD (KDD 2014), pp. 641–650. ACM Press (2014)
13. de Oca, M.A.M., Aydin, D., Stützle, T.: An incremental particle swarm for large-scale continuous optimization problems: an example of tuning-in-the-loop (re)design of optimization algorithms. Soft Comput. **15**(11), 2233–2255 (2011)
14. Downey, R.G., Fellows, M.R.: Fundamentals of Parameterized Complexity. Texts in Computer Science. Springer, London (2013)
15. Fawcett, C., Hoos, H.H.: Analysing differences between algorithm configurations through ablation. In: Proceedings of 10th MIC, pp. 123–132 (2013)
16. Fellows, M.R., Langston, M.A., Rosamond, F.A., Shaw, P.: Efficient parameterized preprocessing for cluster editing. In: Csuhaj-Varjú, E., Ésik, Z. (eds.) FCT 2007. LNCS, vol. 4639, pp. 312–321. Springer, Heidelberg (2007)
17. Gramm, J., Guo, J., Hüffner, F., Niedermeier, R.: Graph-modeled data clustering: exact algorithms for clique generation. Theory Comput. Syst. **38**(4), 373–392 (2005)
18. Grötschel, M., Wakabayashi, Y.: A cutting plane algorithm for a clustering problem. Math. Program. **45**(1–3), 59–96 (1989)
19. Guo, J.: A more effective linear kernelization for cluster editing. Theor. Comput. Sci. **410**(8–10), 718–726 (2009)
20. Guo, J., Hartung, S., Komusiewicz, C., Niedermeier, R., Uhlmann, J.: Exact algorithms and experiments for hierarchical tree clustering. In Proceedings of 24th AAAI. AAAI Press (2010)
21. Hoos, H.H.: Programming by optimization. Commun. ACM **55**(2), 70–80 (2012)
22. Hutter, F., Hoos, H.H., Leyton-Brown, K.: Sequential model-based optimization for general algorithm configuration. In: Coello, C.A.C. (ed.) LION 5 2011. LNCS, vol. 6683, pp. 507–523. Springer, Heidelberg (2011)
23. Niedermeier, R.: Invitation to Fixed-Parameter Algorithms. Oxford University Press, Oxford (2006)
24. Sanders, P., Wagner, D.: Algorithm engineering. It - Inf. Technol. **53**(6), 263–265 (2011)
25. van Zuylen, A., Williamson, D.P.: Deterministic algorithms for rank aggregation and other ranking and clustering problems. In: Kaklamanis, C., Skutella, M. (eds.) WAOA 2007. LNCS, vol. 4927, pp. 260–273. Springer, Heidelberg (2008)

OSCAR: Online Selection of Algorithm Portfolios with Case Study on Memetic Algorithms

Mustafa Mısır[✉], Stephanus Daniel Handoko, and Hoong Chuin Lau

School of Information Systems, Singapore Management University,
Singapore, Singapore
{mustafamisir,dhandoko,hclau}@smu.edu.sg

Abstract. This paper introduces an automated approach called OSCAR that combines algorithm portfolios and online algorithm selection. The goal of algorithm portfolios is to construct a subset of algorithms with diverse problem solving capabilities. The portfolio is then used to select algorithms from for solving a particular (set of) instance(s). Traditionally, algorithm selection is usually performed in an offline manner and requires the need of domain knowledge about the target problem; while online algorithm selection techniques tend not to pay much attention to a careful construction of algorithm portfolios. By combining algorithm portfolios and online selection, our hope is to design a problem-independent hybrid strategy with diverse problem solving capability. We apply OSCAR to design a portfolio of memetic operator combinations, each including one crossover, one mutation and one local search rather than single operator selection. An empirical analysis is performed on the Quadratic Assignment and Flowshop Scheduling problems to verify the feasibility, efficacy, and robustness of our proposed approach.

1 Introduction

We propose in this paper a framework that combines the ideas of *algorithm portfolio* and *online selection*. We call this framework OSCAR (Online SeleCtion of Algorithm poRtfolio). Algorithm selection [1] essentially learns the mapping between instance features and algorithmic performance, and this is usually performed in an offline fashion, as the process is typically very computationally intensive. The learned mapping can be utilized to choose the best algorithms to solve unseen problem instances based on their features. Algorithm portfolio [2,3] treats the algorithm selection problem in a broader perspective. The goal is to construct a diverse suite of algorithms that altogether are capable of solving a wide variety of problem instances, thus reducing the risk of failure. In terms of online algorithm selection, Adaptive Operator Selection (AOS) [4] deals with a single type of operators at a time, performs on-the-fly selection of evolutionary operators. Selecting from the pool of all possible combinations of crossover, mutation, and local search operators might be beneficial as this would capture the correlation among the different types of operators, but it could be

© Springer International Publishing Switzerland 2015
C. Dhaenens et al. (Eds.): LION 9 2015, LNCS 8994, pp. 59–73, 2015.
DOI: 10.1007/978-3-319-19084-6_6

challenging for the AOS methods. Hyperheuristics [5] can be seen as generic online algorithm selection methods that typically make use of a portfolio of very simple algorithms.

This work is motivated by the objective to provide a rich generic algorithm selection framework for solving diverse problem instances of a given target optimization problem. More specifically, we focus our attention on memetic algorithms (MA) [6] that represent a generic evolutionary search technique for solving complex problems [7]. By interleaving global with local search, MA reaps the benefit of the global convergence of the stochastic global search method as well as the quick and precise convergence of the deterministic local search method thereby avoiding the local optimum trap of deterministic search technique and alleviating the slow, imprecise convergence of the stochastic search technique. Like other evolutionary algorithms, however, the efficacy of MA depends on the correct choice of operators and their parameters. Various evolutionary (i.e. crossover, mutation) operators lead to different solution qualities [8]. For constrained problems, the choice of ranking operator is also important [9]. Reference [10] focused on the frequency of the local search, or in other words, whether local search is needed or can be skipped, since it can be expensive computationally, and may cause difficulty in escaping from local optimality (especially when the population diversity is too low such that all individuals reside in the same basin of attraction). All the above works suggest that there is indeed a correlation between a problem instance and the MA configuration that can render efficacious search.

Rather than relying primarily on the personal expertise or simply employing the widely-used ones, automatic selection of the potentially efficacious operators makes MA not only more likely to yield superior performance, but also easier to use, even by those inexperienced users. In our context, an algorithm refers to one combination of evolutionary operators that need to be successively applied in each MA iteration. Dummy operator is introduced for each operator type to cater for the possibility of not using any operator of that type. As shown in Fig. 1, the algorithm portfolio is constructed offline via a series of operations which encompass *feature extraction*, *feature selection*, *algorithm clustering*, and *portfolio generation*. The resulting portfolio is then sent to an online selection mechanism that performs on-the-fly selection of combination of operators in each MA iteration. The efficacy of the proposed framework is then assessed empirically on quadratic assignment problem (QAP) and flowshop scheduling problem (FSP).

The contributions of the work presented in this paper is three-fold:

1. We propose OSCAR, a novel framework which takes the advantage of both the algorithm portfolio and online selection paradigms. To our knowledge, OSCAR is the first online selection of algorithms in a portfolio.
2. We generate problem-independent features for the construction of portfolio, thereby eliminating the necessity of problem domain expertise.
3. We provide a means of identifying reasonable number of sufficiently diverse combinations of operators for the evolutionary algorithm, such as the MA, allowing AOS to capture the correlation among different types of operators.

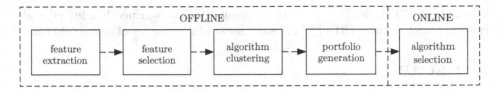

Fig. 1. Workflow of OSCAR

The remainder of the paper is presented as follows. Section 2 reviews related works in the literature. Section 3 introduces OSCAR and explains how it works in detail. Section 4 presents and discusses the experimental results on QAP and FSP. Finally, conclusion and future research directions are given in Sect. 5.

2 Related Works

Algorithm portfolios and (offline) selection have had a long history, and in the following, we review some recent works. SATZilla [11] is a well-known algorithm portfolio selection methodology that is particularly used to solve the SAT problem. It pursues a goal of providing a runtime prediction model for the SAT solvers. A number of problem-specific features for a given SAT instance are used to calculate the expected runtime of each algorithm in the portfolio. Its different versions are consistently ranked among the top portfolio-based solvers in the SAT competitions. 3S [12] utilised the resource constrained set covering problem with column generation to deliver solver schedules. Its superior performance was shown on the SAT domain. A cost-sensitive hierarchical clustering model was proposed in [13]. While the clustering model delivers a selection system, a static solver schedule is generated by 3S. SAT and MaxSAT were used as the application domains. Additionally, a Bayesian model combined with collaborative filtering is introduced to solve the constraint satisfaction and combinatorial auction problems in [14]. Unlike these studies, Hydra [15] addresses algorithm portfolios using parameter tuning. A portfolio is constructed by combining a particular solver with different parameter configurations provided by a version of ParamILS, i.e. FocusedILS [16]. The effectiveness of Hydra was also shown on SAT. Another tool developed for SAT, i.e. SATEnstein [17], targeted the algorithm generation process via tuning. It considers a variety of design elements for stochastic local search algorithms in the form of parameter tuning using ParamILS.

In terms of online algorithm selection, existing studies mostly refer to the terms Adaptive Operator Selection (AOS) [4] and Selection Hyper-heuristics [5]. The main idea is to monitor the search progress while solving a problem instance to immediately make changes on the choice of algorithms. Besides that, the online algorithm selection community deals with the algorithms and problems where solutions can be shared. However, in the case of offline methods, solution sharing can be cumbersome thus usually ignored when multiple algorithms are selected, like CPHydra [18]. Adaptive pursuit [19], multi-armed bandits [4] and reinforcement learning (RL) [20] are some successful examples of online selection.

3   OSCAR

Unlike most existing algorithm portfolio approaches that seek to deliver a portfolio of single solvers, this paper focuses on building a portfolio of algorithm combinations (even though our underlying approach can be used in the context of portfolio of single solvers). Each combination consists of a crossover operator, a mutational heuristic and a local search method. Our goal is to generate a small number of algorithm combinations with diverse performance that can successfully solve a large set of instances from a given problem domain. In order to have such a portfolio, it is initially required to generate a performance database revealing the *behavior* of each combination. Behavior here is denoted as the generic and problem-independent features primarily used in hyper-heuristic studies such as [21]. A class of hyper-heuristics, i.e. selection hyper-heuristics, aims at efficiently managing a given set of heuristics by selecting a heuristic(s) at each decision step. Due to the selection element in hyper-heuristics and their generic nature, we make use of the following features to characterize algorithm combinations for memetic algorithms.

- Number of new best solutions: N_{best}
- Number of improving solutions: N_{imp}
- Number of worsening solutions: N_{wrs}
- Number of equal quality solutions: N_{eql}
- Number of moves: N_{moves}
- Amount of improvement: $\triangle_{imp}$
- Amount of worsening: $\triangle_{wrs}$
- Total spent time: T.

A pseudo-code for OSCAR is presented in Algorithm 1. The process starts by collecting performance data regarding each algorithm combination a_x. The goal here is to perform a *feature extraction* about algorithms. For this purpose, each instance i_y is solved by a memetic algorithm successively using a randomly selected algorithm combination a_x. Algorithm 2 illustrates the basic memetic algorithm implementation. It should be noted that the performance data generation process differs for the cases where offline algorithm selection is applied. In the offline case, each algorithm is separately trained since these algorithms neither interact nor share solutions. Considering that an online selection device is employed and solutions are shared, it is vital to gather the performance data by running all the algorithms while they are selected online and operating on the same solutions.

The corresponding crossover (c_x), mutation (m_x) and local search (l_x) operators of a_x are applied in a relay fashion. The performance data generation process ends after each instance is solved within a given time limit (t_{limit}). The resulting performance data is used to generate features for each algorithm, $F(a_x)$.

Algorithm 1. OSCAR($\mathcal{A}$, $\mathcal{I}_{train}$, $\mathcal{I}_{test}$, $\mathcal{FS}$, $\mathcal{C}$, $\mathcal{OAS}$, $\mathcal{BC}$)

Input : $\mathcal{A}$: an algorithm with multiple operators to choose from, $\mathcal{I}_{train}$: a set
of training instances, $\mathcal{I}_{test}$: a set of test instances, $\mathcal{FS}$: a feature
selection method, $\mathcal{C}$: a clustering algorithm, $\mathcal{OAS}$: an online algorithm
selector, $\mathcal{BC}$: criterion for algorithm comparison

Operator combination $a_x = c_x + m_x + l_x$ where c_x, m_x and l_x refer to crossover,
mutation and local search operators respectively

Performance vector for the algorithm combination a_x on the instance i_y:
$P(a_x, i_y) = \{p_1(a_x, i_y), \ldots, p_k(a_x, i_y)\}$

Feature vector for the algorithm combination a_x:
$F(a_x) = \{p_1(a_x, i_1), \ldots, p_k(a_x, i_m)\}$

Feature extraction

1 $F \leftarrow P = A(.)$ algorithm A is iteratively applied using randomly selected
operator combinations a_x to gather performance data P for generating features
F

Feature selection

2 $F \leftarrow \mathcal{FS}(F)$

Algorithm clustering

3 Cluster algorithm combinations: $\mathcal{C}(A, F)$

Portfolio generation

4 Build portfolio using best algorithm combination from each cluster of C:
$AP = \{cl_1 \rightarrow a, \ldots, cl_t \rightarrow a\}$ w.r.t. $\mathcal{BC}$

Online selection

5 $S_{best} \leftarrow \mathcal{A}(AP, \mathcal{OAS}, \mathcal{I}_{test})$

Algorithm 2. MA(c, m, l)

---

n: population size, k: number of newly produced individuals / solutions at each
generation

1 **Initialisation**: Generate a population of solutions: $P(S_i)$ for $1 \leq i \leq n$
2 **while** !$stoppingCriteria()$ **do**
    $k = 1$
3   **while** $c \leq n_c$ **do**
4   Apply a crossover: $S_{n+k} = c(S_a, S_b)$
5   Apply a mutation method: $S_{n+k} = m(S_{n+c})$
6   Apply a local search operator: $S_{n+k} = l(S_{n+c})$
7          $k++$
    end
8      $updatePopulation(P)$
end

Each feature vector is composed of the normalised versions of the following 7 features for each instance: $f_1 = N_{best}/T$, $f_2 = N_{imp}/T$, $f_3 = N_{wrs}/T$, $f_4 = N_{eql}/T$, $f_5 = \triangle_{imp}/T$, $f_6 = \triangle_{wrs}/T$ and $f_7 = T/N_{moves}$ As a result, each algorithm combination has $\#instances \times 7$ features.

After completing the feature extraction process, a *feature selection* or elimination [22] method is applied. Gini Importance[1] [23] and Gain Ratio[2] [24] were used for feature selection purpose. Gini Importance is mostly used with Random Forests to detect the effective features w.r.t. the given class information. Gain Ratio is a information theoretic measure used to detect the effect of each feature by checking the variations on the values of each feature.

Next, *algorithm clustering* is performed. k-means clustering is applied as the clustering method $\mathcal{C}$ to identify the (dis-)similarity of the algorithm combinations. The best performing algorithm combinations, one from each selected cluster compose the portfolio during the *portfolio generation* process. During this process, the clusters with operator combinations which couldn't find any new best solution are ignored. Of significant importance is that when a cluster manage to find some new best solution, that cluster must be part of the portfolio, no matter how small the cluster may be. Such small cluster may in fact be the special combination that works well only on some very specific problem instances. The best combination for each cluster are then determined w.r.t. $\mathcal{BC}$ which is the number of new best solutions found. The overall procedure is finalised by applying the corresponding memetic algorithm with a given online selection approach $\mathcal{OAS}$ to the test instances $\mathcal{I}_{test}$ during the *online selection* phase. For the experiments, uniform random selection is used as the $\mathcal{OAS}$ option.

4 Computational Results

For the memetic algorithm, the population size is set to 40. As many as 20 new individuals are generated during each generation. 4 crossovers, 1 mutation operator and 3 local search heuristics are the available memetic operators. Since the mutation operator needs a mutation rate to be set, 6 different values are considered: 0.0, 0.2, 0.4, 0.6, 0.8, and 1.0. Setting the mutation rate to zero actually means that the mutation operator is not used. In order to have the same effect for the other two operator types, we added one dummy crossover operator and one dummy local search heuristic. In total, 119 (5 crossovers $\times$ 6 mutations $\times$ 4 local search - 1^3) operator combinations are generated. The details of these memetic operators are given as follows:

– Crossover:
  - *CYCLE* crossover: iteratively construct individuals by taking values from one parent and appointing the location of a next value from the second parent.
  - *DISTANCE_PRESERVING* crossover: outputs an individual where the distance referring to the number of genes assigned to different locations should be the same for the both parents.

---

[1] Using Scikit http://scikit-learn.org.
[2] Using Java-ML http://java-ml.sourceforge.net/.
[3] No crossover + no mutation + no local search case is ignored.

- *ORDER* crossover: a subgroup of genes are taken from one parent and the remaining genes come from the second parent respecting their order.
- *PARTIALLY_MAPPED* crossover: two randomly gene segments swap and partial maps denoting the elements located at common loci are used to change the conflicting genes with the swapped segment.
- Mutation: perturbs a given individual based on a mutation rate
- Local search:
  - *BEST_2_OPT* local search: attempts pairwise swap between 2 loci and applies the one producing best improvement in an iterative manner.
  - *FIRST_2_OPT* local search: attempts pairwise swap between 2 loci in a systematic fashion and applies the first one that produces improvement in an iterative manner.
  - *RANDOM_2_OPT* local search: attempts pairwise swap between 2 loci in a random order and applies the first one that produces improvement in an iterative manner.

For the training phase, t_{limit} is set to 300 s. The testing is performed with the per-instance execution time limit of 30 min for 5 trials. Java on an Intel Core I5 2300 CPU @ 2.80 GHz PC is used for the experiments.

4.1 Quadratic Assignment Problem

The QAP [25] requires the assignment of n facilities to n locations. Equation 1 shows the objective to minimise for the QAP. $f_{\pi_i \pi_j}$ is the flow between the facilities π_i and π_j. π refers to a solution where each element is a facility and the locus of each facility shows its location. d_{ij} is the distance between the location i and j. The objective is to minimise the total distance weighted by the flow values.

$$min \sum_i^n \sum_j^n f_{\pi_i \pi_j} d_{ij} \tag{1}$$

60 QAP instances from QAPLIB [26] were used. 31 instances are selected for training such that we can have enough performance data for each algorithm combination within the aforementioned time limit.

Portfolio Generation. The feature generation process resulted in 217 (31 instances × 7 per instance features) features. The features calculated for each operator combination on each instance is discarded if the number of moves performed is less than 10. After eliminating such features, 182 (26 instances × 7 per instance features) are left for each operator combination. Next, k-means was called with $k = 5$ to detect clusters of operator combinations. The features with this cluster information was considered as a classification problem in order to understand the nature of clusters. For this purpose, a random forests based feature importance evaluation method, i.e. Gini importance [23], is applied.

It revealed that 27 out of 182 features are the ones actually shaping these clusters. In addition, the features $f_1 = N_{best}/T$ and $f_2 = N_{imp}/T$ are from these 27 features for most of the QAP instances.

Besides using these 27 features, the same number of features are taken from the most critical features determined by other feature importance metrics. Table 1 lists the algorithm combination portfolios found using different feature sets provided by the metrics. The general view of these portfolios suggest that it is not always a good idea to keep applying all the three types of memetic operators together. Thus, in certain operator combinations, one or two operator types are missing. DISTANCE_PRESERVING and PARTIALLY_MAPPED crossovers are not included any of the operator combinations of the derived portfolios. Mutation is either ignored or applied with a small rate, i.e. 0.2 and 0.4. Among the local search heuristic, FIRST_2_OPT is detected as the most popular local search method while BEST_2_OPT is never picked. Besides, the portfolio sizes vary between 3 and 4. Considering that $k = 5$, 1 or 2 clusters have no operator combination yielded new best solutions during the training phase. In order to show whether using multiple operator combinations in an online setting is useful, the single best combination is also detected. The single best for the QAP uses CYCLE crossover and FIRST_2_OPT without mutation.

Table 1. Operator combination portfolios determined by OSCAR for the QAP

Feature selection	Algorithm portfolios		
	Crossover	Mutation	Local search
No selection	CYCLE	–	FIRST_2_OPT
	CYCLE	–	RANDOM_2_OPT
	ORDER	0.4	FIRST_2_OPT
	CYCLE	0.2	FIRST_2_OPT
Gini importance	CYCLE	–	FIRST_2_OPT
	CYCLE	–	RANDOM_2_OPT
	–	–	FIRST_2_OPT
Gain ratio	CYCLE	–	FIRST_2_OPT
	CYCLE	–	RANDOM_2_OPT
	–	–	FIRST_2_OPT
	CYCLE	0.2	FIRST_2_OPT

Figure 2 visualises the operator combinations for each operator type to determine what actually shapes these clusters via multidimensional scaling (MDS) [27] with Euclidean distance. These graphs indicate that the operator combinations are grouped particularly in reference to the local search operators. Figure 3 shows the effect of individual performance measures on clustering. The amount of improvement and worsening w.r.t. the total time spent by each operator combination is utilised as the most critical performance measures. The operator combinations' speed, the number of new best solutions and equal quality solutions detected w.r.t. the total time spent by each operator combination are determined as the measures affecting clusters least.

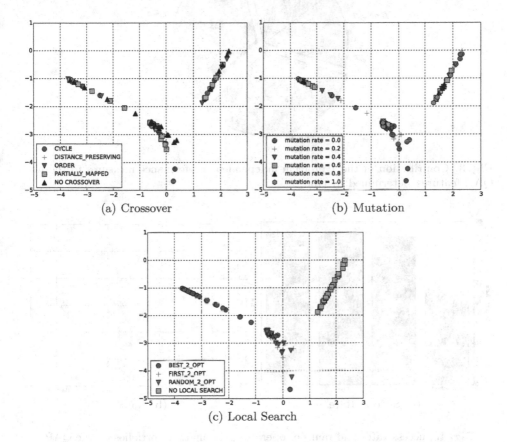

(a) Crossover   (b) Mutation

(c) Local Search

Fig. 2. MDS of operator combinations w.r.t. each operator type for the QAP

Online Algorithm Selection. Figure 4(a) shows the performance of three portfolios together with the Single Best combination when Random is used as online selector, in terms of the success rate (i.e. how many times the best known or optimum solutions are found, expressed in percentage). The results indicate that the single best is able to deliver around 23 % of the best known QAP solutions while OSCAR with different portfolios can find between 36 % and 37 % of the best known solutions. Although Gini and Gain Ratio based portfolios perform slightly better than the case without feature selection, there seems to be of only slight difference. However, when we look at the results closely by considering the solution quality, the performance difference becomes clearer. Figure 4(b) presents box plots indicating the ranks of each tested method. Besides the superior performance of OSCAR against the Single Best in ranking, the portfolio constructed using Gini delivers the best results among the three portfolios.

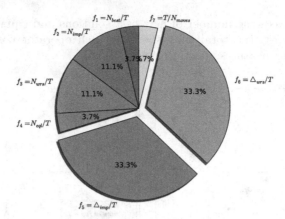

Fig. 3. Contribution of the 7 problem-independent performance measures to the top QAP features, determined by Gini

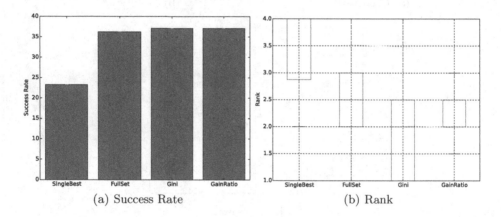

(a) Success Rate   (b) Rank

Fig. 4. Success rates and ranks of operator combination portfolios on the QAP

4.2 Flowshop Scheduling Problem

The Flowshop Scheduling Problem (FSP) is related to the assignment of n jobs to m machines aiming at minimizing the completion time of the last job, i,e. the makespan. The 68 FSP instances from the Taillard FSP benchmarks[4] [28] are used. 41 of these instances are taken as the training instances while the remaining 27 instances are considered as the test set.

Portfolio Generation. The feature generation process provided 287 features (41 instances × 7 per instance features) for each instance. After performing

---

[4] http://mistic.heig-vd.ch/taillard/problemes.dir/ordonnancement.dir/ordonnance ment.html.

k-means clustering with $k = 5$, the Gini importance metric calculated via applying Random Forests indicated that only 29 of these 287 features contributed to the clustering process. Thus, we use 29 as the number of top features to check. This is achieved using the aforementioned importance metrics as we did for the QAP case. Table 2 lists the portfolios of operator combinations derived using each of these importance metrics. Unlike the QAP case, DISTANCE_PRESERVING and PARTIALLY_MAPPED crossovers are also used in the FSP portfolios. For Mutation, higher rates are preferred, i.e. 0.6 and 0.8, or no mutation is applied. RANDOM_2_OPT, here, is as frequently picked as FIRST_2_OPT and BEST_2_OPT is used in one operator combination where DISTANCE_PRESERVING is included. Similar to the QAP portfolios, here each portfolio has either 3 or 4 operator combinations. The single best combination for the FSP applies PARTIALLY_MAPPED crossover, mutation with rate of 0.6 and RANDOM_2_OPT.

Table 2. Operator combination portfolios determined by OSCAR for the FSP

Feature selection	Algorithm portfolios		
	Crossover	Mutation	Local search
No selection	CYCLE	–	FIRST_2_OPT
	CYCLE	–	RANDOM_2_OPT
	DISTANCE_PRESERVING	0.6	BEST_2_OPT
	PARTIALLY_MAPPED	0.6	RANDOM_2_OPT
Gini importance	CYCLE	-	FIRST_2_OPT
	CYCLE	–	RANDOM_2_OPT
	PARTIALLY_MAPPED	0.6	RANDOM_2_OPT
	ORDER	–	FIRST_2_OPT
Gain ratio	PARTIALLY_MAPPED	0.6	RANDOM_2_OPT
	–	0.8	FIRST_2_OPT
	ORDER	–	FIRST_2_OPT

Figure 5 presents the operator combinations w.r.t. their problem-independent features in 2D via MDS. As with the QAP, the local search operators mainly characterise the operator combinations' groups. Figure 6 shows the which individual performance measure is used while clustering. Operator combinations' speed is detected as the major factor. Additionally, the number of new best solutions, worsening solutions and equal quality solutions w.r.t. the total time spent by each operator combination are also highly effective on the clusters. The amount of worsening w.r.t. the total time spent by each operator combination is utilised as the least important performance measure.

Online Algorithm Selection. Figure 7(a) details the performance of 3 portfolios and the single best combination in terms of success rate (i.e. how many

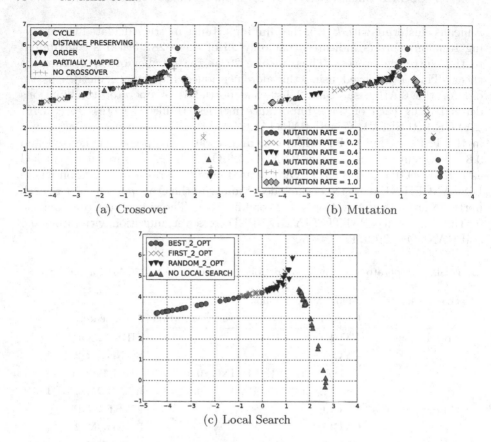

(a) Crossover

(b) Mutation

(c) Local Search

Fig. 5. MDS of operator combinations w.r.t. each operator type for the FSP

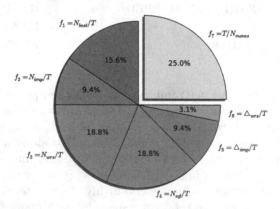

Fig. 6. Contribution of the 7 problem-independent performance measures to the top FSP features, determined by Gini

times the best known or optimal FSP solutions are found, expressed in percentage). The portfolios generated using full feature set and Gain Ratio show similar performance to the single best combination by reaching between 47 % and 49 % of the best known or optimum solutions. However, the portfolio with Gini found around 56 % of the best known solutions as the best tested method. Figure 7(b) presents these results in terms of ranks w.r.t. the solution quality where OSCAR's superior performance can be clearly seen. Among the reported portfolios, the Gini based portfolio reveals the statistically significant best results.

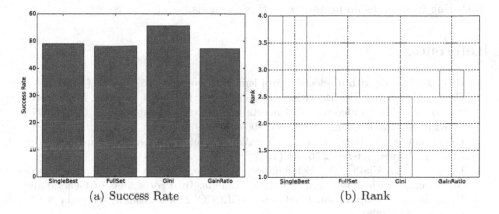

(a) Success Rate   (b) Rank

Fig. 7. Success rates and ranks of operator combination portfolios on the FSP

Overall, the results on both the QAP and the FSP indicate that using multiple operator combinations is profitable when they are selected online. This shows that OSCAR is able to combine the strengths of both offline algorithm portfolios and online algorithm selection in a problem-independent manner. Of particular significance is that the Gini-based portfolio always perform the best.

5 Conclusions

In this paper, we have introduced OSCAR as a framework that performs Online SeleCtion of Algorithm poRtfolio. The algorithm portfolio is constructed offline to determine which combinations of the memetic operators are efficacious for solving certain problem domains. Those combinations in the portfolio are then fetched to some online selection mechanism. This hybridization allows an online selection method to capture the correlation among different types of the memetic operators. This paper presents the first study of such hybridization. Additionally, OSCAR does not require any problem-specific features to generate the portfolio, thereby eliminating the necessity of problem domain expertise.

Empirical assessments on QAP and FSP have demonstrated the efficacy of OSCAR. OSCAR is able to deliver superior performance compared to the single best operator combinations for both problems. This shows that the problem-independent features introduced are practical to differentiate one available operator combination from the others, which eventually lead to an efficient portfolio. Furthermore, the improving performance delivered after feature selection, particularly when Gini importance index is employed, indicates the usefulness of the feature selection part of OSCAR.

Moving forward, the explanatory landscape analysis [29] will be incorporated to extend the algorithm feature space. The multi-objective performance measures shall be studied to build portfolios for multi-objective evolutionary algorithms. An in-depth analysis will be performed to evaluate the performance of different clustering techniques and online selection methods.

References

1. Rice, J.: The algorithm selection problem. Adv. comput. **15**, 65–118 (1976)
2. Gomes, C., Selman, B.: Algorithm portfolio design: theory vs. practice. In: Proceedings of the 13th Conference on Uncertainty in Artificial Intelligence (UAI 1997), Providence/Rhode Island, USA, pp. 190–197 (1997)
3. Huberman, B., Lukose, R., Hogg, T.: An economics approach to hard computational problems. Science **275**, 51 (1997)
4. Da Costa, L., Fialho, A., Schoenauer, M., Sebag, M., et al.: Adaptive operator selection with dynamic multi-armed bandits. In: Proceedings of Genetic and Evolutionary Computation Conference (GECCO 2008), Atlanta, Georgia, USA, pp. 913–920 (2008)
5. Burke, E., Gendreau, M., Hyde, M., Kendall, G., Ochoa, G., Ozcan, E., Qu, R.: Hyper-heuristics: a survey of the state of the art. J. Oper. Res. Soc. **64**, 1695–1724 (2013)
6. Moscato, P., Cotta, C., Mendes, A.: Memetic algorithms. In: Moscato, P., Cotta, C., Mendes, A. (eds.) New Optimization Techniques in Engineering, pp. 53–85. Springer, Heidelberg (2004)
7. Krasnogor, N., Smith, J.: A memetic algorithm with self-adaptive local search: TSP as a case study. In: Proceedings of Genetic and Evolutionary Computation Conference (GECCO 2000), Las Vegas/Nevada, USA, pp. 987–994 (2000)
8. Yuan, Z., Handoko, S.D., Nguyen, D.T., Lau, H.C.: An empirical study of off-line configuration and on-line adaptation in operator selection. In: Pardalos, P.M., Resende, M.G.C., Vogiatzis, C., Walteros, J.L. (eds.) LION 2014. LNCS, vol. 8426, pp. 62–76. Springer International Publishing, Switzerland (2014)
9. Runarsson, T.P., Yao, X.: Stochastic ranking for constrained evolutionary optimization. IEEE Trans. Evol. Comput. **4**, 284–294 (2000)
10. Handoko, S.D., Kwoh, C.K., Ong, Y.S.: Feasibility structure modeling: an effective chaperone for constrained memetic algorithms. IEEE Trans. Evol. Comput. **14**, 740–758 (2010)
11. Xu, L., Hutter, F., Hoos, H., Leyton-Brown, K.: SATzilla: portfolio-based algorithm selection for SAT. J. Artif. Intell. Res. **32**, 565–606 (2008)
12. Kadioglu, S., Malitsky, Y., Sabharwal, A., Samulowitz, H., Sellmann, M.: Algorithm selection and scheduling. In: Lee, J. (ed.) CP 2011. LNCS, vol. 6876, pp. 454–469. Springer, Heidelberg (2011)

13. Malitsky, Y., Sabharwal, A., Samulowitz, H., Sellmann, M.: Algorithm port-folios based on cost-sensitive hierarchical clustering. In: Proceedings of the 23rd International Joint Conference on Artifical Intelligence (IJCAI 2013), pp. 608–614 (2013)
14. Stern, D., Herbrich, R., Graepel, T., Samulowitz, H., Pulina, L., Tacchella, A.: Collaborative expert portfolio management. In: Proceedings of the 24th AAAI Conference on Artificial Intelligence (AAAI 2010), Atlanta/Georgia, USA, pp. 179–184 (2010)
15. Xu, L., Hoos, H., Leyton-Brown, K.: Hydra: automatically configuring algorithms for portfolio-based selection. In: Proceedings of the 24th AAAI Conference on Artificial Intelligence (AAAI 2010), pp. 210–216 (2010)
16. Hutter, F., Hoos, H., Leyton-Brown, K., Stützle, T.: ParamILS: an automatic algorithm configuration framework. J. Artif. Intell. Res. **36**, 267–306 (2009)
17. KhudaBukhsh, A.R., Xu, L., Hoos, H.H., Leyton-Brown, K.: Satenstein: automati-cally building local search sat solvers from components. In: Proceedings of the 21st International Joint Conference on Artificial Intelligence (IJCAI 2009), vol. 9, pp. 517–524 (2009)
18. O'Mahony, E., Hebrard, E., Holland, A., Nugent, C., O'Sullivan, B.: Using case-based reasoning in an algorithm portfolio for constraint solving. In: Irish Confer-ence on Artificial Intelligence and Cognitive Science (2008)
19. Thierens, D.: An adaptive pursuit strategy for allocating operator probabilities. In: Proceedings of the 7th Annual Conference on Genetic and Evolutionary Com-putation (GECCO 2005), pp. 1539–1546. ACM (2005)
20. Nareyek, A.: Choosing search heuristics by non-stationary reinforcement learning. In: Resende, M., de Sousa, J. (eds.) Metaheuristics: Computer Decision-Making, pp. 523–544. Kluwer Academic Publishers, Dordrecht (2003)
21. Mısır, M.: Intelligent hyper-heuristics: a tool for solving generic optimisation prob-lems. Ph.D. thesis, Department of Computer Science, KU Leuven (2012)
22. Guyon, I., Elisseeff, A.: An introduction to variable and feature selection. J. Mach. Learn. Res. **3**, 1157–1182 (2003)
23. Breiman, L.: Random forests. Mach. Learn. **45**, 5–32 (2001)
24. Quinlan, J.R.: C4.5. Programs for Machine Learning. Morgan Kaufmann, San Francisco (1993)
25. Lawler, E.: The quadratic assignment problem. Manag. Sci. **9**, 586–599 (1963)
26. Burkard, R.E., Karisch, S.E., Rendl, F.: Qaplib-a quadratic assignment problem library. J. Global Optim. **10**, 391–403 (1997)
27. Borg, I., Groenen, P.J.: Modern Multidimensional Scaling: Theory and Applica-tions. Springer, New York (2005)
28. Taillard, E.: Benchmarks for basic scheduling problems. Eur. J. Oper. Res. **64**, 278–285 (1993)
29. Mersmann, O., Bischl, B., Trautmann, H., Preuss, M., Weihs, C., Rudolph, G.: Exploratory landscape analysis. In: Proceedings of the 13th Annual Conference on Genetic and Evolutionary Computation, pp. 829–836. ACM (2011)

Learning a Hidden Markov Model-Based Hyper-heuristic

Willem Van Onsem$^{(\boxtimes)}$, Bart Demoen, and Patrick De Causmaecker

Department of Computer Science, KU Leuven,
Celestijnenlaan 200A, 3001 Heverlee, Belgium
willem@gmail.com

Abstract. A simple model shows how a reasonable update scheme for the probability vector by which a hyper-heuristic chooses the next heuristic leads to neglecting useful mutation heuristics. Empirical evidence supports this on the MAXSAT, TRAVELINGSALESMAN, PERMUTATION-FLOWSHOP and VEHICLEROUTINGPROBLEM problems. A new approach to hyper-heuristics is proposed that addresses this problem by modeling and learning hyper-heuristics by means of a hidden Markov Model. Experiments show that this is a feasible and promising approach.

1 Introduction

A *hyper-heuristic* is a problem-independent algorithm that aims to select which heuristic to apply next during an evolutionary process. The aim of the hyper-heuristic is to speed up convergence toward an optimum in an optimization problem.

A hyper-heuristic is thus an optimization problem itself: one aims to optimize the convergence speed by scheduling heuristics appropriately. By problem independent we mean the hyper-heuristic can observe only some properties of the solution, and not the solution itself. Most software packages only allow inspecting the fitness-value.

One traditional approach to hyper-heuristics is to reward a heuristic that has - in the past - improved the solution, and punish the heuristics that computed a worse solution. This is achieved by adapting the probabilities by which a heuristic is chosen. We show with a simple model that such a scheme is doomed to underuse the *bad* heuristics, while they are necessary to find an optimal solution. This slows down convergence to an optimal solution. We report on experiments on four problem classes that confirm this.

Literature proposes several alternatives to avoid the underuse of certain heuristics in a more or less ad-hoc way. We choose for a more radical approach: we propose to model the choices made by a hyper-heuristic as a hidden Markov Model (HMM), and learn this HMM by means of (a sample of) the performance of the individual heuristics on (a sample of) the solution space.

This paper is structured as follows: Sect. 2 defines necessary terminology and concepts. In Sect. 3 we argue why a popular model – the probability vector – will probably fail to increase convergence. This is done both using a model

C. Dhaenens et al. (Eds.): LION 9 2015, LNCS 8994, pp. 74–88, 2015.
DOI: 10.1007/978-3-319-19084-6_7

and empirically. Section 4 introduces our approach to modeling and learning hyper-heuristics: it is based on the Mealy Input-Output hidden Markov model (MIOHMM) also explained there. Results and experiments using this model are reported in Sect. 5. Section 6 concludes and discusses future work.

2 Preliminaries

A single-objective optimization problem consists of an implicit solution space S and a fitness function $f : S \rightarrow \mathbb{R}$. The aim is to find a (pseudo) optimal solution $s^\star$ such that the fitness value $f(s^\star)$ is the infimum of $f(S)$.

An evolutionary algorithm aims to achieve this by applying a chain of heuristics on an initial solution s_0 and returns the best solution encountered so far when the time limit is reached.

A heuristic h is a function $h : S \rightarrow S$ that maps one solution to another solution.

Most heuristics are probabilistic in nature: the generated solution depends on both a solution and the seed of a random number generator. The set of possible outcomes of the heuristic h given the solution s, is called the *neighborhood H* of s, $H(s)$.

The concept of a heuristic can be generalized further: genetic algorithms for instance make use of heuristics that take as input two or more solutions. Such heuristics are called *"crossover" heuristics*. We do not consider such heuristics here.

We consider two types of heuristics: *local search* and *mutation heuristics*. A local search heuristic or *hill climber* is a heuristic h that guarantees that the generated solution is at least as good as the original solution, or more formally $f(h(s)) \geq f(s)$ for all solutions s. Mutation heuristics do not guarantee this behavior.

A *local optimum* with respect to a set of heuristics h_i is a solution s such that for each element s' in the union of the neighborhoods H_i of s, $f(s') < f(s)$ or $s = s'$. Any local search heuristic applied on a local optimum results in the same solution.

3 Modeling Heuristic Behavior with Probability Vectors

Hyper-heuristics [1,2] commonly use a *probability vector* [3] for guiding the selection of the heuristic to apply next.

A probability vector is a list of probabilities that sum up to one and associates elements - in this case heuristics - with probabilities. The well known *roulette wheel selection procedure* [4] can select heuristics proportional to their probability.

A probability vector is trained by updating the weights in function of accumulated empirical evidence. An algorithm that updates the probabilities is called an *update scheme*.

Hyper-heuristic systems use different [1,2] update schemes. We will show that under assumptions stated later, a *reasonable* update scheme eventually makes escaping from a local optimum less probable.

A reasonable update scheme rewards heuristics that produce a better solution, penalizes heuristics that produce a worse solution, and rewards or is neutral to heuristics that produce a new solution with the same quality. The update weights are furthermore monotonic with respect to the absolute difference in fitness value: if the difference increases, the weight either increases or remains the same. A reasonable update scheme is also oblivious to the type of heuristic: e.g. the update strategy does not differ between local search and mutation heuristics. Not all update schemes proposed in literature are reasonable.

3.1   On the Probability to Escape from a Local Optimum

Our claim is that probability vectors eventually antagonize convergence of the evolutionary process given. This is true under a number of reasonable assumptions.

1. The hyper-heuristics runs with both a local search and mutation heuristic: this is true for all hyper-heuristics we are aware off.
2. The heuristics are *stationary*. A heuristic is stationary if the probability of generating a solution only depends on the given current solution and not on other parameters like the elapsed time in the process.
3. It is very unlikely that the result of a mutation heuristic is a local optimum or that a mutation heuristic can improve the result of a local search heuristic. This seems true in practice. For simplicity, we assume here that it is not just very unlikely, but impossible.
4. For a mutation heuristic, the average fitness value of the solution after the application of the mutation heuristic is eventually worse than the fitness value of the solution before the application of the mutation heuristic.

    Experiments on four different problems[1] implemented in the *HyFlex 1.0* [5] framework show that this is true for 12 out of 15 mutation heuristics. Other mutation heuristics are iso-fitting: they produce always solutions that have the same fitness value as the original solution.

Once the evolutionary process is at a local optimum, a local search heuristic cannot generate a solution different from the active solution. Mutation heuristics are thus necessary to escape from a local optimum (assumption 1). Empirical evidence shows that evolutionary algorithms are stuck in a local optimum a significant number of times[2]. The detection of and the escape from local optima should thus be performed as efficiently as possible.

    Figure 1 illustrates with an (in)finite state machine how a generic hyper-heuristic process escapes a local optimum. The nodes represent the possible

---

[1] See goo.gl/vVTZNE for details.
[2] If the heuristics are applied with uniform probability, around 5 % to 20 % of the time, although it strongly depends on the problem. See goo.gl/vVTZNE for empirical evidence.

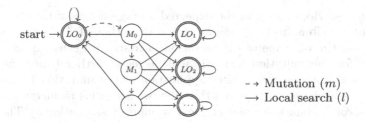

Fig. 1. Representation of the different states in an escaping process.

states of a hyper-heuristic algorithm: they are distributions over the solution space S that represent how probable it is for a solution to be the "active solution". The escaping process starts in a state we represented by LO_0: the state represents the fact that a local optimum is the active solution at that moment. The aim is to get the system in another local optimal solution: other local optima are represented by the states $LO_1, LO_2, \ldots$. All local optima states are marked as "accepting" since they mark the end of an "escaping attempt".

In the initial state of the escaping process, the probability vector is represented by $\langle p_l, p_m \rangle$ with p_l and $p_m = 1 - p_l$ respectively the probability of the local search and mutation heuristic. The expected number of function calls before the mutation heuristic is called is determined by:

$$\sum_{i=0} p_l^i = \frac{1}{1 - p_l} = \frac{1}{p_m}. \tag{1}$$

During these escape attempts local search always generates the same solution: a reasonable update scheme either does nothing with this information or it rewards the local search heuristic l.

If an acceptance scheme is incorporated, this can take even more attempts since rejecting the result of a local search application makes no difference, and rejecting the solution generated by the mutation heuristic only results in more attempts to escape the optimum.

After the mutation heuristic m is eventually applied, the generated solution is worse since otherwise the initial solution would not have been a local optimum. Reasonable update schemes will penalize this with a reduction of the probability.

Since we assume it is impossible that the mutation heuristic produces a local optimum (assumption 3), the process escapes local optimum LO_0 and ends up in state M_0 at the cost of a decrease in the probability of the mutation heuristic. M_0 describes a probability distribution over the possible outcomes of the mutation heuristic. Now two possible scenarios can unfold:

1. The mutation heuristic is called a second time. The process ends up in state M_1 (with a possibly different distribution over the solution space) note that this scenario can be repeated;
2. The local search heuristic is applied and the system ends up in a (possibly different) local optimum. In case we end up in the same local optimum, all invested computational resources are wasted.

In the first scenario, in general the expected average fitness value over the distribution of solutions in M_{i+1} will be worse than that of M_i (assumption 4). Depending on the outcome of the mutation heuristic, we thus expect that the probability for the mutation heuristic will decrease further (since the probability vector is "reasonable"). After application of the mutation heuristic, the algorithm is still in an M-state and thus the two scenarios reemerge.

In the second scenario we reach a local optimum (assumption 3). The probability of the local search heuristic will increase at the expense of the probability of the mutation heuristic, since it is expected that the probability of the local search heuristic will increase. Equation (1) shows that the higher the probability of the local search heuristic, the longer it takes to escape a local optima.

One can describe this phenomena as the fact that the local search heuristic "takes full credit" for the work that was partially carried out by the mutation heuristic: escaping out of a local optimum. Since mutation heuristics are crucial in such process, at least a small probability should be maintained to prevent a hyper-heuristic locking itself in.

As the probability of applying a mutation heuristic decreases, it takes longer to escape from a (new) local optimum. Hence we claim:

Claim. A method using a probability vector with a reasonable update scheme eventually takes more and more time to escape from local optima and thus its convergence speed decreases.

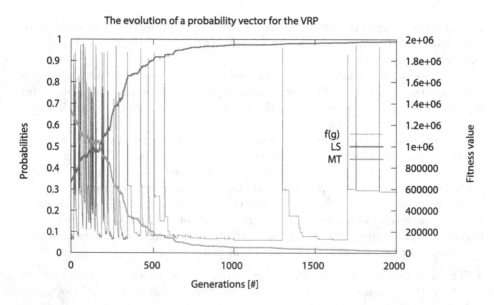

Fig. 2. Evolution of a probability vector solving the VEHICLEROUTINGPROBLEM.

3.2 Empirical Evidence of the Claim

We performed experiments using the *HyFlex 1.0* [5] framework to test whether our claim about probability vectors hold. Note that we proved our claim using just one mutation heuristic and one local search heuristic. In our experiments we used multiple heuristics of both kinds. Several reasonable update schemes were used. The tests were performed on the MAXSAT, PERMUTATIONFLOWSHOP, TRAVELINGSALESMAN and VEHICLEROUTINGPROBLEM problems.

Figure 2 depicts the state of the probability vector at each generation for the VEHICLEROUTINGPROBLEM problem using a constant penalty/reward update scheme.

The thin gray line shows the fitness value of the active solution[3] and thus indicates whether the system is stuck in a local optimum. When the spikes in the fitness value are close together, this indicates that the escape from a local optimum was fast. When the fitness value remains the same during some generations, it indicates a slower escape. It is clear that as the number of generations increases, it becomes harder to escape the local optimum, still the fitness function $f(g)$ does not show that the local optima become significantly better.

This trend is matched by the evolution of the probability of the mutation heuristic, and consequently the local search heuristic: the thick red line indicates the total probability of the three employed local search heuristics. Initially the probability is set to $1/9$ for each heuristic, so the total probability for the local search heuristics is $p_l = 0.333$. As the number of iterations grows, the probability approaches 1 quickly. The thick green line shows the sum of the probabilities of mutation and ruin-recreate heuristics. Since the problem runs with 6 such heuristics, initially the probability is set to $p_m = 0.666$, but it decreases fast below any reasonable probability.

The same effects were observed for all other tested problems and employed update schemes, so we conclude that our claim about probability vectors generally holds.

One can argue that in practice, hyper-heuristics never implement such a "pure" probability vector with a reasonable update scheme. For instance, many approaches use a probability vector per heuristic. This probability vector then determines which heuristic to apply next given the previous heuristic that was called first. We have performed empirical tests on such probability transition matrices as well and the same effects were observed although the convergence of the probability of local search heuristics towards 1 was slower. The reason seems that it takes several generations to update all the elements of the transition matrix.

3.3 Working Around the Problem with Probability Vectors

Hyper-heuristics try to solve the above stated problem in various ways:

Reinforcement learning [6,7] is a state-oriented update scheme with *memory*. It gives credit not only to the item last applied in the sequence, but uses

---

[3] See the right axis for the appropriate unit.

a smoothing off approach where each heuristic in the sequence receives credit: the more recent the heuristic was called, the more the heuristic is rewarded. The rewards are given in the context of an implementation-defined state. A first limitation to this approach is that a programmer must find a good way to define states that can only depend on the observed fitness values. Furthermore reinforcement learning sometimes tend to reward items in a sequence that have nothing to do with the result: if multiple local search heuristics were applied in the evolutionary chain, they are all rewarded for delivering the same solution.

AdapHH [8,9] solves the problem using a *tabu search* [10] approach where heuristics that take a significant amount of time without generating a better solution are tabued. Since local search heuristics take in general more time than mutation heuristics, local search heuristics will get tabued more often. Mutation heuristics can get tabued as well resulting in a potential lower convergence rate. Since eventually the heuristics are untabued again, such algorithms have a more stable performance.

Finally, *Iterative Local Search* [11] interleaves mutation heuristics with local search heuristics and applies pairs of a mutation and local search heuristic. If such move generates a better solution, both heuristics are rewarded. A potential pitfall with this approach is that it can take more than one mutation heuristic application to get out of a local optimum. Extensions on iterative local search exist that take this into account.

We think that it might also be worthwhile to experiment with an update scheme that penalizes heuristics that generate the same (quality) solution as the given one. As a result, in a long sequence of local search heuristics, these local search heuristics will become less favorable and the evolutionary process can escape from the local optimum.

However, we think that it might be better to abandon probability vectors altogether.

4   From Probability Vectors to Hidden Markov Models

Although in the previous section we showed that a probability vector cannot learn well heuristic behavior, we think that probabilistic reasoning is a promising way to reason about heuristic behavior. The missing aspect in many implementations is maintaining a "state". This state is incorporated in a hyper-heuristic by the notion of the active solution.

We already discussed that reinforcement learning maintains states, but it is up to the programmer to decide what these states represent. In this section we discuss an approach were the semantic interpretation of states is left to a learning algorithm. This makes the algorithm more flexible.

We first discuss hidden Markov models and their application to hyper-heuristics in Sect. 4.1. We then show how we can learn heuristic behavior using such models in Sect. 4.2. In Sects. 4.3 and 4.4, we propose methods that decide which heuristic to apply next in an evolutionary process and forgetting learned behavior in favor of new experience.

4.1 Hidden Markov Models

Definition 1 (Hidden Markov Model (HMM)). *A hidden Markov Model is a 3-tuple $\langle \pi, A, B \rangle$ with π a probability n-vector, A an $n \times n$ transition matrix and B an $n \times m$ emission matrix. Each row of A and B are probability vectors.*

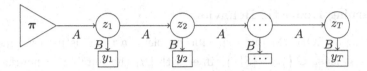

Fig. 3. A Markov process described by a hidden markov model.

A hidden Markov model describes a Markov process as depicted on Fig. 3: a probabilistic function that varies in time $Y : \mathbb{N} \to O : t \mapsto Y(t)$ with $O = \{o_1, o_2, \ldots, o_m\}$ a finite set of possible *observations*. This is done by considering a set of *"hidden"* states $\{s_1, s_2, \ldots, s_n\}$. Given the system is in state $z_t = s_i$ at time step t, a_{ij} describes the probability of the system being in state $z_{t+1} = s_j$ at time step $t+1$. The hidden states cannot be observed directly. At each time step t the active state z_t emits an observation. The probability of emitting observation o_k is defined by b_{ik}. The initial probability vector π element for index i is defined as the probability of generating the corresponding solution as initial solution in the evolutionary process.

For the purpose of this paper, we take as observations the set of fitness values of the solutions. Since a hyper-heuristic can only inspect directly the fitness value of a solution, this is our only possibility.

Depending on which "behavior" we want to model, we can generate a set of observations O. Since the number of solutions in a combinatorial optimization problem is finite, the domain of behavioral aspects attached to the solution space is finite as well. From a hyper-heuristic point of view, the only reasonable behavioral property we can extract from a solution is its fitness value. Given the set of all possible fitness values O, the emission probability b_{ik} is set to 1 if the solution corresponding to hidden state s_i has the fitness value represented by o_k and 0 otherwise.

The above discussed model shows that we can model the behavior of a single stationary heuristic with a HMM. The aim of a hyper-heuristic however is to determine which heuristic to apply next in a set of *multiple* heuristics. In order to model multiple heuristics, we use the concept of an Input-Output hidden Markov model [12].

Definition 2 (Input-Output Hidden Markov Model (IOHMM)). *An Input-Output hidden Markov model is a 3-tuple $\langle \pi, A, B \rangle$ with π a probability n-vector, A an $l \times n \times n$ transition matrix and B an $n \times m$ emission matrix. Each matrix A_i is a transition matrix as defined in Definition 1.*

An IOHMM considers not only a set of observations O but an input alphabet Σ as well. In the case of a hyper-heuristic the alphabet consists out of the set of heuristics one can apply. Depending on the input h_i, a different transition matrix A_i is applied. One can generate an IOHMM model for a set of heuristics H and an initializer analogue to a HMM, but the process of calculating the transition matrices A_i is repeated for each heuristic.

4.2 Learning Heuristic Behavior

Constructing an IOHMM for a specific problem instance is useless: first of all it requires at least $\mathcal{O}\left(|\mathcal{H}| \cdot |\mathcal{S}|^2\right)$ time, with $|\mathcal{H}|$ the number of heuristics and $|\mathcal{S}|$ the number of solutions of the problem instance, to generate an IOHMM for a problem instance. Thus, it is easier to enumerate the entire solution space in search for the global optimum.

An evolutionary process generates empirical evidence: a list of tuples containing both the heuristic that was called and the resulting behavior (i.e. the fitness value of the generated solution). The well known BAUM-WELCH algorithm uses the *Expectation-Maximization* methodology to learn values for π, A and B such that probability of generating a sequence like the empirical evidence is maximized for an a priori determined number of hidden states n. This is the best we can hope given we cannot make any assumptions regarding how the heuristics work.

The algorithm runs in $\mathcal{O}\left(t \cdot \left(n^2 + n \cdot m\right)\right)$ with t the number of data points, n the number of hidden states and m the number of possible observations. This is the time complexity of one step in the expectation-minimization process: it is possible that multiple iterations are necessary before the model parameters $\langle \pi, A, B \rangle$ converge towards a local optimum[4]. At this point we realized that the number of observations and hidden states is too large to learn a model effectively, so they must be reduced.

By reducing the number of hidden states, the hidden states no longer represent solutions, but distributions over the set of solutions. Each distribution marks solutions that show, according to the BAUM-WELCH algorithm, similarly with respect to the observations. Since the algorithm aims to maximize the probability of the observed data, solutions will be grouped if one or more heuristics behave similarly on both solutions. The number of hidden states can be a limiting factor: if the heuristic behavior is complex, it requires more hidden states. As the amount of empirical evidence grows, one can increase the number of hidden states to increase the quality of the model.

We also reduced the set of observations (in the context of a hyper-heuristic, the set of possible fitness values). If an IOHMM considers the set of all possible fitness values, the model has no means to generalize heuristic behavior and the model would have a hard time learning that a local search heuristic applied to one local optimum would generate the same solution, regardless of the fitness value of the first solution.

---

[4] Experiments show that such local optimum is nearly always near the global optimum, although sequences can be derived that are hard to learn.

Reasoning about the difference between two fitness values is therefore a better decision: it enables the model to learn that if there is no difference between the initial and final solution of a local search heuristic, there never will be any in the future.

Considering "differences" between two fitness values as the observation set leads to an inconsistency: differences between fitness values of two solutions do not correspond to a single solution. Heuristics can produce different fitness differences for the same solution.

We can solve this issue by squaring the number of hidden states: in that case each hidden state represents a tuple containing the old and the new solution. In that case the hidden Markov model has a "memory"[5] of 1 time step. The computational effort invested in learning how to handle such memory will however increase significantly: the number of parameters to learn is now $n^4 + n^2 \cdot m$ with n the number of original hidden states and m the number of "difference" observations.

One can "pre-encode" the use of memory using a *Mealy Input-Output hidden Markov Model*: a IOHMM where the observed difference depends on both the solution and the heuristic applied on that solution.

Definition 3 (Mealy Input-Output Hidden Markov Model (MIO-HMM)). *A* Mealy Input-Output hidden Markov model *is a 3-tuple* $\langle \pi, A, B \rangle$ *with* π *a probability n-vector, A an $l \times n \times n$ transition matrix and B an $l \times n \times m$ emission matrix. Each matrix A_i is a transition matrix and each matrix B_i is an emission matrix as defined in Definition 1.*

The observation y_t no longer depends on current hidden state z_t, but on the previous hidden state z_{t-1} and the input token x_t. The BAUM-WELCH algorithm can be modified such that it trains a MIOHMM in the same time complexity as training a hidden Markov model. Figure 4 depicts the evolution of a MIOHMM in time.

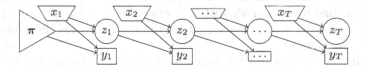

Fig. 4. The markov process described by a mealy input-output hidden markov model.

For our experiments, we reduced the number of observations to three: *better*, *worse* and *same*.

---

[5] This in contrast with the *Markov assumption* that states that a Markov process has no memory.

4.3   Selecting a Heuristic

Based on the information collected, compressed and stored by the MIOHMM, one needs to decide which heuristic to apply next in the evolutionary process. An advantage of a hidden Markov model is that it can calculate the distribution over the hidden states at any point in the process. Based on the earlier historical evidence and the model itself, one can predict with which probability a heuristic will produce a better, equivalent or worse solution, this of course given the model is correct.

Our hyper-heuristic submitted to the *CheSC 2014*[6] challenge used the following selection procedure: we designed a desired emission probability vector with values:

$$\boldsymbol{d} = \langle d_+, d_=, d_- \rangle = \langle 0.6, 0.05, 0.35 \rangle \tag{2}$$

At each decision point the heuristic behavior is predicted. The heuristic for which the dot-product between $\boldsymbol{d}$ and the predicted output is maximized is selected as the next heuristic.

We did not perform any tuning on the $\boldsymbol{d}$ vector: the vector merely favors generating a better solution over a worse solution and a worse solution over an equivalent solution. Since we are only interested in the heuristic that maximizes the dot-product we think this is a robust metric: a small difference in the $\boldsymbol{d}$-vector will typically only lead to a different decision on rare occasions.

The selection procedure is still a weak spot in our hyper-heuristic algorithm.

4.4   Forgetting Learned Experience

Since the number of empirical samples keeps increasing, adapting our model to the latest measuring point would require increasing computational effort at each time step. By considering only a time window of samples, we set a threshold on the maximum amount of effort spend on improving the learned model.

Since the observed data originates in many cases from the same region in the search space, the MIOHMM will learn a model aligned to this region. By considering a time frame, our algorithm has the ability to forget past experience that would make the model less suited for the challenges for the evolutionary process at that moment.

The BAUM-WELCH algorithm tends to stick with an earlier learned model. For instance, transition probabilities close to 0.0 require many iterations to increase to a significant level. Since the transition matrices of learned hidden Markov models tend to be sparse learning a better model can be hard.

We solved this by adding additional noise to the matrices: small probabilities were added to or subtracted from the elements from transition matrices. This noise can be seen as "forgetting" what has been learned in favor of accepting new experience. *Markovitch* [13] argues that forgetting is a vital point in learning that many algorithms tend to ignore.

---

[6] See Sect. 5.3.

5   Results

In this section we show that local search heuristics can be learned perfectly, the effect of the number of hidden states on the model quality and the hyper-heuristic performance in practice.

5.1   Local Search Heuristics

An encouraging theoretical result of the use of hidden Markov models, is that the model can easily learn the behavior of local search heuristics using two hidden states. The two states are called the *improvement state* s_i and the *non-improvement state* s_n.

Fig. 5. A model of a local search heuristic requires two hidden states.

The probability of a better solution $(+)$ in the first state is 100% as is the probability of generating an equal solution $(=)$ in the second. The transition probability of s_n to itself is $p_{nn} = 100\%$ as well since our local search heuristic reached an optimum.

In case the *local search heuristic* guarantees a local optimum after one function application (as is the case in *HyFlex 1.0* [5]), the transition from the improvement state s_i to the optimum state is 100%. In case it can take an undetermined number of applications of the heuristic, the probability is $p_{in} = 1/\lambda$ with λ the average number of consecutive improvements until an optimum is reached. The transition probability from the improvement state to itself is defined by $p_{ii} = 1 - p_{in}$.

If the result is guaranteed to be a local optimum, this behavior can be learned from three observations. Otherwise it requires a sequence of heuristic applications until an optimum is reached to estimate p_{in} effectively. The precision of p_{in} increases with $\mathcal{O}\left(1/\sqrt{k}\right)$ with k the number of sequences of the local search heuristic that end up in a local optimum.

5.2   Number of Hidden States Versus Model Quality

An advantage of the hidden Markov model approach is that the learning component acts rather independent from the decision component[7]. This allows one to analyze whether the MIOHMM is capable of modeling the heuristic behavior correctly. Sometimes machine learning algorithms fail to improve the overall model quality: some problem instances can be modeled better at the expense of others so that we end up with a zero-sum result.

---

[7] The decision component will however have an impact on the generated evidence.

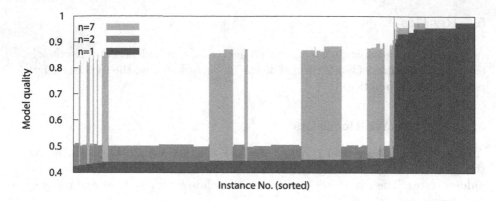

Fig. 6. The effect of the number of hidden states on the model quality.

We performed a batch of experiments in which we iterated over all possible MAX3SAT problems with 8 variables and 4 clauses. With symmetry breaking, this results in 199'057 unique problems. For each problem, we tried to learn the behavior of three low level MAX3SAT heuristics with a MIOHMM for a varying number of hidden states using data collected exhaustively over the entire solution space.

The quality of the learned model was evaluated by calculating the average number of times the model could predict the result of heuristic application in an evolutionary process correctly. Although we argued that the selection procedure is still a weak spot in our approach, the more accurate a model can predict the outcome of a heuristic, the better the decision a hyper-heuristic can make.

Figure 6 shows the results obtained with 1, 2 and 7 hidden states. As the number of hidden states increases, the achievable quality of a MIOHMM is guaranteed to increase[8]. Since there is no inherent order in MAX3SAT problem instances, we ordered the problem instances on increasing model quality of a MIOHMM with 1 hidden state. The regions in green and yellow show the increase of model quality compared to a MIOHMM with less hidden states.

The graph shows that as the number of hidden states increases, the model quality of certain chunks of problem instances increases significantly. For some instances, the learned model predicts the behavior correctly in more than 80 % of the cases.

The results might seem not that impressive, but note that a completely random selection would result in a model quality of 0.333. Moreover in this experiment we aimed to learn the heuristic behavior over the entire search space.

Since this data is not available in a real hyper-heuristic process, the hyper-heuristic will learn based on evidence of "local" data and thus specialize in the active region.

---

[8] Since the BAUM-WELCH algorithm is a heuristic learning algorithm, this is not guaranteed, but we never encountered such an example.

To the best of our knowledge, this is the first experiment performed with hyper-heuristics where a problem instance set is exhaustively enumerated. These experiments are an indication that one hyper-heuristic is more suited than another to learn generic heuristic behavior given the learning algorithm can be separated from the decision algorithm. The full batch of experiments is available at goo.gl/vVTZNE.

5.3 Hyper-heuristic Performance

Our hyper-heuristic based on MIOHMM was submitted to the *CheSC 2014* [14,15] competition for the parallel track. Parallelization was performed by multiple threads, learning and selecting based on the same model. No a priori knowledge about the heuristics[9] was added and the parameters as described by Eq. (2) were not fine-tuned.

The algorithm outperformed the only competitor, the *Evolving Tree Hyper-Heuristic (ETH)* [16] for the PROBESELECTIONPROBLEM and MULTI-DIMENSIONALKNAPSACKPROBLEM problem, but achieved worse results for the VEHICLEROUTINGPROBLEM. For more details, visit goo.gl/IQZ1bj.

6   Conclusion and Future Work

We have shown that the probability vector approach has problems learning heuristic behavior.

As an alternative, we have modeled a hyper-heuristic as an IOHMM. To effectively learn such model, its number of observations and hidden states must be reduced. This can result in a model with lower quality. Whether this is a serious issue depends on the problem at hand.

An additional advantage of our approach is that at each point in time, one can measure how well the model is trained with respect to historical data. The model can learn from multiple sources concurrently and thus has a benefit with respect to parallelization as was shown on the *CheSC 2014* competition.

On the whole we think our results are promising and warrant further investigation. In particular, further research must study alternatives for reducing the observation set. Also the selection procedure can be improved. It is possible to encode a priori known aspects about heuristics into our model such that well known aspects should not be learned.

Acknowledgements. This research is funded by the *Institute for Innovation through Science and Technology (IWT)* under grant 131′751.

References

1. Khamassi, I.: Ant-Q hyper heuristic approach applied to the cross-domain heuristic search challenge problems. In: CHeSC 2011 (2011)

---

[9] For instance, how a hyper-heuristic can model a local search heuristic.

2. McClymont, K., Keedwell, E.: A single objective variant of the online selective Markov chain Hyper-heuristic (MCHH-S). In: CHeSC 2011 (2011)
3. Wikipedia: Probability vector – wikipedia, the free encyclopedia (2014). Accessed 24 Sep 2014
4. Lipowski, A., Lipowska, D.: Roulette-wheel selection via stochastic acceptance. Phys. A Stat. Mech. Appl. **391**(6), 2193–2196 (2012)
5. Ochoa, G., et al.: HyFlex: a benchmark framework for cross-domain heuristic search. In: Hao, J.-K., Middendorf, M. (eds.) EvoCOP 2012. LNCS, vol. 7245, pp. 136–147. Springer, Heidelberg (2012)
6. Watkins, C.J.C.H.: Learning from delayed rewards. Ph.D. thesis, King's College, Cambridge, UK, May 1989
7. Gaspero, L.D., Urli, T.: A reinforcement learning approach for the cross-domain heuristic search challenge. In: CHeSC 2011 (2011)
8. Mısır, M., De Causmaecker, P., Vanden Berghe, G., Verbeeck, K.: An adaptive hyper-heuristic for CHeSC 2011. In: CHeSC 2011 (2011)
9. Mısır, M., Verbeeck, K., De Causmaecker, P., Vanden Berghe, G.: An intelligent hyper-heuristic framework for CHeSC 2011. In: Hamadi, Y., Schoenauer, M. (eds.) LION 2012. LNCS, vol. 7219, pp. 461–466. Springer, Heidelberg (2012)
10. Glover, F., Hanafi, S.: Tabu search and finite convergence. Discrete Appl. Math. **119**(1–2), 3–36 (2002)
11. Burke, E.K., Curtois, T., Hyde, M.R., Kendall, G., Ochoa, G., Petrovic, S., Rodríguez, J.A.V., Gendreau, M.: Iterated local search vs. hyper-heuristics: towards general-purpose search algorithms. In: IEEE Congress on Evolutionary Computation, pp. 1–8. IEEE (2010)
12. Bause, F.: Input-output hidden markov models for the aggregation of performance models. Sfb Teilprojekt M and Ls Informatik Iv and Modellierung Grosser and Sfb Teilprojekt M SFB559-03010, Universität Dortmund, Dortmund, July 2003
13. Markovitch, S., Scott, P.D.: The role of forgetting in learning. In: Laird, J.E. (ed.) ML, pp. 459–465. Morgan Kaufmann, Ann Arbor (1988)
14. Asta, S., Özcan, E., Parkes, A.J.: Batched mode hyper-heuristics. In: Nicosia, G., Pardalos, P. (eds.) LION 7. LNCS, vol. 7997, pp. 404–409. Springer, Heidelberg (2013)
15. Van Onsem, W., Demoen, B., De Causmaecker, P.: HHaaHHM: hyper-heuristics as a hidden Markov model. In: Proceedings of the Cross-domain Heuristic Selection Competition 2014, April 2014
16. Pihera, J., Musliu, N.: ETHH - evolving tree hyper-heuristic. In: Proceedings of the Cross-domain Heuristic Selection Competition 2014, April 2014

Comparison of Parameter Control Mechanisms in Multi-objective Differential Evolution

Martin Drozdik[✉], Hernan Aguirre, Youhei Akimoto, and Kiyoshi Tanaka

Interdisciplinary Graduate School of Science and Technology,
Shinshu University, Nagano, Japan
martin@iplab.shinshu-u.ac.jp

Abstract. Differential evolution (DE) is a powerful and simple algorithm for single- and multi-objective optimization. However, its performance is highly dependent on the right choice of parameters. To mitigate
this problem, mechanisms have been developed to automatically control
the parameters during the algorithm run. These mechanisms are usually
a part of a unified DE algorithm, which makes it difficult to compare
them in isolation. In this paper, we go through various deterministic,
adaptive, and self-adaptive approaches to parameter control, isolate the
underlying mechanisms, and apply them to a single, simple differential
evolution algorithm. We observe its performance and behavior on a set
of benchmark problems. We find that even the simplest mechanisms can
compete with parameter values found by exhaustive grid search. We
also notice that *self-adaptive* mechanisms seem to perform better on
problems which can be optimized with a very limited set of parameters.
Yet, *adaptive* mechanisms seem to behave in a problem-independent way,
detrimental to their performance.

Keywords: Differential evolution · Multi-objective optimization · Parameter control · Comparative study

1 Introduction

Differential evolution (DE) [14] is a simple to understand, but nevertheless powerful optimizer. However, its performance is highly sensitive to the choice of
parameters. Moreover, this dependency changes from problem to problem. Selection of well performing fixed parameters for a particular optimization problem
is a relatively little understood subject, especially in the multi-objective realm.
This motivated many researchers to develop techniques to set the parameters
automatically, during the run of the DE algorithm.

According to the taxonomy in [5], parameter setting techniques are divided
into *parameter tuning*, which happens before the run, and *parameter control*,
which happens during the run. The former is a subset of a larger field, called

M. Drozdik—The work of Martin Drozdik has been supported by the Ministry of
Education, Culture, Sports, Science, and Technology of Japan.

C. Dhaenens et al. (Eds.): LION 9 2015, LNCS 8994, pp. 89–103, 2015.
DOI: 10.1007/978-3-319-19084-6_8

algorithm configuration, which is itself a deeply researched subject [8]. In this work however, we study only parameter control mechanisms.

Parameter control (PC) techniques are further divided into *deterministic, adaptive,* and *self-adaptive. Deterministic* techniques apply the parameters according to a given rule, while ignoring any feedback from the search process. *Adaptive* techniques continually update their parameters using feedback from the population. *Self-adaptive* techniques attach different parameters to each individual. These parameters undergo mutation and recombination along with the individuals. Better parameter values lead to individuals with a higher chance to survive and therefore have a higher chance to propagate to the next generation.

Each mentioned paradigm of parameter control is represented by numerous algorithms in the literature. One of the first attempts to control parameters in DE is the (multi-objective) SPDE algorithm [1] belonging to the self-adaptive category. An adaptive mechanism based on population diversity for both single- and multi-objective DE has been proposed by Zaharie in [19]. The use of fuzzy controllers to adapt the parameters has been proposed by Liu et al. [11] The SaDE algorithm [15], originally proposed for single-objective DE, adapts the used DE strategies as well as the parameters. SaDE, which is an adaptive algorithm according to our classification, has been generalized to multi-objective realm and subsequently improved to OW-MOSaDE [6]. A comparison of *single-objective* adaptive and self-adaptive methods is presented in [2] and in [3].

A typical modern multi-objective algorithm is in fact an orchestra of sub-algorithms, each playing its own instrument. There is a sub-algorithm to initialize the population, a sub-algorithm to select individuals for reproduction, a sub-algorithm to maintain diversity, and so on. Various techniques for parameter control are usually published as a part of a unified production-ready algorithm. Apart from the parameter control mechanism, this algorithm usually has its own sub-algorithms to perform tasks *not* related to parameter control. These sub-algorithms usually vary from algorithm to algorithm and make the comparison of algorithms difficult, since it is not clear if the difference in performance should be attributed to the parameter control mechanism itself, or to the difference in sub-algorithms. For example, to estimate diversity of an individual, the OW-MOSaDE algorithm [6] uses the harmonic average distance measure, while the JADE2 algorithm [21] uses the product of distances. In order to isolate these effects, we implement all the parameter control methods within a simple multi-objective DE algorithm DEMO [16].

In this paper, we want to find out if some parameter control paradigm is inherently better in terms of performance and whether the parameter control mechanisms can find favorable parameters in problems which can be successfully optimized only with a limited set of parameters. We are also interested in finding an explanation of the observed performance. We do this by observing the evolution of parameters used by the parameter control methods throughout the optimization process. For this paper, we tried to choose representative examples from each group. We compare one deterministic, three adaptive, and four self-adaptive methods. Some of the methods we present here are originally used only for single-objective optimization, but they can be easily adopted to multi-objective optimization, which we do in this paper.

We conclude, that *self-adaptive* methods are the most robust methods, while performing on a par with the best fixed parameter settings. We found out that adaptive methods may have big problems to find good parameters. Moreover, they seem to adapt their parameters in patterns *independent* of the problem.

However, our conclusions come from empirical results with a single (DEMO) algorithm. It is possible, that applying the studied PC mechanisms to a different algorithm may yield very different results. Such was the case in [13] where the authors studied parameter control in ant colony optimization algorithms.

In the following section we introduce the various mechanisms we use in this work. In Sect. 3 we explain the details of our experimental setup. In Sect. 4 we discuss and interpret the empirical results and we conclude in Sect. 5.

2   Approaches to Parameter Control in DE

In this section we describe DE in more detail and introduce its parameters. Then we introduce the mechanisms that we examine in this paper.

2.1   Differential Evolution Parameters

The fundamental principle of DE is to create new individuals by adding scaled *differences* of individuals to each other. Let $P = \{X_1, ..., X_{NP}\}$, where $X_i = (x_{i,1}, ..., x_{i,n}) \in \mathbb{R}^n$, be the population. In its most basic form, DE traverses through P, attempting to improve each individual X_{target} by generating a *new* individual X_{trial} in the following way: First, three distinct individuals, $X_{r_1}, X_{r_2}, X_{r_3}$ are chosen from P. Then a *scaled* difference of two of these individuals is added to the third one and an intermediate individual X_{mutant} is created:

$$X_{\text{mutant}} := X_{r_1} + \text{F}(X_{r_2} - X_{r_3}). \tag{1}$$

The scaling factor F is the first parameter of DE. Then the X_{trial} is generated by randomly inheriting variables from either X_{mutant} or from X_{target}. One variable with a randomly chosen index inv is automatically inherited from X_{mutant} to avoid generating a copy of X_{target}. This is described in (2), where $\text{rand}_U(0, 1)$ is a generator of uniformly randomly distributed numbers in $[0; 1]$.

$$x_{\text{trial},i} := \begin{cases} x_{\text{mutant},i} & \text{if } \text{rand}_U(0, 1) < \text{Cr or } i = inv \\ x_{\text{target},i} & \text{else} \end{cases} \tag{2}$$

The number Cr in (2) is the second parameter of DE and it is called the *crossover probability*. Cr controls the proportion of variables that are perturbed in an incumbent individual X_{target} to create a new individual. When Cr = 0, only one variable changes at a time, hence Cr = 0 is well suited for *separable* problems.

Very significant work on understanding the theoretical properties of F and Cr has been done by Zaharie in [18]. An empirical analysis has been performed by Kukkonen in [10]. The population size NP is also considered a parameter of DE, and there have been attempts to adapt the population size as well [17], but in this paper we restrict ourselves to parameters F and Cr. Moreover, strategies to generate X_{trial}, different than the one in (1) and (2) have been proposed, but in this work we shall consider *only* the default strategy. Next, we present all the parameter control mechanisms that we consider in this study.

2.2 Deterministic Mechanism for Parameter Control

The MDDE algorithm [22] initializes the parameters as relatively big values F_0, Cr_0, to prevent premature convergence. Then it monotonically decreases them with respect to the generation g, in a geometric sequence, according to:

$$F_g := F_0 \exp(-a_0 \frac{g}{g_{\text{max}}})$$
$$Cr_g := Cr_0 \exp(-a_1 \frac{g}{g_{\text{max}}}),$$

where g_{max} is the maximum number of generations.

2.3 Adaptive Mechanisms for Parameter Control

JADE2. The adaptive mechanism in the JADE2 algorithm generates new values of F and Cr for each new X_{trial}. If a particular X_{trial} Pareto dominates the X_{target}, the combination of F and Cr which generated the X_{trial} is recorded as a *successful* one. The values of F are generated from a Cauchy distribution with median μ_F and scale $\gamma = 0.1$, the values of Cr from a normal distribution with mean μ_{Cr} and $\sigma = 0.1$. At the end of each generation, the parameters of these distributions are updated using the following rules:

$$\mu_F := (1 - c)\mu_F + c.\text{avg}_L(F_s)$$
$$\mu_{Cr} := (1 - c)\mu_{Cr} + c.\text{avg}_A(Cr_s),$$

where $c \in [0; 1]$ is a learning factor, $\text{avg}_L(F_s)$ is the Lehmer mean of all successful F's and avg_A is the arithmetic mean of successful Cr's in the previous generation. In our experiments we used $c = 0.1$, as recommended by the authors.

OW-MOSaDE. Objective-wise MOSaDE [6] attempts to learn which value of Cr is good for a *particular objective*. For each objective $f_i \in (f_1, \ldots, f_m)$ OW-MOSaDE holds one value of $\mu_{i,Cr}$. These values are updated at the end of each generation if the X_{trial} generated by a particular Cr improves objective f_i. In addition, a master μ_{Cr} is updated if *all* objectives are improved simultaneously. At each generation, one of these $m + 1$ values is randomly chosen to serve as the mean of a normal random distribution with $\sigma = 0.1$, which is sampled to generate the values of Cr. That is, each generation the algorithm concentrates on either one randomly chosen objective or attempts to improve all objectives at once. As opposed to JADE2, there is no learning factor, but the successful values of Cr are retained for lp generations, where $lp = 50$ is a learning period. The value F is not adapted, but generated randomly from a fixed set of normal distributions for each individual.

Control of Diversity Adaptation Algorithm (PDCaDE). Zaharie discovered a simple algebraic relationship between the expected variance of the DE population before and after the generation of new individuals [18]. Based on these results, she developed an algorithm which monitors the variance of the population in the decision space and alters the parameters according to this relationship, so that the variance of the population decreases in a specified, steady manner throughout the entire run. The motivation is to prevent premature convergence and to use the allocated budget of generations evenly.

The algorithm does not have a specific name, so we call it *Population Diversity Control Adaptive DE (PDCaDE)* in the rest of this article. PDCaDE introduces a new parameter γ, which we held constant at $\gamma = 1.25$ for all our experiments. This value was determined by some limited tuning, since the author does not provide a recommendation for multi-objective problems.

2.4 Self-adaptive Mechanisms for Parameter Control

The main idea behind self adaptive mechanisms is that each individual carries the set of parameters *by which it was created*. This way, if an individual is created by a good set of parameters and survives into the next generation, the parameters it carries survive too. Conversely, bad parameter combinations get pruned away.

In all self-adaptive DE mechanisms considered in this paper, the principle is the same. New individuals X_{trial} are generated using (1) and (2), where the F and Cr are not fixed, but replaced by F_{trial} and Cr_{trial}. These values are generated on the spot and then carried by the newly generated X_{trial}. Let us denote by F_i and Cr_i the parameter values carried by individual X_i. Then the methods to generate F_{trial} and Cr_{trial} can be described by simple equations in Table 1.

3 Experimental Design

3.1 Algorithmic Framework

Algorithm 1 shows the unified framework used to compare the selected parameter control mechanisms. The lines that apply *only* to self-adaptive mechanisms are highlighted in *yellow*, while the ones that apply *only* to adaptive mechanisms are highlighted in *purple*.

If we want to draw conclusions about PC mechanisms in general using this methodology, we rely on the following assumption: Let A, B be two PC mechanisms and X, Y be two DE algorithms. If $X(A)$ (algorithm X with mechanism A) is better than $X(B)$ in some regard, then $Y(A)$ is better than $Y(B)$. Surely the validity of this assumption depends on many factors. Some research in this direction has been done in [13]. Since all our experiments are performed within this single algorithm, the most important task for future work is to explore the validity of our assumption.

Some methods have their own parameters, which we held constant at the values recommended by their authors. Some methods also use several strategies to generate new individuals, but in this work we limited ourselves to the default strategy described in Eqs. (1) and (2).

Table 1. Summary of used *self-adaptive* mechanisms

Name	Main formula	Additional parameters
SPDE [1]	$F_{trial} := rand_N(0,1)$ $Cr_{trial} := Cr_{r_1} + rand_N(0,1)(Cr_{r_2} - Cr_{r_3})$	$Cr_{init} := rand_N(\mu, \sigma)$ $\mu = 0.5, \sigma = 0.15$
jDE [3]	$F_{trial} := \begin{cases} rand_U(0.1, 1.0) & \text{if } rand_U(0,1) < \tau_1 \\ F_{target} & \text{else} \end{cases}$ $Cr_{trial} := \begin{cases} rand_U(0.1, 1.0) & \text{if } rand_U(0,1) < \tau_2 \\ Cr_{target} & \text{else} \end{cases}$	$\tau_1 = 0.1$ $\tau_2 = 0.1$
DEMOwSA [20]	$F_{trial} = \frac{F_i + F_{r_1} + F_{r_2} + F_{r_3}}{4} e^{\tau rand_N(0,1)}$ $Cr_{trial} = \frac{Cr_i + Cr_{r_1} + Cr_{r_2} + Cr_{r_3}}{4} e^{\tau rand_N(0,1)}$	$\tau = \frac{1}{\sqrt{2n}}$
SAMDE [12]	$F_{trial} = F_{r_1} + F'(F_{r_2} - F_{r_3})$ $Cr_{trial} = Cr_{r_1} + F'(Cr_{r_2} - Cr_{r_3})$	$F' := rand_U(0.7, 1.0)$
$rand_N(\mu, \sigma)$ - generator of normal random numbers		
$rand_U(a, b)$ - generator of uniform random numbers		

Algorithm 1. Adaptive and self-adaptive DEMO [16] algorithm

---

1 initialize $P = \{X_1, ..., X_{NP}\}$ uniformly randomly in the decision space
2 initialize F and Cr generators
3 initialize values of F_i and Cr_i for $i = 1, \ldots, NP$
4 **for** generation := 1 *to* G_{max} **do** *Evolutionary loop*
5   **for** target := 1 *to* NP **do** *Generational loop*
6   generate F_{trial} and Cr_{trial}
7   compute F_{trial} and Cr_{trial} using Table 1
8   generate X_{trial} using F_{trial} and Cr_{trial} from (1) and (2)
9   attach F_{trial} and Cr_{trial} to X_{trial}
10   project X_{trial} to decision space
11   **if** X_{target} *dominates* X_{trial} **then**
12   discard X_{trial}
13   **else if** X_{trial} *dominates* X_{target} **then**
14   replace X_{target} with X_{trial}
15   **else if** X_{target} *and* X_{trial} *are mutually non-dominated* **then**
16   add X_{trial} to the end of the population
17   **end**
18   update success memories
19   **end**
20   update parameter generators
21   Trim P to size NP using non-dominated sorting [16] and MNN diversity [9]
22 **end**

Table 2. Characteristics of the selected WFG problems

	WFG4	WFG6	WFG7	WFG9
Separable	yes	no	yes	no
Unimodal	no	yes	yes	no

3.2   Problems

WFG Problems. To test the mechanisms in various conditions, we chose a subset of the WFG [7] test suite with the same concave Pareto front and all possible combinations of separability and modality characteristics. These problems are summarized in Table 2. We held the number of variables fixed at 10 and performed tests for 2 and 3 objectives.

Quadratic Problems. As we shall later see, even the non-separable multi-modal WFG problems can be successfully optimized using many combinations of fixed parameters. To test the ability of parameter control mechanisms to solve challenging problems we developed a scalable problem, that can be solved by relatively few combinations of F and Cr, called Q. The problem Q consists of m functions: $Q = (q_1, \ldots, q_m)$. Each function is a quadratic form $q_m(X) = (X - c_m)D_m(X - c_m)^T$ where

$$D_1 = \operatorname{diag}(1, 2, 4, \ldots, 2^{n-1}),$$
$$D_2 = \operatorname{diag}(2^{n-1}, 1, 2, \ldots, 2^{n-2}),$$
$$\ldots$$

and the vectors c_q are generated uniformly randomly in a unit sphere. The resulting problem is then rotated in the decision space around all $n - 2$ rotation subspaces by 45 degrees.[1] Moreover, the population for this problem is generated randomly uniformly in a sphere of radius 2^{10} which is shifted from the origin in a random direction by 2^{14}. In this work, we explore the Q problem for 2, 3, and 4 objectives, while the number of variables remains fixed at 10.

3.3   Observed Statistics

We are interested in the *performance* of the various methods as well as in their *behavior*. To measure the performance, we use the hypervolume [23] metric, since it measures both convergence and diversity of the resulting Pareto front approximation. As a reference point for both types of problems we first construct the hyperbox which contains the entire true Pareto front and add a unit vector to its upper corner.

In order to simplify the interpretation of the hypervolume, we normalize it by dividing it by the maximal attainable hypervolume in the case of WFG problems, and by the volume of the hyperbox between the origin and the reference point

---

[1] Details on this methodology can be found in [4].

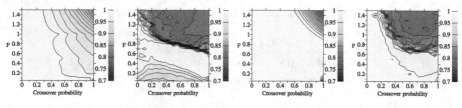

(a) WFG4 (S-MM)   (b) WFG6 (S-UM)   (c) WFG7 (NS-UM)   (d) WFG9 (NS-MM)

Fig. 1. Average normalized hypervolume for 2 objectives

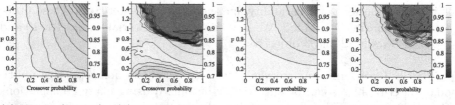

(a) WFG4 (S-MM)   (b) WFG6 (S-UM)   (c) WFG7 (NS-UM)   (d) WFG9 (NS-MM)

Fig. 2. Average normalized hypervolume for 3 objectives

for the Q problem. This way we know that the maximal attainable normalized hypervolume, corresponding to complete convergence is 1.

In order to observe the *behavior* of the mechanisms, we log each combination of F and Cr that the algorithm uses in one generation.

4   Results and Discussion

4.1   Parameter Tuning

For each problem we performed a preliminary tuning of the F and Cr parameters by grid search. We explored the ranges F $\in$ [0.05; 1.5] and Cr $\in$ [0; 1] with a resolution of 0.05. For each combination of parameters, we ran 10 independent runs of Algorithm 1 with fixed parameters. The average normalized hypervolume from this tuning is presented in the form of heat-maps, with hot colors meaning good performance. The tuning results for the WFG problems are in Figs. 1 and 2. In each figure we see a bright, L-shaped region of favorable values. Some theoretical explanation of the shape of this region can be found in [10] and [4].

4.2   Parameter Control on WFG Problems

For each of the studied methods we ran 50 independent runs with a fixed population size (NP) of 500 individuals. Each run was limited by 500 generations. The average normal hypervolume along with the standard deviation across the 50 runs is presented in Table 3. The value of normalized hypervolume at the

Table 3. Average normalized hypervolume for the WFG problems

		2 objectives			
		WFG4	WFG6	WFG7	WFG9
	start	0.774 (1.2e-02)	0.618 (1.5e-02)	0.706 (8.6e-03)	0.666 (2.2e-02)
	ideal	0.999 (6.9e-05)	0.999 (5.9e-04)	0.999 (3.5e-04)	0.996 (1.4e-03)
	MDDE	0.998 (3.4e-04)	0.820 (8.6e-04)	0.999 (7.5e-06)	0.905 (8.3e-02)
adaptive	JADE2	**0.999** (7.2e-06)	**0.992** (4.4e-03)	0.999 (1.0e-05)	**0.996** (8.7e-04)
adaptive	OW-MOSaDE	0.997 (6.6e-04)	0.975 (6.4e-03)	**0.999** (9.1e-06)	0.993 (2.0e-03)
adaptive	PDCaDE	0.998 (1.7e-03)	0.980 (7.1e-03)	0.999 (4.9e-05)	0.993 (2.2e-03)
self-adaptive	DEMOwSA	0.998 (2.9e-04)	**0.991** (3.8e-03)	0.999 (2.1e-05)	0.989 (3.5e-03)
self-adaptive	jDE	**0.999**(7.5e-06)	0.985 (1.7e-02)	**0.999** (1.4e-05)	**0.996** (1.0e-03)
self-adaptive	SAMDE	0.999 (8.0e-06)	0.980 (1.4e-02)	0.999 (1.5e-05)	0.995 (9.5e-04)
self-adaptive	SPDE	0.999 (1.1e-05)	0.970 (1.1e-02)	0.999 (8.8e-06)	0.996 (4.7e-04)
		3 objectives			
		WFG4	WFG6	WFG7	WFG9
	start	0.669 (1.9e-02)	0.563 (8.9e-03)	0.646 (1.0e-02)	0.575 (2.0e-02)
	ideal	0.987 (2.2e-04)	0.983 (1.6e-03)	0.988 (1.1e-04)	0.976 (2.9e-03)
	MDDE	0.978 (6.7e-04)	0.807 (3.6e-02)	0.986 (1.4e-04)	0.892 (8.4e-02)
adaptive	JADE2	**0.982** (4.3e-04)	**0.977** (3.5e-03)	0.978 (4.5e-04)	**0.965** (1.9e-03)
adaptive	OW-MOSaDE	0.968 (1.3e-03)	0.971 (8.3e-03)	0.977 (5.2e-04)	0.961 (1.6e-03)
adaptive	PDCaDE	0.974 (2.6e-03)	0.966 (8.4e-03)	**0.979** (9.6e-04)	0.962 (2.2e-03)
self-adaptive	DEMOwSA	0.972 (1.0e-03)	0.970 (1.6e-03)	0.975 (9.3e-04)	0.959 (1.8e-03)
self-adaptive	jDE	0.982 (7.0e-04)	0.967 (1.3e-02)	**0.981** (5.3e-04)	**0.968** (2.7e-03)
self-adaptive	SAMDE	0.983 (5.8e-04)	0.968 (8.8e-03)	0.977 (5.4e-04)	0.963 (1.5e-03)
self-adaptive	SPDE	**0.985** (6.6e-04)	0.964 (1.1e-02)	0.980 (3.8e-04)	0.965 (1.6e-03)

start of the run is denoted as *start*. For each problem, based on the initial tuning, we constructed an *ideal* set of fixed parameters and ran the algorithm for 50 independent runs with these settings. Within the group of adaptive methods and the group of self-adaptive methods we marked the highest value in bold.

We can see that both adaptive and self-adaptive methods are performing almost on a par with the ideal parameter set. The only exception is the deterministic MDDE algorithm, which shows significant problems for the non-separable WFG6 and WFG9 problems.

For each method, we plot the path of the average used F and Cr with respect to the generation. We call this plot the *trajectory* of that method. The averaged (over the 50 runs) trajectories of adaptive methods along with the MDDE method are plotted in Fig. 3 and the trajectories of the self-adaptive methods

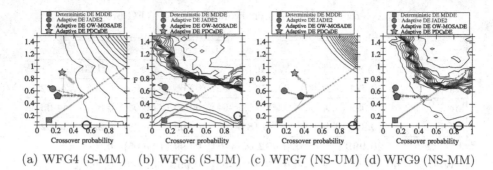

(a) WFG4 (S-MM)   (b) WFG6 (S-UM)   (c) WFG7 (NS-UM)   (d) WFG9 (NS-MM)

Fig. 3. [Cr; F] trajectories of adaptive methods for 2 objectives

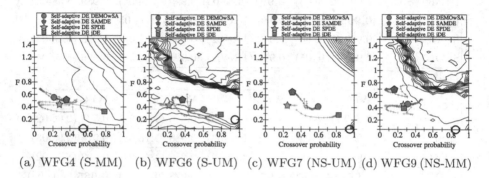

(a) WFG4 (S-MM)   (b) WFG6 (S-UM)   (c) WFG7 (NS-UM)   (d) WFG9 (NS-MM)

Fig. 4. [Cr; F] trajectories of self-adaptive methods for 2 objectives

are in Fig. 4. The small crosses are plotted for each 10 generations and the *final* reached value is marked by a large *symbol*. The optimal value of F and Cr is marked by a *black circle*. Moreover, all graphs contain the contour lines of the average normalized hypervolume obtained by parameter tuning.

It is immediately clear that all trajectories have different starting points. This is because each PC mechanism has its own way of initialization. Next, we see that each *adaptive* method seems to behave the same way across all observed problems. That is, both JADE2 and OW-MOSaDE aim for the lower values of Cr, while not adapting F much and PDCaDE seems to always converge to the same point, regardless of where the optimal parameter combination is. Conversely, the self-adaptive methods behave differently on each problem.

The situation is very similar for 3 objectives. The trajectories for the adaptive and deterministic methods in Fig. 5 seem to behave indifferently to the problem and to the number of objectives. On the other hand, the behavior of self-adaptive methods in Fig. 6 depends on the problem. Looking at the results of parameter tuning in Figs. 1 and 2 we see a possible explanation. The heat-maps of normalized hypervolume for problems WFG4 and WFG6 have more structure than those of WFG7 and WFG9. The contour lines are more evenly distributed, which may help the algorithms find favorable parameter values. Conversely, the

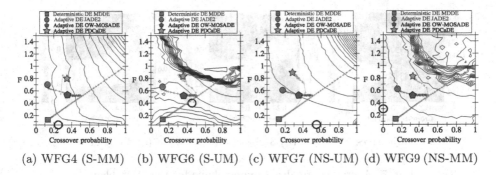

(a) WFG4 (S-MM)   (b) WFG6 (S-UM)   (c) WFG7 (NS-UM)   (d) WFG9 (NS-MM)

Fig. 5. [Cr; F] trajectories of adaptive methods for 3 objectives

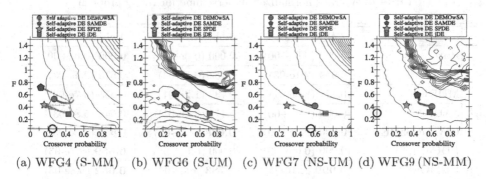

(a) WFG4 (S-MM)   (b) WFG6 (S-UM)   (c) WFG7 (NS-UM)   (d) WFG9 (NS-MM)

Fig. 6. [Cr; F] trajectories of self-adaptive methods for 3 objectives

heat-maps for WFG7 and WFG9 have large plateaus associated with favorable parameters, separated by steep cliffs from plateaus with bad parameters. Consequently we see that on WFG4 and WFG6 problems, the trajectories of self-adaptive methods aim correctly for the more favorable regions, while on the WFG7 and WFG9 problems, the behavior seems more random.

4.3 Q Problems

The performance heat-maps for the Q problems are in Fig. 7. The contrast with the data for WFG in Figs. 1 and 2 is immediately visible. The area of favorable parameter combinations is relatively small. Moreover, the favorable area is surrounded by steep cliffs. Even a small change in one parameter may mean the difference between a successful convergence and total failure. On such hard problems, the difference in performance of parameter control methods becomes apparent. The averages and standard deviations of 50 independent runs for 500 generations with a population size of 500 individuals are presented in Table 4.

On the 2-objective Q problem *all* the adaptive methods fail completely. Out of 50 runs, not one of them approached close enough to the Pareto front. Some minor success has been achieved by the deterministic MDDE method, but the best performers are the *self-adaptive* methods. On the 3-objective problem, OW-MOSaDE catches up, while the other adaptive methods are lagging. For the

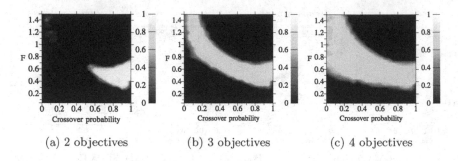

(a) 2 objectives   (b) 3 objectives   (c) 4 objectives

Fig. 7. Average normalized hypervolume for the Q problem

Table 4. Average normalized hypervolume for the Q problem

		2 objectives	3 objectives	4 objectives
	start	0.000 (0.0e+00)	0.000 (0.0e+00)	0.000 (0.0e+00)
	ideal	0.999 (2.1e-05)	0.783 (6.7e-04)	0.673 (2.4e-03)
	MDDE	0.128 (2.8e-01)	0.732 (1.2e-01)	0.653 (5.7e-03)
adaptive	JADE2	0.000 (0.0e+00)	0.175 (2.8e-01)	0.648 (5.8e-03)
	OW-MOSaDE	0.000 (0.0e+00)	**0.668** (1.1e-01)	0.654 (4.3e-03)
	PDCaDE	0.000 (0.0e+00)	0.000 (0.0e+00)	**0.659** (8.7e-03)
self-adaptive	DEMOwSA	**0.999** (2.1e-05)	**0.783** (6.8e-04)	**0.652** (5.8e-03)
	jDE	0.745 (4.0e-01)	0.433 (3.7e-01)	0.651 (6.1e-03)
	SAMDE	0.994 (1.5e-02)	0.778 (2.0e-03)	0.638 (7.6e-03)
	SPDE	0.548 (4.6e-01)	0.640 (2.4e-01)	0.643 (6.7e-03)

4-objective problem, the performances even out. It may seem counterintuitive that increasing the number of objectives makes the parameter control easier, but looking at Fig. 7 we see that the more objectives the Q problem has, the bigger is the set of favorable parameter combinations.

The trajectories of the *adaptive* mechanisms in Fig. 8 again seem to be very similar for the 2, 3 and 4 objective Q problems. Disturbingly, they resemble those of the WFG problems. The PDCaDE algorithm seems to always converge to Cr = 0.4 and F = 0.8. The OW-MOSaDE cannot adapt the distribution of the F parameter and invariably pushes the value of Cr down. This makes sense for the WFG problems, but it is counterproductive for the 2-objective Q problem. The JADE2 mechanism seems to be lured towards small values of Cr even more. Both JADE2 and OW-MOSaDE try to adapt the parameters by learning which parameters generate individuals which dominate another individual. This suggests that for each problem the parameters with this property are similar and that this property does not guarantee good performance. Of course, a more detailed and rigorous investigation is suggested as future work.

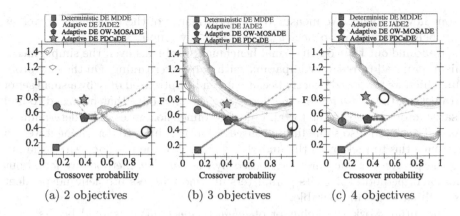

Fig. 8. [Cr; F] trajectories of adaptive methods for the Q problem.

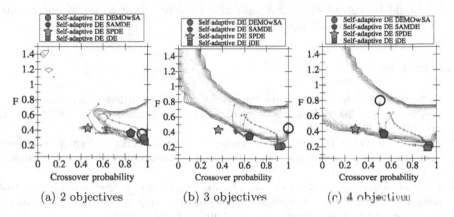

Fig. 9. [Cr; F] trajectories of self-adaptive methods for the Q problem.

The behavior of self-adaptive mechanisms in Fig. 9 is completely different. On the 2-objective problem, all self-adaptive mechanisms achieve at least half of the possible hypervolume. This is even true for the SPDE mechanism, which does *not* find the area of favorable parameter combinations. It seems that since the parameters of SPDE are generated randomly, favorable parameter combinations arise often enough to converge partially. The adaptive algorithms also generate their parameters randomly, but the centers of the random distributions from which these parameters are generated are shifting in the wrong direction.

5   Conclusion

In this paper we compared various deterministic, adaptive, and self-adaptive mechanisms of parameter control in multi-objective differential evolution. We isolated the mechanisms and applied them to a single multi-objective algorithm. We then tested this algorithm on a set of known benchmark problems as well as one new problem. We measured the performance of these methods as well as their behavior in terms of *which parameters* they found.

We found out that on the usual benchmark problems even the simple mechanisms can lead to results comparable with parameter tuning. On the new problem, which we proposed exactly because it can be optimized only by a small set of parameters, the *self-adaptive* methods were the only ones that managed to find a satisfactory Pareto front for all objective dimensionalities. The deterministic method achieved also some limited success, but it is hard to determine if we can attribute this to luck or to the underlying quality of the method. After examining the progress of the parameters used by the *adaptive* methods we found out that each method evolves its parameters in a more or less problem independent way, which seems undesirable.

For future work the behavior of *adaptive* mechanisms should be first confirmed to exist in other contexts, and if so, to be examined in detail and its cause should be established. It would also be interesting to see if our results hold for more modern DE algorithms.

References

1. Abbass, H.A.: The self-adaptive Pareto differential evolution algorithm. In: Proceedings of the 2002 Congress on Evolutionary Computation, CEC 2002, vol. 1, pp. 831–836 (May 2002)
2. Brest, J., Bošković, B., Greiner, S., Žumer, V., Maučec, M.S.: Performance comparison of self-adaptive and adaptive differential evolution algorithms. Soft Comput. **11**(7), 617–629 (2007)
3. Brest, J., Greiner, S., Boskovic, B., Mernik, M., Zumer, V.: Self-adapting control parameters in differential evolution: a comparative study on numerical benchmark problems. IEEE Trans. Evol. Comput. **10**(6), 646–657 (2006)
4. Drozdik, M., Tanaka, K., Aguirre, H., Verel, S., Liefooghe, A., Derbel, B.: An analysis of differential evolution parameters on rotated bi-objective optimization functions. In: Dick, G., Browne, W.N., Whigham, P., Zhang, M., Bui, L.T., Ishibuchi, H., Jin, Y., Li, X., Shi, Y., Singh, P., Tan, K.C., Tang, K. (eds.) SEAL 2014. LNCS, vol. 8886, pp. 143–154. Springer, Heidelberg (2014)
5. Eiben, A.E., Hinterding, R., Michalewicz, Z.: Parameter control in evolutionary algorithms. IEEE Trans. Evol. Comput. **3**(2), 124–141 (1999)
6. Huang, V.L., Zhao, S.Z., Mallipeddi, R., Suganthan, P.N.: Multi-objective optimization using self-adaptive differential evolution algorithm. In: IEEE Congress on Evolutionary Computation, CEC 2009, pp. 190–194 (May 2009)
7. Huband, S., Hingston, P., Barone, L., While, L.: A review of multiobjective test problems and a scalable test problem toolkit. IEEE Trans. Evol. Comput. **10**(5), 477–506 (2006)
8. Hutter, F., Hoos, H.H., Leyton-Brown, K., Sttzle, T.: ParamILS: an automatic algorithm configuration framework. J. Artif. Int. Res. **36**(1), 267–306 (2009)
9. Kukkonen, S., Deb, K.: A fast and effective method for pruning of non-dominated solutions in many-objective problems. In: Runarsson, T.P., Beyer, H.-G., Burke, E.K., Merelo-Guervós, J.J., Whitley, L.D., Yao, X. (eds.) PPSN 2006. LNCS, vol. 4193, pp. 553–562. Springer, Heidelberg (2006)

10. Kukkonen, S., Lampinen, J.: An empirical study of control parameters for the third version of generalized differential evolution (GDE3). In: IEEE Congress on Evolutionary Computation, pp. 2002–2009 (2006)
11. Liu, J., Lampinen, J.: A fuzzy adaptive differential evolution algorithm. Soft Comput. **9**(6), 448 (2005)
12. Pedrosa Silva, R.C., Lopes, R.A., Guimares, F.G.: Self-adaptive mutation in the differential evolution. In: GECCO, GECCO 2011, pp. 1939–1946. ACM, New York (2011)
13. Pellegrini, P., Sttzle, T., Birattari, M.: A critical analysis of parameter adaptation in ant colony optimization. Swarm Intell. **6**(1), 23–48 (2012)
14. Price, K., Storn, R.M., Lampinen, J.A.: Differential Evolution: A Practical Approach to Global Optimization. Springer, New York (2005)
15. Qin, A.K., Huang, V.L., Suganthan, P.N.: Differential evolution algorithm with strategy adaptation for global numerical optimization. IEEE Trans. Evol. Comput. **13**(2), 398–417 (2009)
16. Robič, T., Filipič, B.: DEMO: differential evolution for multiobjective optimization. In: Coello Coello, C.A., Hernández Aguirre, A., Zitzler, E. (eds.) EMO 2005. LNCS, vol. 3410, pp. 520–533. Springer, Heidelberg (2005)
17. Teo, J.: Exploring dynamic self-adaptive populations in differential evolution. Soft Comput. **10**(8), 673–686 (2006)
18. Zaharie, D.: Critical values for the control parameters of differential evolution algorithm. In: Proceedings of MENDEL 2002 (2002)
19. Zaharie, D.: Control of population diversity and adaptation in differential evolution algorithms. In: Matousek, R., Osmera, P. (eds.) Proceedings of Mendel 2003, 9th International Conference on Soft Computing, pp. 41–46, Brno, Czech Republic (jun 2003)
20. Zamuda, A., Brest, J., Boskovic, B., Zumer, V.: Differential evolution for multi-objective optimization with self adaptation. In: IEEE Congress on Evolutionary Computation, CEC 2007, pp. 3617–3624 (Sept 2007)
21. Zhang, J., Sanderson, A.C.: JADE: adaptive differential evolution with optional external archive. IEEE Trans. Evol. Comput. **13**(5), 945–958 (2009)
22. Zhang, M., Zhao, S., Wang, X.: Multi-objective evolutionary algorithm based on adaptive discrete differential evolution. In: IEEE Congress on Evolutionary Computation, CEC 2009, pp. 614–621 (May 2009)
23. Zitzler, E.: Evolutionary algorithms for multiobjective optimization: methods and applications. Ph.D. thesis, Computer Engineering and Network Laboratory, Swiss Federal Institute of Technology (ETH), Zurich, Switzerland (1999)

Genetic Programming, Logic Design and Case-Based Reasoning for Obstacle Avoidance

Andy Keane[✉]

Faculty of Engineering and the Environment, University of Southampton,
Southampton SO17 1BJ, UK
ajk@soton.ac.uk

Abstract. This paper draws on three different sets of ideas from computer science to develop a self-learning system capable of delivering an obstacle avoidance decision tree for simple mobile robots. All three topic areas have received considerable attention in the literature but their combination in the fashion reported here is new. This work is part of a wider initiative on problems where human reasoning is currently the most commonly used form of control. Typical examples are in sense and avoid studies for vehicles – for example the current lack of regulator approved sense and avoid systems is a key road-block to the wider deployment of uninhabited aerial vehicles (UAVs) in civil airspaces.

The paper shows that by using well established ideas from logic circuit design (the *"espresso"* algorithm) to influence genetic programming (GP), it is possible to evolve well-structured case-based reasoning (CBR) decision trees that can be used to control a mobile robot. The enhanced search works faster than a standard GP search while also providing improvements in best and average results. The resulting programs are non-intuitive yet solve difficult obstacle avoidance and exploration tasks using a parsimonious and unambiguous set of rules. They are based on studying sensor inputs to decide on simple robot movement control over a set of random maze navigation problems.

Keywords: Decision tree · Data mining · Feature engineering · Classification · Algorithm construction

1 Introduction

Genetic Programming (GP) is the basic building block of the methods proposed here. GP was initiated in the early 1990s, when Koza applied genetic algorithms (GAs) to the evolution of computer programs (Koza 1992). GP extends the use of GAs to evolve structures of significant complexity which demand sophisticated adaptive plans to improve their performance. Since that time very many papers have used the basic ideas proposed by Koza to study a bewildering array of problems, see for example the review by Espejo *et al.* (2010) of the applications

© Springer International Publishing Switzerland 2015
C. Dhaenens et al. (Eds.): LION 9 2015, LNCS 8994, pp. 104–118, 2015.
DOI: 10.1007/978-3-319-19084-6_9

of GP in classification and for an application in robotics the work of (Seo *et al.* 2010). In this paper GP is used to evolve control programs for a simple mobile robot equipped with adjacency touch sensors and the ability to move forward or rotate 90° to the left or right. This is a very basic problem but one that serves to illustrate the ideas being proposed and some of the shortcomings of a straightforward application of GP to obstacle avoidance programs.

In a conventional GP approach a series of random initial programs are used to seed the GP process and these are then evolved using the standard operations of crossover and mutation under the impact of selection pressure to produce improved programs to tackle a set of trial tasks. The basic problem with this simple naïve use of GP is that the decision trees evolved are generally difficult to interpret and can rapidly become very large and cumbersome. When designs become large they then become more costly to evaluate and this slows the whole search process down. Moreover, it is quite common to find that evolutionary improvement can stall after a short period of initial improvement unless recourse is made to very large population sizes. The whole aim of the present work is to bring to bear ideas from logic design and case-based reasoning to deal with these issues.

This paper is laid out as follows: in section two we briefly introduce the path planning task being studied before going on to describe the basic ideas of how GP systems can be applied to this problem in section three. In section four we provide some illustrative results obtained using a naïve GP system, while in section five we show how the *espresso* algorithm can be used to simplify evolved program structures, speeding up the search process and improving outcomes. We then show how the *espresso* logic tool can be applied to yield an improved GP process in section six before drawing our conclusions in section seven.

2   Path Planning

In path planning and obstacle avoidance two basic approaches can be adopted: either an initial exploration phase is carried out to map the robot's world, following which planning decision can be made using the derived map, possibly with map updates or, alternatively, decisions must be based on current sensor inputs and possibly previous readings and actions without a world map. When dealing with significantly changing or new environments such as for UAV control, it is clear that the off-line building of world maps of other UAVs and obstacles, prior to decision making is not appropriate. Therefore we consider here the problem of planning based on sensor inputs. Moreover, to simplify the problems being studied, we restrict ourselves to problems where decisions must be based solely on current information states. Thus the task faced by the robot is: given a mission to accomplish and current sensor readings, should the robot move forwards, turn right or turn left. The mission considered is the commonly used one of visiting as much of the available world as possible in the least number of moves while not colliding with obstacles, see for example Bearpark and Keane (2008).

Such a task can be carried out using a range of programming structures but that most similar to the approach adopted by human navigators is a form of case-based reasoning: i.e., a decision tree where a series of sensor predicates are tested one at a time until a match is found following which a pre-defined action is performed. Moreover, while human operators can be inconsistent in the actions they take when presented with sets of sensor inputs, it is important when considering problems subject to regulatory approval, such as vehicle navigation, that predictable behaviour is adopted. For example "rule-of-the-road" requirements on cars and aircraft lay down the expected behaviour of operators when confronted with certain known scenarios, i.e., turn to the right when confronted with an oncoming vehicle so as to pass port side to port side. The great attraction of such case-based approaches is that their logic can be studied and understood, often exhaustively, for all possible scenarios; see for example Weng et al. (2009). Our aim here is to try and build a GP system that produces well-structured case-based programs for dealing with the robot task planning problem studied. This is not as trivial as might at first be thought if the full power of the evolutionary operators of cross-over and mutation is not to be curtailed.

The key step proposed here is to adopt ideas developed for logic circuit design to allow a GP system to automatically build the case-based program structures desired. In a previous paper it was shown that the convergence of GP in producing robot path planning programs could be improved by re-writing the programs being developed in case-based form during the evolutionary search process (Bearpark and Keane 2008). In that paper arbitrary Reverse Polish Notation (RPN) control programs were re-written as large single case statements before being re-inserted into the GP system, something the authors termed conversion to "canonical form". The process used to carry out this conversion was not straightforward or simple to implement. Here a similar approach is adopted but program re-writing is accomplished by borrowing tools from VLSI circuit design. VLSI logic circuit design commonly involves millions of logic gates and these are routinely simplified using well established methods. The most well-known of these is the so called "*espresso*" algorithm originated at the University of California, Berkeley, now made available as part of the Octtools[1] package (McGeer et al., 1993). VLSI logic circuit simplification essentially boils down to taking a truth table that maps input states to output states for the proposed design and producing a new, simplified, design that produces the same truth table. *Espresso* does this in a very efficient way giving a high quality, if not always perfect, reduction in the number of logic gates needed to create a given truth table. Here we use the same algorithms to re-write the RPN structures being evolved by GP. Since the resulting structures are almost always shorter, often massively so, their evaluation may then be carried out much more quickly, allowing either faster or more exhaustive searches to be run, see for example Moraglio et al. (2012). The reduced program structures also fundamentally change the actions of mutation and crossover on a population of designs. As far as can be found by searches of ISI publication data-bases, this is the first time that the *espresso* algorithm has been applied to a GP system in this way.

---

[1]   http://embedded.eecs.berkeley.edu/pubs/downloads/octtools/.

3   Genetic Programming and Robot Control

A Genetic Algorithm (GA) uses techniques based on the natural evolution of species to gradually improve the quality of the data structures that it produces until an optimal solution is found, or the run is terminated. Generally, an initial set of possible solutions is chosen randomly to form the first generation. The GA performs genetic operations on the population, to produce another generation, and the process is repeated for a number of generations. The principal genetic operations are those of selection and crossover. The selection process measures the success of each member of the population in performing the allotted task, and selects the better members as candidates for a mating pool, here of the same size as the population. On average, high-scoring members of the population appear multiple times in the mating pool, while low-scoring members are not selected. Consequently the average quality of the population increases with each generation. A new generation is evolved by the crossover operation using the concepts of sexual reproduction. Further genetic operations may be performed, particularly mutation, in which randomly chosen genetic material is removed from a child and randomly generated material inserted in its place. In a GP system the data structures are computer algorithms, normally represented as trees in which the nodes are function nodes, representing a sub-routine in the algorithm, or terminal nodes, representing constants or variables defined by the algorithm. A simple example is shown in Fig. 1. The algorithm is 'executed' by a depth-first traverse of the tree, starting from the root node, searching for function nodes and their operands. An examination of nodes 2, 3 and 4 yields the logical value *true* or the logical value *false*. A *true* result causes the traverse to continue by examining node 5 and its operands. A *true* value for Z causes the execution of action A while a *false* value executes B. In the case where the expression X AND Y is *false*, the algorithm executes C. This program may be written in Reverse Polish Notation (RPN) form as: C, B, A, Z, IF, Y, X, AND, IF. To encode this for GP each operator, state and action is given a numerical code and the RPN re-written as a numerical string. When carrying out mutation and crossover care must be taken to ensure that only syntactically correct strings are produced. Thus each operator has a fixed number of operands (its cardinality) and each operator must have the correct type(s) of operands (i.e., AND, OR and NOT operands require state inputs and generate state outputs while an IF

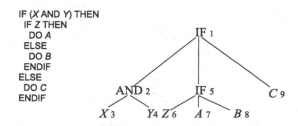

Fig. 1. A sample of control pseudo-code and the equivalent tree.

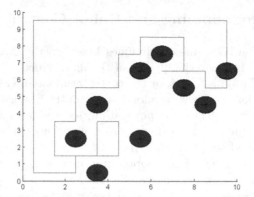

Fig. 2. The robot in its world (coloured red) with obstacles coloured blue and a possible path (the red solid line) (Color figure online).

requires either one state input and two action outputs or one state input and two state outputs and generates either an action or a state output, respectively – in this example X, Y and Z are states and A, B and C are actions).

Our GP system evolves algorithms that enable a software agent to solve basic problems in spatial exploration. An example is shown in Fig. 2. An enclosed 10 by 10 space has fixed internal obstacles coloured blue while the software robot itself is coloured red. The robot can either move forward one space, turn left or turn right before reassessing its environment. Here the robot is required to visit as many cells as possible using only a sense of touch, i.e., knowing only which of the eight adjacent cells is occupied. The robot thus has eight state inputs and three action outputs while the red line indicates a possible path. The score achieved by the robot is the number of squares visited divided by the number of vacant squares in the world. By testing a robot over a set of maze problems with random obstacle numbers and positions an average performance score can readily be computed – if the collection of random obstacle sets (mazes) is stored and used repeatedly, that score is stationary and can be used in a GP process to evolve better control programs. Notice that because the robot is blind and has no memory it can only navigate by touch and so all competitive navigation schemes involve moving towards an obstacle and then moving from obstacle to obstacle using these obstacles to aid navigation (precisely the sort of groping manoeuvre familiar to humans entering a darkened room). In an empty world a

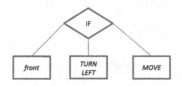

Fig. 3. A simple wall follower tree – this design scores 0.2761 when averaged over 50 of the test mazes.

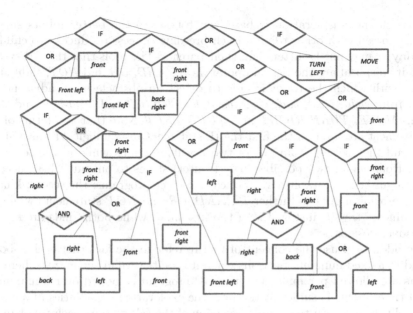

Fig. 4. A tree of 47 elements evolved by naïve GP after 100 generations – in RPN this is *MOVE, turn left, front, front right, front, front, left,* OR, IF, OR, *front right, right, back,* AND, *right,* IF, *front, front left, left,* OR, IF, OR, *front right, front, front, left,* OR, IF, OR, *front right, back, right, front left,* IF, *front, front left, front right, front right,* OR, *right, back,* AND, *right,* IF, OR, IF, OR, IF – this design scores 0.3922 when averaged over 50 mazes.

simple thought experiment reveals that the best logic that can be achieved is to move to the edge of the world and then circumnavigate the edge either clockwise or anticlockwise, giving a maximum score in a ten by ten world of 44/100 (if one starts with one's back to the wall and walks forward to the opposite wall and then moves around the perimeter). The tree in Fig. 3 does just this and will cause the robot to move in the direction in which it is facing until it reaches the wall when it will turn left and follow the wall in an anticlockwise direction. This tree can be encoded in RPN as: *MOVE, TURN LEFT, front,* IF. As already noted, execution of the program requires the identification of a path from the root node to a terminal node by a depth-first traversal of the tree. Such a path is determined by the internal logical operators and the values of their operands. If the terminal node of a path is a sensor value, the traversal continues by providing this value to its parent operator. If the terminal node is an action, the action is taken and the traversal is terminated. This changes the relationship between the robot and its environment and the control program is executed again from the root node. The purpose of the GP system used here is to supply each robot in the population with an algorithm to guide it in solving the set of maze problems. The fitness of the algorithm is measured by the score achieved by the robot averaged over a set of 50 mazes. Selection for the mating pool that will produce the next generation is based on fitness, so the better robots survive and may breed with other robots. Each breeding process produces two children who may or may not be fitter than their parents. The tools that the GP system has at its disposal are the logical operators *IF*, *AND*, *OR* and *NOT*, eight state sensors with which the robot is able to detect an obstacle in an adjacent cell (*front, front-right, right, back-right, back, back-left, left* and *front-left*) and three actions *MOVE*, *TURN RIGHT* and *TURN LEFT*. Note that two forms of the *IF* statement are used: in the first the IF statement tests a logical (sensor state) input and then selects from one of two possible actions while in the second it selects from one of two possible logic (sensor) states to output. These need to be distinguished to establish syntactically correct programs: the second kind of *IF* can provide inputs to further *IF*, *AND*, *OR* and *NOT* statements while the first cannot – the *IF* in Fig. 3 is of the first kind, while both kinds are seen in subsequent designs.

We additional use the concept of fuel to refer to the fact that each robot is limited to a fixed number of actions when it gets its turn in the mazes – here 100 actions are allowed per maze, i.e., the GP program is looped over a maximum of 100 times – this is sufficient to allow the exploration capabilities of a design to be fully assessed. In the simplest version of the GP system, each robot in the population has sole occupancy in the maze while its fitness is measured. The mazes each have on average 10 obstacles but varying between as few as five and as many as 15.

4   Some Illustrative Results

If we run a simple GP process to design control programs for this problem, permitting crossover between any valid position and mutation across any node or leaf (ensuring syntactically correct outcomes) it is possible to rapidly evolve the simple wall following design of Fig. 3 into more powerful forms. Figure 4 shows a typical program structure while Fig. 5 shows its path around a typical world. As can be seen from Fig. 4 a highly complex and non-intuitive program structure has evolved. This particular run of the naïve GP used a population size of 100 with 100 generations, each member of the population being trialled over 50 obstacle courses at each generation. The initial population is seeded with the structure from Fig. 3 plus 99 random, but syntactically valid, designs. The GP uses roulette wheel selection, 40 % cross-over probability and 20 % mutation probability (n.b., this latter probability is the probability that a member of the population undergoes a single mutation operation as opposed to the quantity of genetic material being modified). The final score for this design averaged over 50 mazes is 0.3922 as compared to 0.2761 for the original wall follower design. Not only are the programs evolved by the naïve GP often cumbersome, their structure makes them increasingly difficult to improve on using just selection, cross-over and mutation – note the duplication and occasionally redundant elements in Fig. 4 (where identical sensor inputs are ORed together – highlighted). This was the observation made by Bearpark and Keane referred to earlier (Bearpark and Keane 2008).

If this basic GP process is repeated 100 times using different random number sequences in the GP (with a fixed set of mazes) the results of Figs. 6 and 7 are produced – the best, mean and standard deviation scores are 0.4547, 0.4111 and 0.0306, respectively. The elapsed CPU time required to carry these 100 searches using a single CPU was 5.464E6 s. It is this performance that any new algorithm should seek to improve on.

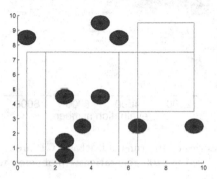

Fig. 5. The naïve GP evolved robot of Fig. 4 in its world (coloured red) with obstacles coloured blue and resulting path (the red solid line) (Color figure online).

5   Using *espresso* to Simplify Logic Trees – Case Statements

To make improvements to the design process we observe that our robot control program is a way of mapping a set of eight observable states that can be either *true* or *false* into one of three action outcomes. This is entirely similar to the task performed by VLSI logic circuits, although it is noted that VLSI designs most commonly work with two output states (0 or 1) though −1, 0, 1 output devices are also possible. We therefore proceed as follows: first a given control program is interrogated by exhaustively tabulating all possible output values for the 2^8 possible combinations of sensor input states; second the resulting truth table is passed to a VLSI logic circuit simplification routine (here a version of *espresso*) and third the resulting reduced logic table is then used to construct a case-based robot control program where each unique input state needed to recreate the program action is mapped to exactly one element in the case statement. The resulting decision tree is (a) much more highly structured, (b) generally more compact, (c) therefore faster to execute, (d) more readily understood by a human reader and e) more useful for subsequent use by the GP system since cross-over can now move entire sets of case elements between members of its population. The reduction in program size significantly speeds up the whole GP process while allowing it to produce better designs, typically halving evaluation times. Since VLSI logic programs are so highly optimized the cost of thus rewriting population members is trivially small as compared to evaluating designs

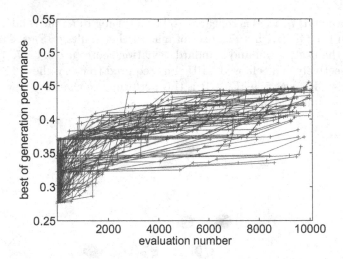

Fig. 6. Optimization histories of 100 runs of the basic GP process with a population of 100 members each being used for 100 generations – plots stop when the search stalls.

over multiple mazes. The only subtlety required to carry out this process is to adapt the essentially two output level tools available online to the task of dealing with more actions, here three. The use of the *espresso* algorithm in this context represents a considerable advance over the custom coded approach adopted earlier (Bearpark and Keane 2008), allowing faster and more compact designs to be produced. Figure 8 illustrates the *espresso* processed version of Fig. 4: it is apparent that the simplified representation is much easier to comprehend, and that its form better reflects the abilities of the robot as it tackles the different problems encountered in its exploration. Essentially it is a case-based system: each rule tackles a problem case faced by the robot. These rules may be seen to take the following actions based on four distinct cases (sensor feature sets):

1. If *front* turn *left*;
2. Else if *front right* and not *left* turn *left*;
3. Else if *front right* and *right* and not *back* and not *front left* turn *left*;
4. Else if *right* and *back* and not *left* and not *front left* turn *left*;
5. Else *move*.

It is by no means obvious that this refined structure generates the same operational behaviour as the original one but in fact they generate identical tables.

As already noted the process used here to simplify logic trees involves first converting any given control program into its equivalent truth table. Since the robot has eight proximity sensors there are $2^8 = 256$ possible combinations of input sensor states. Each state is presented, one at a time, to the control logic using the same code that is used to simulate the software robot and the resulting action noted – this can take one of three values: turn left, move or turn right, which are then represented as -1, 0 and 1. Table 1 shows part of the truth table for a robot program. Truth tables of this kind cannot be presented

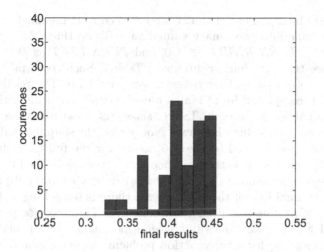

Fig. 7. Histogram showing variations of final results for the optimization histories of Fig. 6 – the best, mean and standard deviation scores are 0.4547, 0.4111 and 0.0306, respectively.

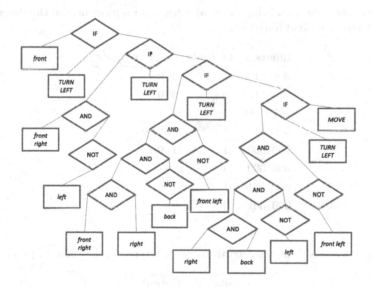

Fig. 8. The *espresso* simplified tree of 32 elements derived from Fig. 4 – in RPN this is *MOVE, TURN LEFT, front left*, NOT, *left*, NOT, *back, right*, AND, AND, AND, IF, *TURN LEFT, front left*, NOT, *back*, NOT, *right, front right*, AND, AND, AND, IF, *TURN LEFT, left*, NOT, *front right*, AND, IF, *turn left, front*, IF – this design scores 0.3922 when averaged over 50 mazes.

directly to *espresso* as it is designed to work with binary valued inputs and binary valued outputs – while the robot sensor map is a binary one, the action list is not. *Espresso* does, however, permit multiple function outputs providing each one is only binary. Thus the next step is to convert the single three valued output in the table into two binary valued ones. To do this a *MOVE* output is encoded as [0 0], *TURN RIGHT* as [1 0] and *TURN LEFT* as [0 1], while [1 1] is not used, see the right hand columns of Table 1. Such truth tables can then be presented to *espresso* and rewritten in compact form. The resulting output will show which compound input states must be explicitly dealt with and what the appropriate action is for each. Table 2 shows the re-written table for the full truth table from which Table 1 is drawn. Notice that the simplified table contains only seven rows as compared to the 256 needed for the full table derived from the initial program. Note also the presence of "–" symbols in the table which are "don't care" symbols meaning the relevant line can be used for multiple matching entries, e.g., the final line of the table says if there is something in front of the robot and no other line has fired then turn left. This re-write process works for almost all input truth tables but it does permit an output action of [1 1] which has no meaning for a three action problem. *Espresso* rule re-writing can lead to such outcomes because it is designed for logic circuits where there is no concept of one-at-a-time testing of input states and thus multiple input states

Table 1. Part of a truth table for a typical robot control program and the three valued outputs converted to dual binary outputs

Inputs	Outputs	Binary outputs	
00001100	0	0	0
00001101	0	0	0
00001110	1	1	0
00001111	1	1	0
00010000	1	1	0
00010001	-1	0	1
00010010	1	1	0
00010011	-1	0	1

Table 2. *espresso* simplified truth table for a typical robot control program.

Inputs	Outputs
0001- - -0	10
0-00- -1-	10
00011- - -	10
- - -10- -1	01
- -1-0111	01
-101- -1-	01
1- - - - - - -	01

can sometimes be handled by concatenation. If such output lines do occur in the re-written tables the re-write operation is abandoned – such events are very rare. The final step in the process is to convert tables like that of Table 2 back into programmatic form. This is readily accomplished by creating a case statement where each line of the truth table maps to one case which is then terminated with a final action of *MOVE*. Each individual case is simply a sequence of sensor readings unioned together with AND statements if the sensor column shows a 1 and AND NOT statements if the sensor column shows a zero with the relevant action as specified in the output line, otherwise control passes to the next line in the table. It will be obvious to the reader that the truth table in Table 2 maps to the case structure in Fig. 6. Notice that the re-written statements do not make use of the OR statement at all. Because of this fact it is important not to over use the *espresso* algorithm as it can lead to a loss of genetic diversity.

6   Results: The Impact of *espresso* on the GP Process

To make use of the *espresso* capability an additional re-write operator must be added to the GP system. Re-writing 20% of the population after the actions of cross-over and mutation in every generation has been found to work effectively with the problem being studied here, though the results are not particularly sensitive to this setting – values ranging from 10% to 40% work similarly well. If this enhanced GP process is repeated 100 times using different random number sequences in the GP (with the same fixed set of mazes as used before) the results of Figs. 9 and 10 are produced – the best, mean and standard deviation scores are now 0.5133, 0.4151 and 0.0319, respectively as compared to the previous

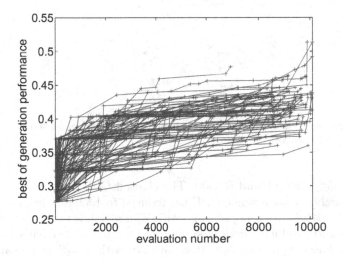

Fig. 9. Optimization histories of 100 runs of the *espresso* enhanced GP process with a population of 100 members each being used for 100 generations – plots stop when the search stalls.

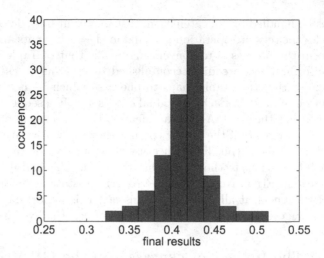

Fig. 10. Histogram showing variations of final results for the optimization histories of Fig. 9 – the best, mean and standard deviation scores are 0.5133, 0.4151 and 0.0319, respectively.

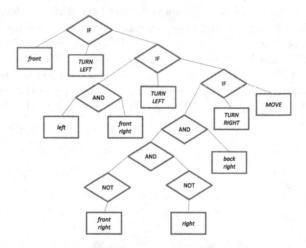

Fig. 11. A tree evolved by the *espresso* enhanced GP – this design scores 0.5133 when averaged over 50 mazes.

results of 0.4547, 0.4111 and 0.0306. The elapsed CPU time required to carry these 100 searches using a single CPU has reduced to 4.329E6 s from the previous run time of 5.464E6 s, a saving of some 20%. The search is faster, achieves better peak results and produces broadly similar average results (Welch's t-test shows the *espresso* based approach is better but only with a 36% significance level). While these gains are not overwhelming, they are very useful, particularly the improvement in peak performance. Figure 11 shows the best structure evolved with this enhanced approach and Fig. 12 shows its path through the same maze as illustrated in Fig. 5. Its control program is appealingly simple, having the desired case-based structure and now with only three distinct cases (a compact set of useful features has been engineered):

1. If *front* turn *left*;
2. Else if *left* and *front right* turn *left*;
3. Else if not *front right* and not *right* and *back right* turn *right*;
4. Else *move*.

It also contains the important capability of being able to turn right or left. The logic of the final rule-base is, however, still not completely obvious and it is not at all clear that a human navigator would develop such an approach. Indeed the problem being studied here is extremely difficult for humans to write successful decision trees for. This is the great attraction of computer generated control programs – powerful yet readily studied logics can be produced without needing any special insights into the problem being confronted.

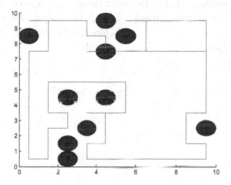

Fig. 12. The *espresso* enhanced GP evolved robot of Fig. 11 in its world (coloured red) with obstacles coloured blue and resulting path (the red solid line) (Color figure online).

7 Conclusions

This paper has shown how tools developed for VLSI logic circuit design can be used to improve the performance of a naïve GP system in producing a simple robot path planning task. The key observation is that restructuring GP derived programs as case statements not only helps improve subsequent understanding of the programs but also aids the GP system in developing them – the GP runs faster and it produces better designs through cross-over. Moreover logic simplification tools are extremely robust and powerful and so have negligible impact on the cost of evolving new structures. Here an eight sensor, three action system is explored but the ideas presented are capable of dealing with arbitrary numbers of inputs and outputs by adopting suitable encoding approaches to switch between decision trees, Reverse Polish Notation (RPN) and truth tables.

References

Bearpark, K., Keane, A.J.: Canonical representation in genetic programming. In: Parmee, I.C. (ed.) Proceedings of the Conference on Adaptive Computing in Design and Manufacture ACDM08, Bristol (2008)

Espejo, P.G., Ventura, S., Herrera, F.: A Survey on the application of genetic programming to classification. IEEE Trans. Syst. Man Cybern. 40(2), 121–144 (2010)

Koza, J.: Genetic Programming: on the Programming of Computers by Means of Natural Selection. MIT Press, Cambridge (1992)

McGeer, P.C., Sanghavi, J.V., Brayton, R.K., Sangiovanni-Vicentelli, A.L.: ESPRESSO-SIGNATURE: a new exact minimizer for logic functions. IEEE Trans. Very Large Scale Integr. (VLSI) Syst. 1(4), 432–440 (1993)

Mencar, C., Castiello, C., Cannone, R., Fanelli, A.M.: Design of fuzzy rule-based classifiers with semantic cointension. Inf. Sci. 181(20), 4361–4377 (2011)

Moraglio, A., Krawiec, K., Johnson, C.G.: Geometric semantic genetic programming. In: Coello, C.A.C., Cutello, V., Deb, K., Forrest, S., Nicosia, G., Pavone, M. (eds.) PPSN 2012, Part I. LNCS, vol. 7491, pp. 21–31. Springer, Heidelberg (2012)

Seo, K., Hyun, K.S., Goodman, E.D.: Genetic programming-based automatic gait generation in joint space for a quadruped robot. Adv. Robot. 24(15), 2199–2214 (2010)

Weng, M., Wei, X., Qu, R., Cai, Z.: A path planning algorithm based on typical case reasoning. Geo-spat. Inf. Sci. 12(1), 66–71 (2009)

Minimizing Total Tardiness on Identical Parallel Machines Using VNS with Learning Memory

Eduardo Lalla-Ruiz[1](✉) and Stefan Voß[2]

[1] Department of Computer and Systems Engineering, University of La Laguna,
San Cristóbal de La Laguna, Spain
elalla@ull.es
[2] Institute of Information Systems, University of Hamburg, Hamburg, Germany
stefan.voss@uni-hamburg.de

Abstract. Minimizing total tardiness on identical parallel machines is an $\mathcal{NP}$-hard parallel machine scheduling problem that has received much attention in literature due to its direct application to real-world applications. For solving this problem, we present a variable neighbourhood search that incorporates a learning mechanism for guiding the search. Computational results comparing with the best approaches for this problem reveals that our algorithm is a suitable alternative to efficiently solve this problem.

1 Introduction

Minimizing total tardiness on identical parallel machines (referred to as $P||\sum T_j$ in standard machine scheduling terminology) assumes a set of n jobs $J = \{1, ..., n\}$ to be processed on m identical parallel machines $M_1, ..., M_m$. Each job j has a processing time $p_j > 0$ and a due date d_j. All jobs are available at time zero, and no job pre-emption is allowed. The tardiness T_j of job j is given by $T_j = max(C_j - d_j, 0)$, where C_j is the completion time of j. The objective of this problem is to find a schedule that minimizes the total tardiness $\sum_{j=1}^{n} T_j$. Recent references include [1, 2, 6].

The $P||\sum T_j$ has been widely studied in the literature. Recent studies are the following. Biskup *et al.* [1] present a comparison of the existing heuristic algorithms for solving this problem. Moreover, they propose a new heuristic approach to solve the problem. This heuristic provides better results in terms of solution quality than the other heuristics. Tanaka and Araki [6] propose a new branch and bound algorithm with a lagrangian relaxation for computing the lower bounds. They propose a set of problem instances to assess their approach. Niu *et al.* [4] propose a Clonal Selection Particle Swarm Optimization (CSPSO) for this problem. The authors evaluate their proposal over the instances proposed in [6]. Deng *et al.* [2] propose a Hybrid Differential Evolution algorithm (HDDE) and also used the problem instances presented in [6]. The computational experience reported in their work shows that HDDE outperforms CSPSO for the

© Springer International Publishing Switzerland 2015
C. Dhaenens et al. (Eds.): LION 9 2015, LNCS 8994, pp. 119–124, 2015.
DOI: 10.1007/978-3-319-19084-6_10

small-sized problem instances. Moreover, they compare HDDE with branch and bound (B&B) for the large-sized instances and indicate that HDDE improves the performance of B&B in terms of computational time.

In this work, we present a Variable Neighbourhood Search with a learning mechanism (VNS-L) for solving $P||\sum T_j$. Our approach combines a variable neighbourhood search with restarting strategy which exploits the use of a memory for learning from past solutions. The goal of this work is to assess the performance of this idea as well as to provide high-quality solutions in short computational times. In doing so, the performance of the VNS-L is evaluated using the well-known benchmark suite proposed by Tanaka and Araki [6] and comparing its results with those given by some arguably best algorithms reported in the literature.

The remainder of this paper is organized as follows. Section 2 describes the VNS-L proposed to address $P||\sum T_j$. Afterwards, the performance of our algorithm is analyzed in realistic scenarios proposed in literature; see Sect. 3. Finally, Sect. 4 provides the main conclusions extracted from the work and suggests several directions for further research.

2 Variable Neighbourhood Search with Learning

Variable Neighbourhood Search (VNS) is a well-established meta-heuristic that systematically exploits the idea of changing neighbourhoods during the search [3]. VNS relies only on the best solutions currently known to center the search. In this regard, the information collected during the search relative to previous good solutions or their characteristics is forgotten. In order to address this deficiency and take advantage of that information, a Variable Neighbourhood Search with a Learning Mechanism is proposed. It is a hybridization within a Multi-Start strategy (MS) embedded with a learning mechanism for taking advantage of the information obtained during the VNS process using a memory structure. Within the VNS-L template we use the exploitation capabilities of the VNS and the exploration capabilities provided by MS as it gives the ability of re-starting the search. As noted by [5], incorporating memory structures into re-starting processes improves their performance.

The learning mechanism within VNS-L is based on (i) a frequency based memory structure, termed as M, with the aim of collecting promising solution features found during the search and (ii) a solution generation procedure. The memory structure, M, is composed of solution features as follows. Consider a set of solutions Λ. Each solution $x \in \Lambda$ has associated a set $C(x) = \{(i,j)\}$ of features. That is, a job j is the i-th job served in the whole schedule. The memory has a matrix structure with dimension $n \times n$, where the rows represent the jobs and the columns the service order. Each time the VNS improves the best solution known within the search, the information related to the solution features is updated, for keeping track of these matches. For example, in case the solution structure considered for updating the memory includes the feature (i,j), i.e., job j is the i-th job served in the schedule, then its corresponding memory position M_{ij} is updated.

Algorithm 1. VNS-L pseudocode

```
1  iter = 1
2  Initialize M
3  while iter ≠ iter_max do
4  │   x ← Solution Generation Procedure using M
5  │   k = 1
6  │   while k ≠ k_max do
7  │   │   Shaking:
8  │   │   Choose a random neighbour x' ∈ N₁(x, k)
9  │   │   Improvement phase:
10 │   │   while stopping criterion is not met do
11 │   │   │   a) Reinsertion move over x' → x''
12 │   │   │   b) Interchange move over x'' → x'''
13 │   │   │   if x''' is better than x' then
14 │   │   │   │   x' ← x'''
15 │   │   Solution assessment:
16 │   │   if x''' is better than x then
17 │   │   │   x ← x'''
18 │   │   if x''' is better than x_best then
19 │   │   │   Update memory M using β parameter
20 │   │   │   x_best ← x'''
21 │   │   │   k = 1
22 │   │   else
23 │   │   │   Update memory M using γ parameter
24 │   │   │   k = k + 1
25 │   iter = iter + 1
26 return x_best
```

The way a memory position M_{ij} is updated is as follows: $M_{ij} = (M_{ij} + 1) \cdot \beta$. The parameter $\beta \geq 1$ is used so that when the memory is updated, those solution features that have been part of the best known solutions more often have greater significance. On the other hand, in VNS-L we keep track of the worst solution obtained during the local search process. In case we are not able to improve the disturbed solution, we update the memory according to $M_{ij} = (M_{ij} + 1) \cdot \gamma$. The parameter $\gamma < 1$ is used so that when the memory is updated, those solution features affected will have less significance.

VNS-L as shown in Algorithm 1 uses a finite set of neighbourhoods based on (a) reinsertion-move, $N_1(x, k)$, namely k jobs are removed from a machine m and reinserted in another machine m', where $m \neq m'$, and (b) interchange-move $N_2(x)$, that consists of exchanging a job j assigned to machine m with a job j' assigned to machine m', where $m \neq m'$. For any given k the application of these neighbourhood structures is performed sequentially, i.e., firstly the reinsertion-move is applied and thereafter the interchange-move. The shaking process allows

Table 1. Computational results for the 2250 instances proposed by [6]. Note that only the computational times (measured in seconds) are reported since all the approaches reach the optimal solutions

Instance		HDDE		B&B		VNS-L	
n	m	Avg. t(s.)	Max. t(s.)	Avg. t(s.)	Max. t(s.)	Avg. t(s.)	Max. t(s.)
20	2	0.01	0.30	0.43	1.41	0.06	0.12
	3	0.02	0.41	0.30	4.00	0.07	0.13
	4	0.03	1.14	0.15	4.03	0.08	0.13
	5	0.02	0.17	0.08	0.47	0.08	0.13
	6	0.02	0.38	0.05	0.36	0.09	0.16
	7	0.02	0.27	0.04	0.36	0.09	0.17
	8	0.02	0.63	0.03	0.28	0.10	0.18
	9	0.01	0.52	0.11	8.81	0.10	0.19
	10	0.01	0.88	0.03	0.33	0.12	0.18
25	2	0.02	0.45	1.16	13.47	0.12	0.22
	3	0.06	1.92	19.85	757.28	0.15	0.32
	4	0.08	1.22	47.47	4148.98	0.17	0.28
	5	0.13	5.97	14.15	1534.72	0.17	0.43
	6	0.12	4.00	0.37	27.22	0.24	0.45
	7	0.10	1.52	0.16	2.84	0.28	0.58
	8	0.18	13.78	0.11	0.89	0.30	0.54
	9	0.06	0.72	0.18	12.78	0.32	0.57
	10	0.03	0.28	0.07	2.81	0.33	0.59
		0.05	1.92	4.71	362.28	0.16	0.30

to escape from those local optima found along the search by using the reinsertion-move. Once the search process in the VNS-L is over, the information stored in M is used by the solution generation procedure for re-starting the VNS-L. To do so, a roulette wheel mechanism using the information stored in M is applied to generate a job order sequence. Then, the first job in that sequence will be assigned to the machine which adds the minimum tardiness completion time to the solution.

3 Computational Results

This section is devoted to present the computational experiments carried out in order to assess the performance of the proposed algorithm. For this purpose, we use a set of 2250 instances provided by [6]. All the computational experiments reported in this work are conducted on a computer equipped with an Intel 3.16 GHz and 4 GB of RAM. We run 20 executions of VNS-L with the following parameter values: $iter_{max} = 15$, $k_{max} = 3$, $\beta = 1.2$, $\gamma = 0.95$, and M initialized as the one-matrix.

Table 1 shows the average computational results provided by (i) the best approximate approach reported in the literature based on a Hybrid Discrete Differential Evolution Algorithm, HDDE [2], (ii) the best exact approach based on a Branch and Bound, B&B [6], and (iii) our VNS-L. HDDE and B&B were executed on an Intel 3.2 GHz with 512 MB of RAM by [2]. The first columns correspond to the sizes of the instance sets. Since all the sets are composed of 125 instances each and the optimal solution values are known, in the tables we only report the average computational time values since the three methods always obtain the optimal solution values.

As can be seen in the table, VNS-L maintains a consistent temporal performance during the search. VNS-L reduces the maximum required time in 85 % and 99.92 % of the cases in comparison to HDDE and B&B, respectively. In this regard, it should be noted that, on average, there is not much difference between the average and maximum running performance of VNS-L (about 0.15 s). This gives a sense of the temporal performance of VNS-L, which makes it suitable when addressing related problems, solving larger instances or tackling integrated problem schemes where this problem is involved.

4   Conclusions and Further Research

In this work, the problem of minimizing total tardiness on identical parallel machines $(P||\sum T_j)$ has been addressed. In order to solve it, a Variable Neighbourhood Search with a Learning mechanism (VNS-L) is proposed. According to our computational experience over a well-known set of instances proposed in the literature, our algorithm shows a competitive performance in terms of solution quality and computational time. In this regard, it exhibits a similar performance by means of average and maximum computational time, which makes it suitable for those environments where the expected computational time may not vary from one instance to another within the same scenario dimensions.

As further research, we are going to assess the contribution of VNS-L for different configurations of its learning parameters as well as to determine an adaptive method to parameterize it according to given problem instances.

Acknowledgements. This work has been partially funded by the European Regional Development Fund, the Spanish Ministry of Economy and Competitiveness (project TIN2012-32608). Eduardo Lalla-Ruiz thanks the Canary Government for the financial support he receives through his doctoral grant.

References

1. Biskup, D., Herrmann, J., Gupta, J.N.D.: Scheduling identical parallel machines to minimize total tardiness. Int. J. Prod. Econ. **115**(1), 134–142 (2008)
2. Deng, G., Zhang, K., Gu, X.: A hybrid discrete differential evolution algorithm to minimise total tardiness on identical parallel machines. Int. J. Comput. Integr. Manuf. **26**(6), 504–512 (2013)

3. Hansen, P., Mladenovic, N., Moreno Pérez, J.A.: Variable neighbourhood search. Ann. Oper. Res. **175**, 367–407 (2010)
4. Niu, Q., Zhou, T., Wang, L.: A hybrid particle swarm optimization for parallel machine total tardiness scheduling. Int. J. Adv. Manuf. Technol. **49**(5–8), 723–739 (2010)
5. Stützle, T.: Local search algorithms for combinatorial problems. Darmstadt University of Technology. Ph.D. thesis (1998)
6. Tanaka, S., Araki, M.: A branch-and-bound algorithm with lagrangian relaxation to minimize total tardiness on identical parallel machines. Int. J. Prod. Econ. **113**(1), 446–458 (2008)

Dynamic Service Selection with Optimal Stopping and 'Trivial Choice'

Oliver Skroch[(⊠)]

Faculty of Computer Science,
Darmstadt University of Applied Sciences,
64295 Darmstadt, Germany
Oliver.skroch@h-da.de

Abstract. Two different strategies for searching a best-available service in adaptive, open software systems are simulated. The practical advantage of the theoretically optimal strategy is confirmed over a 'trivial choice' approach, however the advantage was only small in the simulation.

Keywords: Intelligent optimization technique · Optimal stopping algorithm · Run-time self-adaptive software · Simulation results

1 Introduction

Service model paradigms for intelligent software systems include the idea of dynamically adapting service invocations as needed, which aims at overcoming certain limitations that constrain traditional software systems. In distributed system architectures, services are implemented through well defined component interfaces. The discovery and selection of suitable services is a decisive development step in composing larger distributed systems [1, 2].

Initial service composition takes place at design time and guides *static* software architectures. But in adaptive, *dynamic* systems services can be recomposed later at run-time. In *closed* architectures these run-time alternatives are predefined at design time and built into the system. But useful external services that were unknown at design time may later become available. Specifically, open and uncontrolled software platforms (the Internet) may provide numerous suitable service candidates. *Open* systems can benefit from opportunistically performing parts of their functionality also through such externally hosted service alternatives.

This practical challenge opens up an interesting field for the application of advanced optimization techniques. The opportunistic run-time search for service alternatives on open platforms is challenging for a number of reasons which are discussed in Sect. 3. Therefore, intelligent optimization strategies are required to improve the related run-time adaptation process. Next to heuristic approaches, there is also an interesting, exact algorithmic meta strategy which can be considered: the identification of the best moment when to stop the search. Optimal stopping theory has been applied for this purpose already in the past. This paper compares the advantages of an optimal stopping approach with a trivial but suitable approach.

C. Dhaenens et al. (Eds.): LION 9 2015, LNCS 8994, pp. 125–130, 2015.
DOI: 10.1007/978-3-319-19084-6_11

The rest of the paper is organized as follows. Section 2 explains the background and presents related work. Section 3 presents the application scenario and the related stopping algorithm. Section 4 describes the simulation experiments and their results, studying the algorithm and its behavior in different settings. Section 5 briefly summarizes and concludes the paper.

2 Background and Related Work

Building larger distributed systems by *composition* implies matching well-defined supplied service interfaces with equally well-defined requested service interfaces [1].

Supplied service interfaces might be available from other components which could well be located externally on the Internet or on other open platforms. Adaptive systems with run-time recomposition capabilities can opportunistically make use of such externally provided services and search, before the invocation of a built-in service, for a better external service [2].

Validated approaches of such self-adaptive systems reach back at least to the Viable Systems Model from the 1960s [3]. More recent discussions include for instance [4, 5] and a particular research focus deals with functionally equivalent Web services of differing quality, e.g. in [6, 7].

When discovering a suitable service option on an uncontrolled open platform, the final decision has to be made on the spot, to either accept or reject, because recalls are inadmissible on open platforms. At first glance it seems that in this situation, regardless of the decision method, the probability to make a right decision approaches zero when the number of available options grows. But surprisingly, *optimal* decision strategies are known from mathematical statistics. The complex problem class relevant for this paper's application scenario has been described already in [8, 9], and within the framework of stopping Markov chains in [10]. The so-called $1/e$ stopping law that is applied in the simulation is described and proven in [11].

Optimal stopping as a meta strategy was repeatedly proposed in computer science with focus on evolutionary algorithms, and recent discussions also include, for example, areas such as multi-criteria optimization [12] or local search heuristics [13].

3 Application Scenario and Stopping Algorithm

Recomposing service invocations at run-time with opportunistic search attempts on the open Internet is particularly challenging for three main reasons. *First*, the number of suitable service alternatives which are available on the Internet is unknown. There could be very many such services, so exhaustive search is out of the question. *Second*, a realistic application scenario will have practical limits as to the acceptable run-time delay for end-to-end processing, and consequently also for each single service invocation. To meet these two challenges, the time frame t which is available for an opportunistic search preceding a service invocation is restricted to a hard limit in the application scenario.

Invoking publicly available services from open platforms raises yet a *third* issue. Services that are hosted externally are not controlled by the self-adaptive system. One important consequence is that any externally provided service alternative can be unavailable or changed in the next moment. Therefore it is not possible to memorize external options and, after further unsuccessful search, get back and use a service that was found previously. This implies that the final decision whether to invoke a certain external service must be made straight away when discovering it.

Next to heuristic approaches, these conditions allow an exact algorithmic search optimization meta strategy through the application of an optimal stopping rule which cuts off the search at the best possible point. The further simulation of a respective stopping algorithm is guided by this scenario.

The solution approach taken in mathematical statistics basically defines a *waiting time* as the time up to which all discovered services are observed without accepting, while the value of the leading service is remembered. The intention is to choose the best service and it can be shown that there exists an optimal waiting time w^* maximizing the success probability for the optimal choice when the first leading option arriving after w^* is accepted, if there is one, and all options are rejected if there is none. The only waiting time policy with the best possible success probability is determined by the so-called $1/e$ law of optimal stopping [11] and the well-known value $1/e \approx 0.3679$ is the (asymptotically) best possible lower bound. Cf. [8, 11] for the details and proofs.

The optimal point when to stop the search for further service alternatives in our application scenario can be determined by directly applying the $1/e$ law.

The corresponding stopping algorithm identifies externally provided service options suitable for matching with the internal composition's requested service and initially rejects all options, while memorizing the best option found yet. But as soon as a proportion of t/e of the predefined time frame t has passed the next leading service option is chosen, if there is one within the predefined time period. Otherwise, if no more option that is better than all previous ones shows up until the time frame t expires, no choice is made, the internal composition remains in place and the built-in service is invoked. Since t is predefined in the application scenario, t/e is constant. As quoted, the method yields the best possible lower probability bound for the optimal choice and always stops within the predefined time frame.

Since $1/n < 1/e$ for all $n \geq 3$, the algorithm should outperform any other strategy already with three or more suitable service option alternatives.

4 Simulation Experiments

To investigate into the algorithm and its behavior, several simulation experiments were conducted. Uniformly distributed integer service utility values U between U_b (best) and U_w (worst) were randomly assigned to the compositions with external service options. The internal composition was given fixed service utility value of U_s (static), with $U_b < U_s < U_w$ to exclude trivial cases. Smaller utility values were considered to be better. The maximum run-time slot was set to t s and n differently suited Web service options were available for each invocation.

The simulation comprised 3000 experiments for two different settings. To compare the developed algorithm, each experiment was additionally run also for a *trivial choice strategy* following the obvious rule: identify suitable, externally provided service options, choose the first option with $U < U_s$ and, if no such option is found by the end of the time frame, invoke the internal service. This rule of choice fulfills the application scenario restraints described in 3 and is an intuitive and very simple, trivial alternative approach. The optimal stopping policy is compared to the trivial choice and, implicitly, to a non-adaptive static system (Table 1).

Table 1. Simulation results for the optimal stopping strategy, compared to a trivial choice approach and (implicitly) to a non-adaptive system where U_s equals 3 or 30. Cell values originate from 3000 conducted experiments for each setting. Column headings denote the settings for (t, n, U_s, U_w).

Trivial choice	(20, 10, 3, 9)	(100, 100, 30, 99)
Best	961	92
Fallback	89	2
Avg. services	2.8	3.2
Avg. run-time	0.0102	0.0124
Avg. utility	1.06	14.67
First	*850*	*883*
Optimal stopping	*(20, 10, 3, 9)*	*(100, 100, 30, 99)*
Best	714	340
Fallback	1204	1149
Avg. services	5.3	21.3
Avg. run-time	0.0180	0.0780
Avg. utility	1.94	14.62
Ref. services	*6.92*	*28.32*

The first simulation was set to $t = 20$ ms, $n = 10$ service options, $U_s = 3$, $U_b = 0$ and $U_w = 9$. The optimal stopping (lower half of Table 1) resulted in a total average improvement to 1.94 utility points (35 %) over the assumed static system, and was outperformed by the trivial choice (upper half of Table 1) with an average improvement to 1.06 utility points (65 %). Remember that the objective of optimal stopping is to optimize $P(U_b)$ and not $E(U)$. Over all 3000 experiments, and related to all $n = 10$ options, trivial choice selected the best service in 961 cases (32.0 %) and in 850 cases the very first discovered service was chosen. This was better than optimal stopping with 714 optimal hits in 3000 experiments.

At first glance, the optimal stopping results seemingly yield a success rate of only 0.238 (714/3000) while the $1/e$ law predicts 0.3679. But the assumed number n of provided services is required for the implementation of the simulation framework only. We remember that the amount of services available on the Internet is unknown in the application scenario, and its distribution is unknown in the $1/e$ law. The corresponding total number of service options for a correct calculation therefore is not n but the number of reference services that would have been considered by the optimal stopping algorithm if it would have used the full available time slot (approximate values are shown in the last line of the lower half of Table 1, note that this number is not larger than n). Scaling the values over all experiments yields an average hit rate of 0.3706 which differs by only 0.7 % from the theoretic prediction, possibly due to rounding errors and timing imprecisions in the simulation.

The other simulation was run with $n = 100$ provided service options, $U_b = 0$ and $U_w = 99$. With optimal stopping, the value of U_s affects the achieved utility only but is not used in the algorithm itself. Still U_s is a relevant parameter for the trivial choice strategy and therefore mentioned.

With these settings, simulation results and comparisons enable a first look at the benefits of the optimal stopping approach. Already at these settings, optimal stopping made the best choice in 340 cases, nearly 4 times as often as trivial choice. Also the average achieved utility with optimal stopping is already slightly better as compared to trivial choice. Trivial choice searches through 3.2 service options only and, at a run-time delay of 12.4 ms, utilizes less than an eighth of the available time slot of 100 ms. Optimal stopping makes much better use of the available time slot by taking 78 % of it with a run-time delay of 78 ms. With the run-time in optimal stopping being about 6.3 times longer as in trivial choice, optimal stopping consequently makes better use of the search space, too, and considers 21.3 services, which is nearly 7 times more in comparison.

5 Summary and Conclusion

An optimal stopping algorithm was compared with a trivial choice approach in simulation experiments with two different settings. The bottom line benefits of the optimal approach were confirmed. However, the advantage was small in the simulation. Both optimal stopping and trivial choice clearly outperformed an assumed static architecture. Further simulation experiments with other settings, for example longer time frames and/or more available services, could give deeper insight into the optimal stopping advantages above trivial choice.

References

1. Atkinson, C., Bunse, C., Groß, H.-G., Kühne, T.: Towards a general component model for web-based applications. Ann. Softw. Eng. **13**, 35–69 (2002)
2. Gamble, M.T., Gamble, R.: Monoliths to mashups: increasing opportunistic assets. IEEE Softw. **25**, 71–79 (2008)
3. Beer, S.: Brain of the Firm, 2nd edn. Wiley, Chichester (1981)
4. Allen, R.B., Douence, R., Garlan, D.: Specifying and analyzing dynamic software architectures. In: Astesiano, E. (ed.) ETAPS 1998 and FASE 1998. LNCS, vol. 1382, pp. 21–37. Springer, Heidelberg (1998)
5. Skroch, O., Turowski, K.: Optimal stopping for the run-time self-adaptation of software systems. J. Inf. Optim. Sci. **31**, 147–157 (2010)

6. Hwang, S.-Y., Lim, E.-P., Lee, C.-H., Chen, C.-H.: Dynamic web service selection for reliable web service composition. IEEE Trans. Serv. Comput. **1**, 104–116 (2008)
7. Yau, S.S., Ye, N., Sarjoughian, H.S., Huang, D., Roontiva, A., Baydogan, M.G., Muqsith, M.A.: Toward development of adaptive service-based software systems. IEEE Trans. Serv. Comput. **2**, 247–260 (2009)
8. Lindley, D.V.: Dynamic programming and decision theory. Appl. Stat. **10**, 39–51 (1961)
9. Presman, E.L., Sonin, I.M.: The best choice problem for a random number of objects. Theor. Probab. Appl. **17**, 657–668 (1972)
10. Dynkin, E.B., Juschkewitsch, A.A.: Sätze und Aufgaben über Markoffsche Prozesse. Springer, Heidelberg (1969)
11. Bruss, F.T.: A unified approach to a class of best choice problems with an unknown number of options. Ann. Probab. **12**, 882–889 (1984)
12. Skroch, O.: Multi-criteria service selection with optimal stopping in dynamic service-oriented systems. In: Janowski, T., Mohanty, H. (eds.) ICDCIT 2010. LNCS, vol. 5966, pp. 110–121. Springer, Heidelberg (2010)
13. Bontempi, G.: An optimal stopping strategy for online calibration in local search. In: Coello, C.A. (ed.) LION 2011. LNCS, vol. 6683, pp. 106–115. Springer, Heidelberg (2011)

A Comparative Study on Self-Adaptive Differential Evolution Algorithms for Test Functions and a Real-World Problem

Shota Eguchi[1], Yuki Matsugano[1], Hirokazu Sakaguchi[1], Satoshi Ono[1](✉),
Hisato Fukuda[2], Ryo Furukawa[3], and Hiroshi Kawasaki[1]

[1] Graduate School of Science and Engineering, Kagoshima University,
1-21-40 Korimoto, Kagoshima 890-0065, Japan
ono@ibe.kagoshima-u.ac.jp
[2] Graduate School of Science and Engineering, Saitama University,
255 Shimo-okubo, Sakura-ku, Saitama 338-8570, Japan
[3] Gradualte School of Information Sciences, Hiroshima City University,
3-4-1, Ozuka-Higashi, Asa-Minami-Ku, Hiroshima 731-3194, Japan

Abstract. This paper compares novel self-adaptive Differential Evolution algorithms (SADEs) on noisy test functions to see how robust the algorithms are against noise in fitness function. This paper also compares the performance of SADEs on real-world problems that estimates Bidirectional Reflectance Distribution Function properties of 3D objects.

Keywords: Self-adaptive differential evolution · Noisy problem · Bidirectional reflectance distribution function · Real-world application

1 Introduction

Differential Evolution (DE) [8] is a simple, easily implemented algorithm with few operators and predetermined control parameters. Various self-adaptive DE (SADE) algorithms have been proposed: jDE [1], Adaptive DE with an optional external archive (JADE) [12], Success-History Based Adaptive Differential Evolution (SHADE) [9], and so on.

Most real-world problems have some degree of uncernity, such as noisy fitness functions and dynamic environments that change optima positions [4]. Problems involving fitness noise may result in incorrect search directions. The search performance of DE is more affected by noisy fitness than other simple EC algorithms [5]. Although sampling fitness values several times for each individual might be a general approach to address noisy fitness [2,6], it requires a large number of fitness evaluations.

In this study, we clarify SADEs search performance on noisy functions to demonstrate their performance in realistic situations. The test functions used in this study involve dynamic Gaussian noise on fitness values. In addition to benchmark functions, we apply SADEs to a real-world computer vision problem,

© Springer International Publishing Switzerland 2015
C. Dhaenens et al. (Eds.): LION 9 2015, LNCS 8994, pp. 131–136, 2015.
DOI: 10.1007/978-3-319-19084-6_12

estimation of Bidirectional Reflectance Distribution Function (BRDF) properties of 3D objects. Its dimensionality is not high; however, static noise exists in the fitness function, and this problem is globally multimodal.

Experimental results on noisy test functions show that novel SADEs have significantly higher performance than DEs that target noisy problems. In addition, the results of the BRDF estimation problem show difference in solution quality and convergence speed between SADEs.

2 Self-Adaptive Des

Self-Adaptive Differential Evolution (jDE) [1] is a simple self-adaptation method in which each vector (individual) has its own scale factor F_i and crossover rate CR_i. With the probability of τ_1 and τ_2 in each generation, jDE randomly changes F_i and CR_i within the specified ranges as $F_{i,g+1} = F_l + F_u \cdot rnd_1$ and $CR_{i,g+1} = rnd_3$, respectively, where $rnd_j (j \in \{1, 2, 3, 4\})$ are uniform random values ranging in $[0, 1]$. τ_1 and τ_2 are probabilities changing F_i and CR_i, respectively. F_l and F_u determines the range of scale factor values. Since appropriate values of F_l and F_u are experimentally clarified, jDE does not require control parameter adjustment except population size N.

JADE is an SADE algorithm which has an archive to store inferior solutions and control parameter adaptation. The archive stores target vectors which are not selected due to the lower fitness than trial vectors. Although, like jDE, each individual in JADE has its own crossover rate CR_i and scale factor F_i, they are updated as $F_i = rndc_i(\mu_F, 0.1)$ and $CR_i = rndn_i(\mu_{CR}, 0.1)$, where $rndc_i$ and $rndn_i$ are Cauchy and Gaussian distributions, $\mu_F = (1 - c) \cdot \mu_F + c \cdot mean_L(S_F)$, $\mu_{CR} = (1 - c) \cdot \mu_{CR} + c \cdot mean_A(S_{CR})$, and $mean_L$ denotes Lehmer mean.

JADE also utilizes a strategy called current-to-pbest, where a base vector is selected from the top $p\%$ of the current population by random. Difference vector $v_{i,g}$ is defined as $v_{i,g} = x_{i,g} + F_i(x_{best,g}^p - x_{i,g}) + F_i(x_{r1,g} - \tilde{x}_{r2,g})$.

SHADE is a variant of JADE which uses different parameter adaptation mechanism based on a history. The historical memories M_{CR} and M_F store mean successful parameter values of CR and F for each generation, and produce various parameter configurations which can be switched during the search. When updating F and CR, SHADE weights arithmetic and Leamer means in JADE i.e., $mean_a$ and $mean_L$, are replaced by $mean_{WA}(S_{CR}) = \sum_{k=1}^{|S_{CR}|} w_k \cdot S_{CR,k}$ $(w_k = \Delta f_k / \sum_{k=1}^{|S_{CR}|} \Delta f_k)$ and $mean_{WL}(S_F) = \sum_{k=1}^{|S_F|} w_k \cdot S_{F,k}^2 / \sum_{k=1}^{|S_F|} w_k \cdot S_{F,k}$.

Although SHADE uses current-to-pbest/1 mutation strategy as well as JADE does, SHADE automatically determines p as $p_i = rnd[p_{min}, 0.2]$ ($2/N$ is recommended for p_{min}) while JADE requires p to be determined by a user.

SHADE with Linear Population Size Reduction (L-SHADE) [10] is an improved version of SHADE, which linearly decreases the population size as the search progresses. Population size in the next generation N_{g+1} is calculated as $N_{g+1} = round$ $[N_{FE} \times (N^{min} - N^{init})/N_{FE}^{max} + N^{init}]$, where N^{min} and N^{init} denote the minimum and initial population sizes, N_{FE}^{max} and N_{FE} denote the maximum and current function evaluation counts.

Table 1. Control parameter configurations

DE	$F = 0.7$, $CR = 0.7$
jDE	$F_l = 0.1$, $F_u = 0.9$, $\tau_1 = 0.1$, $\tau_2 = 0.2$
NADE	$CR = 0.7$, $\{sampling\ number\} = 30$
MUDE	$F_l = 0.1$, $F_u = 0.9$, $\tau_1 = 0.1$, $\tau_2 = 0.2$, $p_G = 0.04$, $p_H = 0.07$, $\{initial\ range\ of\ SFGSS\} = [-1,1]$, $\{samping\ number\} = 40$, $h = 0.1$, $\{FE\ limits\ in\ SFHC\} = 100$
JADE	$\{archive\ size\} = N$, $p = 0.05$, $c = 0.1$
SHADE	$\{archive\ size\} = N$, $\{memory\ size\} = D$
L-SHADE	$\{archive\ size\} = 2N$, $\{memory\ size\} = D$, $\{final\ population\ size\} = 4$
DECC	$\{number\ of\ divisions\} = 10$

Yang *et al.* proposed cooperative coevolution based DE algorithm named DECC which utilizes problem decomposition strategies and DE with self-adaptive neighbourhood search (SaNSDE) [11]. In DECC, first, an n-dimensional objective vector is decomposed into several s-dimensional subcomponents, and then evolve each of them with SaNSDE.

3   Experiments on Noisy Test Functions

To clarify the performance of novel SADEs, JADE, SHADE, L-SHADE, and DECC were compared with Canonical DE[1], jDE, and DEs targeting noisy functions: Noise Analysis Differential Evolution (NADE) [2] and Memetic for Uncertainties DE (MUDE) [6]. Control parameters were configured as shown in Table 1. Strategy of all algorithms except JADE, SHADE, and L-SHADE was unified to DE/rand/1/bin. Population size N was set to 100, and the maximum fitness evaluations (FEs) was to 100,000.

In the first four functions, Sphere, Rosenbrock, Rastrigin, and Griewank, we added Gaussian noise with zero mean and standard deviation of $0.04C$ [2], where $C = \frac{1}{N_c} \sum_{i=1}^{N_c} f(x_i^c) - \bar{f}$, N_c is a sampling number set to 100, x_i^c is i-th individual that is sampled randomly, and $\bar{f}$ is the function value of the global optimum. Note that C does not change during the search, meaning that the more significantly the noise affects the search performance as the search progresses [2]. The last two functions $F4$ and $F17$ in CEC 2005 benchmark problem [3] involve the noise that gradually decreases as the objective function value gets smaller. Dimensions of all the tested functions were set to 50.

Figure 1 shows the fitness transitions of the best solutions in the population. The results were averaged over 50 runs. Note that, in Fig. 1, the function values of the best solutions without noise were plotted.

Overall, JADE and SHADE showed better performance than other algorithms. L-SHADE had a direct influence of noise, and its search performance

---
[1] We represent canonical DE as "DE".

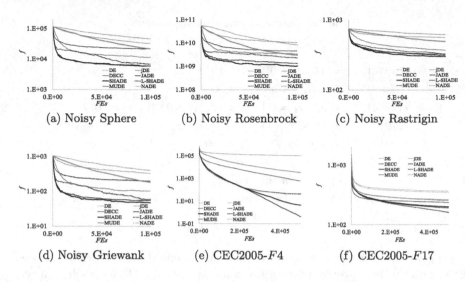

(a) Noisy Sphere   (b) Noisy Rosenbrock   (c) Noisy Rastrigin

(d) Noisy Griewank   (e) CEC2005-F4   (f) CEC2005-F17

Fig. 1. Experimental results on noisy functions.

significantly deteriorated, meaning that the population reduction must be tuned for noisy problems. DECC showed good convergence in noisy Rastrigin functions; however, DECC was slower than jDE and other SADEs in other functions since DECC targets higher dimensional problems. jDE followed JADE and SHADE, and was less subject to noise.

The above results demonstrated SADEs good performance compared to MUDE and NADE targeting noisy functions and the robustness of JADE, SHADE and jDE against fitness noise.

4   BRDF Estimation Problems

Bidirectional Reflectance Distribution Function (BRDF) is a method to describe a light reflection model on a material surface [7]. This paper proposes a method which simultaneously optimizes parameters in combination of Lafortune model for specular reflection and Lambertian BRDF model for diffuse reflection. In addition, it is assumed that RGB components have the same specular reflection model. Simultaneous estimation of the above model is difficult because one of Lafortune model parameters corresponds to an exponent part of the mode equation. This continuous optimization problem is globally multimodal because a certain reflection model can be represented by various combinations of Lafortune and Lambertian model parameters. This problem also involves static noise because a solution is evaluated by comparing the images rendered according to the solution and the actually photographed images.

When assuming that the number of lobes would be one, parameters for each pixel to be adjusted are seven: diffuser reflection parameters ρ_{dR}, ρ_{dG}, ρ_{dB} for color channel in addition to specular parameters $C_{x,1}$, $C_{y,1}$, $C_{z,1}$, and n_1. Lafortune model $f_r(\cdot)$ is modeled by $f_r(\boldsymbol{\omega_i}, \boldsymbol{\omega_r}) = \sum_i \sum_{\mathcal{D} \in \{x,y,z\}} [C_{\mathcal{D},i} \omega_{i\mathcal{D}} \omega_{r\mathcal{D}}]^{n_i}$

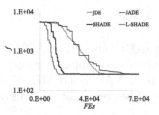

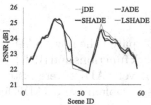

(a) JADE   (b) LSHADE

Fig. 2. Fitness transitions in BRDF estimation.

Fig. 3. PSNR for each scene.

Fig. 4. Rendering result examples (Color figure online).

where ω_i and ω_r are lighting and viewing directions, and i denotes the number of lobes. When assuming just one lobe, four parameters, $C_{x,1}$, $C_{y,1}$, $C_{z,1}$, and n_1, synthesizes the model.

An objective function E which should be minimized is defined by squared error between estimated and observed intensities for all sampled pixels as follows

$$E = \sum_j \sum_{\mathcal{C} \in \{R,G,B\}} (I_{\mathcal{C}(j)} - f_{r\mathcal{C}}(\omega_{i,j}, \omega_{r,j}))^2 \qquad (1)$$

where j denotes index of sampled pixel, $f_{r\mathcal{C}}$ ($\mathcal{C} \in \{R, G, B\}$) is estimated intensities for each RGB color, and $I_{\mathcal{C}}$ is observed intensities.

Experiments with a glazed pottery object were conducted to compare the performance of jDE, JADE, SHADE, and L-SHADE. Their control parameters were configured as follows: in jDE, $F_l = 0.1$, $F_u = 0.9$, $\tau_1 = 0.1$, and $\tau_2 = 0.1$. In JADE, SHADE, and L-SHADE, the archive size and memory size was set to $2N$ and n, respectively. JADE uses p and c with 0.05 and 0.01, respectively. In L-SHADE, final population size was set to four. In all algorithms, population size N was set to 70, and the maximum FEs was set to 7.0×10^4. BRDF models on all 199,838 points on the object surface were estimated individually. Photographed images of one viewpoint and 60×11 (horizontally 60 and vertically 11) light source directions were used for fitness calculation.

Figure 2 shows the example fitness transitions of the best solutions on a certain point of the object surface. L-SHADE was the fastest to converge, and SHADE followed L-SHADE. JDE and JADE were almost the same speed to find the best solutions. Figure 3 shows Peak Signal-to-Noise Ratio (PSNR) for 49 scenes with varying light directions. From the viewpoint of PSNR, the output image quality was high in the order of JADE, jDE, SHADE, and L-SHADE. Significant differences could be seen between JADE and SHADE and between JADE and L-SHADE by paired t-test with 95 % confidence interval. Figure 4 shows example rendering results estimated by JADE and L-SHADE. Rendering results by L-SHADE involved artiacts (white pixels). The above experimental results show that JADE's robustness against problem properties.

5   Conclusions

This paper experimentally compared the novel SADEs: JADE, DECC, SHADE, L-SHADE with noisy benchmark functions and BRDF estimation problem. Experimental results showed that novel SADEs were faster than NADE and MUDE which were specialized in noisy problems, and JADE found bettern solution than SHADE and L-SHADE in BRDF estimation.

Acknowledgements. Part of this work was supported by SCOPE (142110001) of MIC, Japan.

References

1. Brest, J., Greiner, S., Boskovic, B., Mernik, M., Zumer, V.: Self-adapting control parameters in differential evolution: a comparative study on numerical benchmark problems. IEEE Trans. Evol. Comput. **10**(6), 646–657 (2006)
2. Caponio, A., Neri, F.: Differential evolution with noise analyzer. In: Giacobini, M., Brabazon, A., Cagnoni, S., Di Caro, G.A., Ekárt, A., Esparcia-Alcázar, A.I., Farooq, M., Fink, A., Machado, P. (eds.) EvoWorkshops 2009. LNCS, vol. 5484, pp. 715–724. Springer, Heidelberg (2009)
3. Gonzalez-Fernandez, Y., Soto, M.: cec2005benchmark: Benchmark for the CEC 2005 Special Session on Real-Parameter Optimization, R package version 1.0.4 (2015). http://CRAN.R-project.org/package=cec2005benchmark
4. Jin, Y., Branke, J.: Evolutionary optimization in uncertain environments – a survey. IEEE Trans. Evol. Comput. **9**(3), 303–317 (2005)
5. Krink, T., Filipic, B., Fogel, G.: Noisy optimization problems - a particular challenge for differential evolution? In: Congress on Evolutionary Computation, CEC 2004, vol. 1, pp. 332–339, June 2004
6. Mininno, E., Neri, F.: A memetic differential evolution approach in noisy optimization. Memetic Comput. **2**, 111–135 (2010)
7. Nicodemus, F.E., Richmond, J.C., Hsia, J.J., Ginsberg, I.W., Limperis, T.: Geometric Considerations and Nomenclature for Reflectance. National Bureau of Standards, Washington (1977)
8. Storn, R., Price, K.: Differential evolution a simple and efficient heuristic for global optimization over continuous spaces. J. Global Optim. **11**, 341–359 (1997)
9. Tanabe, R., Fukunaga, A.: Success-history based parameter adaptation for differential evolution. In: 2013 IEEE Congress on Evolutionary Computation (CEC), pp. 71–78, June 2013
10. Tanabe, R., Fukunaga, A.: Improving the search performance of shade using linear population size reduction. In: 2014 IEEE Congress on Evolutionary Computation (CEC), pp. 1658–1665, July 2014
11. Yang, Z., Tang, K., Yao, X.: Large scale evolutionary optimization using cooperative coevolution. Inf. Sci. **178**(15), 2985–2999 (2008). nature Inspired Problem-Solving
12. Zhang, J., Sanderson, A.: Jade: adaptive differential evolution with optional external archive. IEEE Trans. Evol. Comput. **13**(5), 945–958 (2009)

Empirical Analysis of Operators for Permutation Based Problems

Pierre Desport[✉], Matthieu Basseur, Adrien Goëffon, Frédéric Lardeux, and Frédéric Saubion

LERIA, University of Angers, Angers, France
pierre.desport@univ-angers.fr

Abstract. This paper presents an analysis of different possible operators for local search algorithms in order to solve permutation-based problems. These operators can be defined by a distance metric that define the neighborhood of the current configuration, and a selector that chooses the next configuration to be explored within this neighborhood. The performance of local search algorithms strongly depends on their ability to efficiently explore and exploit the search space. We propose here a methodological approach in order to study the properties of distances and selectors in order to buildtheir performances operators that can be used either for intensification of the search or for diversification stages. Based on different observations, this approach allows us to define a simple generic hyperheuristic that adapt the choice of its operators to the problem at hand and that manages their use in order to ensure a good trade-off between intensification and diversification. Moreover this hyperheuristic can be used on different permutation-based problems.

1 Introduction

Many optimization problems can be modeled as permutation problems (e.g., flowshop, traveling salesman or quadratic assignment problems). Dedicated efficient solving methods have been proposed for these problems, but their performance often depend on the considered instances and use most of the time ad-hoc heuristics and/or techniques. Therefore, given a new permutation problem a non-expert user would hardly be able to design a solving algorithm using components that she/he could re-use from previous experiments. A recent trend in optimization consists in promoting more autonomous techniques for the design of search algorithms [1], either by automating the tuning of their parameters [2], by dynamically controlling their behaviour [3] or by automating their design [4]. Focusing on this latest aspect, hyperheuristics [5] is a generic paradigm that can be used to manage a set of efficient heuristics in order to solve a sufficiently large set of problem instances, without *a priori* knowledge. The main principle of hyperheuristics is to adapt the solving process to the given instance at hand. Considering a set of possible heuristics that can be applied on a given problem, one may select the best heuristics with regards to the characteristics and properties of this problem and/or change heuristics during the solving process,

© Springer International Publishing Switzerland 2015
C. Dhaenens et al. (Eds.): LION 9 2015, LNCS 8994, pp. 137–150, 2015.
DOI: 10.1007/978-3-319-19084-6_13

according to the current state of the search. Of course, this high level management of the solving heuristics requires to define sufficiently general heuristics and to gather pertinent information on them. In this paper, our focus on permutation problems allows us to consider a sufficiently large family of problems which can be handled by a set of common search operators (from now on we use the generic term operator for our basic search heuristics). These operators, defined in a local search fashion [6], aim at selecting the next configuration of the search space that will be examined by an incremental search process. Such an operator can easily be defined by (1) a notion of neighborhood of the current configuration, thanks to a distance measure, and (2) a selection function within this neighborhood. One aim of this work is thus to carefully study the behaviour of these search operators with regards to different instances of well known permutation problems. Note that different studies on distances for permutation problems have been conducted [7,8]. Here we focus on how operators, which use such distances, can be compared and efficiently chosen with regards to a given instance of a problem. Note that the dynamic control of operators in local search has been studied in [9]. Our approach can be more related to landscape analysis [10]. An analysis of the correlations between the two basic above-mentioned components of operators would help us to better understand their characteristics and to define a simple hyperheuristic to manage them.

This paper is organized as follow. Basic notions on permutations are recalled in Sect. 2. Main concepts concerning local search for combinatorial optimization problems are presented in Sect. 3. The next sections are devoted to the analysis of search landscapes induced by the static structures of the problems, as well as the operational behaviour of local search operators. A simple hyperheuristics is then defined in Sect. 6 in order to illustrate how previous observations can be used to improve solving algorithms.

2   Basic Notions for Permutations

2.1   Permutations

A permutation of n elements is an arrangement of these n objects sorted in a specific order where they appear only once. The group of permutations [11] can be defined as the group of bijections from X to X where X is a non-empty finite set. Let $[n]$ be a set of objects $[n] = \{1..n\}$. A permutation π is a bijective assignment on $[n]$ such that $\pi(i)$ is the element at position i in the permutation π and $pos_\pi(i)$ is the position of the element i in the permutation π. $\Pi([n])$ is the set of permutations on $[n]$, whose cardinality is thus $n!$. Given $[n] = \{1, 2, 3, 4\}$, $\pi = (1, 2, 3, 4)$ or $\pi = (2, 1, 4, 3)$ are two possible permutations.

Identity. Let $\pi \in \Pi([n])$, the identity permutation I is defined as the permutation that assigns each element of π to itself, i.e. $\forall i \in \{1..n\}, I(i) = i$.

Product of permutations. Let $\pi_1, \pi_2 \in \Pi([n])$, $\pi_1(i) * \pi_2(i) = \pi_2(\pi_1(i))\forall i \in [n]$. Note that $\pi_1 * \pi_2 \neq \pi_2 * \pi_1$. The neutral element of the product is I (i.e., $\forall \pi, \pi * I = I * \pi = \pi$).

Inverse. The inverse permutation can be defined using the identity permutation π^{-1} can be constructed from π using the property: $\pi^{-1}(i) = pos_\pi(i)$. For instance, if $\pi = (2, 3, 4, 1)$, then $\pi^{-1} = (4, 1, 2, 3)$.
Let us recall now some properties of permutations.

Adjacency. Given a permutation $\pi \in \Pi([n])$, two elements i and j are adjacent in π if $|pos_\pi(i) - pos_\pi(j)| = 1$.

Longest Increasing Sequence (LIS). Given a permutation π, the longest increasing subsequence $LIS(\pi)$ corresponds to the longest subsequence of elements of π that are sorted in ascending order. For instance, for $\pi = (1, 3, 2, 4)$ the longest increasing subsequences are $(1, 2, 4)$ and $(1, 3, 4)$.

Longest Common Subsequence (LCS). The longest common subsequence of two permutations $\pi_1, \pi_2 \in \Pi([n])$ is $LCS(\pi_1, \pi_2) = \{i \in \{1..n\} | p_1(i) = p_2(i)\}$. For instance, with $\pi_1 = (2, 4, 3, 1, 5, 6)$ and $\pi_2 = (1, 2, 3, 5, 4, 6)$, $LCS(\pi_1, \pi_2) = \{2, 3, 5, 6\}$.

2.2   Distance on Permutations

Table 1 presents different distance indicators which will be considered in the rest of the paper. The diameter of a distance measure represents the maximal distance between all permutations. Note that, as an example, the last column corresponds to the distance between the two following permutations (# is the cardinality function):

- $\pi_1 = (1, 2, 3, 4, 5, 6, 7, 8, 9, 10, 11, 12, 13, 14, 15)$ and,
- $\pi_2 = (15, 3, 1, 11, 2, 7, 9, 10, 4, 14, 6, 12, 5, 13, 8)$.

Note that the interchange distance requires a function, computing the number of permutation cycles which composes a permutation [8].

Table 1. Distances for permutations.

Distances	Formula (π_1 and π_2 are permutations)	Diameter	Ex.
Hamming	$\#\{i|i \in \{1..n\}, \pi_1(i) \neq \pi_2(i)\}$	n	14
Adjacency	$n - 1 - \#\{1 \leq i|adj_{\pi_2}(\pi_1(i), \pi_1(i+1))\}$	$n - 1$	13
Position	$\sum_{i=1}^{n} |pos_{\pi_1}(i) - pos_{\pi_2}(i)|$	$2\lceil n/2 \rceil \lfloor n/2 \rfloor$	62
Lee	$\sum_{i=1}^{n} min(|\pi_1(i) - \pi_2(i)|, n - |\pi_1(i) - \pi_2(i)|)$	$n(n/2)$	48
Swap	$\#\{(i,j)|1 \leq i < j \leq n, pos_{\pi_2}(\pi_1(j)) < pos_{\pi_2}(\pi_1(i))\}$	$n(n-1)/2$	44
Interchange	$n - c(\pi_1^{-1} * \pi_2)$	$n - 1$	12
Ulam	$n - length(LIS(\pi_1^{-1} * \pi_2))$	$n - 1$	9
Insertion	$n - length(LCS(\pi_1, \pi_2))$	$n - 1$	8

3   Optimization Problems and Local Search

Let us define the components of a local search algorithm in the context of solving optimization problems [6].

3.1   General Definitions

Optimization Problem. An optimization problem is a pair $(\mathcal{S}, f)$ where $\mathcal{S}$ is a search space whose elements represent solutions (or configurations) of the problem and $f : \mathcal{S} \to \mathbb{R}$ is an objective function. An optimal solution (for maximization problems) is an element $s^* \in \mathcal{S}$ such that $\forall s \in \mathcal{S}, f(s^*) \geqslant f(s)$.

Local Search. Given an optimization problem, a local search (LS) process consists in starting from an initial configuration and in applying repeatedly basic move operators in order to reach an optimal solution. The trace obtained by such a search process is usually called a search path. An operator is thus a function that returns the next configuration for building the search path. In its simplest form, a move performs the selection of the next configuration to be explored within the neighborhood of the current configuration. A generic and basic outline of an LS metaheuristic is the application of an operator in a simple loop as illustrated in Algorithm 1 where the `SpecificAction()` method represents a step specific to the type of metaheuristic used such as a perturbation (e.g., Iterated Local Search [6]) or enforcing prohibitions (e.g., Tabu Search).

$s \leftarrow$ initial configuration;
$s^* \leftarrow s$;
while *end condition not met* **do**
> $s \leftarrow op(s)$;
> **if** $eval(s) < eval(s^*)$ **then** $s^* \leftarrow s$;
> SpecificAction();

end
return s^*

Algorithm 1. Algorithmic outline of an LS metaheuristic for minimization

Neighborhood. Let $\mathcal{S}$ be the search space of candidate solutions. A neighborhood relation is an irreflexive binary relation $\mathcal{N} \subseteq \mathcal{S}^2$ over the search space. In most cases, the relation is also symmetric.

Search Paths. Given a neighborhood relation $\mathcal{N}$, the set of search paths is defined as $\mathcal{P}_\mathcal{N} = \{s_1 \cdots s_n \in \mathcal{S}^* \mid \forall i > 1, (s_{i-1}, s_i) \in \mathcal{N}\}$, where $\mathcal{S}^*$ represents the set of words constructed over $\mathcal{S}$. Therefore any pair (s, s') of elements of $\mathcal{S}$, such that[1] $(s, s') \in \mathcal{N}^+$, defines an equivalence class over the set $\mathcal{P}_\mathcal{N}$ which

---

[1] $\mathcal{N}^+$ is the transitive closure of $\mathcal{N}$.

corresponds to all paths that link s to s'. This subset is denoted by $\mathcal{P}_\mathcal{N}(s, s')$. In most cases, the neighborhood should be complete, i.e. $\forall s, s' \in S, \mathcal{P}_\mathcal{N}(s, s') \neq \emptyset$.

Distances Induced by Neighborhood. The neighborhood relation defines the structure of the search space. The distance between s and s' can therefore be defined as $d_\mathcal{N}(s, s') = min_{p \in \mathcal{P}_\mathcal{N}(s, s')}|p|$, where $|p|$ is the classic word length and $d_\mathcal{N}(s, s) = 0$. If d define a distance, then $\mathcal{N}$ is necessarily symmetric. Note that a neighborhood induces a distance on the search space but, conversely, a distance on the search space can easily be used to define a neighborhood.

Local Search Operators. An operator is defined by two main components: neighborhood and selection process. A selector is a function that performs a selection over a neighborhood, eventually guided by the ordering $<$, and is defined as $\sigma : S \times 2^{S^2} \to S$ (here the selection returns only one neighbor), such that $(s, \sigma(s, \mathcal{N})) \in \mathcal{N}^=$ (the reflexive closure of S in order to include identity). Note that the selectors may include randomization for computing their results (e.g., random choice of a neighbor). Let us consider three classic different selectors:

- *First improve* (select the first improving neighbor from a randomly ordered set of neighbors),
- *Best improve* (select the best neighbor),
- *Random choice* (randomly chosen neighbor).

An operator is fully defined by a pair $(\mathcal{N}, \sigma)$.

Search Landscape. The search landscape is usually defined by the search space and the objective function that should be maximized (without loss of generality). The ordering relation $<$ over S corresponds to the order induced by the fitness function of the problem.

Operational Landscape. The operational structure of local search is defined by the possible moves in the search landscape according to a neighborhood relation. Again, we consider the paths induced by an operator $o = (\mathcal{N}, \sigma)$:

$$\mathcal{P}_o = \bigcup_{n>1} \{s_1 \cdots s_n \in S^* | \forall i > 1, s_i = \sigma(s_{i-1}, \mathcal{N})\}$$

Here, we should note that we only have the inclusion $\mathcal{P}_o \subseteq \mathcal{P}_\mathcal{N}$, since some neighborhood paths cannot be necessarily constructed by the operators as soon as it includes a selection process among the neighbors. Moreover, if there exists a path in $\mathcal{P}_o$ linking s to s', there does not necessarily exist a path from s' to s. Therefore, due to this non symmetric aspect of operators, it is not obvious to use a simple distance over the paths created by the operators. Now we may handle multiple move operators local search by composing neighborhood relations and selectors.

3.2 Permutations Based Problems

According to the previous definitions, a permutation based problem is a problem such that $\mathcal{S} = \Pi([n])$. Many combinatorial optimization problems can be formulated as permutation-based problems, most of them being NP-hard. In this section we recall three well-known permutation problems that will be used in our experiments.

Quadratic Assignment Problem. Quadratic assignment problem (QAP) [12] models a facilities location problem. The objective of QAP is to assign n facilities to n locations in order to minimize the assignment cost. It may be formalized as follows:

- let f_{ij} the flow between facilities i and j,
- let d_{ij} the distance between locations i and j,
- minimize: $\sum\limits_{i,j=1}^{n} f_{ij} d_{\pi(i)\pi(j)}$.

Flowshop Problem. The flowshop problem [13] is a scheduling problem where the goal is to find the best planning to achieve n jobs on m different machines, minimizing the makespan (total completion time), considering the following constraints :

- All jobs must be processed by all machines,
- A machine can deal with only one job at any time t,
- A job can be processed only by one machine at any time t.

Here we consider that the processing order of the jobs on the machine is always the same. The goal is to find a permutation π representing a processing order of the jobs that minimizes the makespan function $C_{max} = max\{s_{iM} + p_{iM} | i \in [1..n]$ and $M \in [1..m]\}$ where:

- p_{ij} is the time for the machine i to process the job i,
- s_{ij} is the starting time of the job i on the machine j.

Traveling Salesman Problem. The traveling salesman problem (TSP) [14] consists in finding the shortest path in order to visit n cities without visiting any city twice. This constraint is easily enforced by using permutations to represent configurations of the problem. For a permutation π, each element is a city. Given a matrix D such that d_{ij} corresponds to the distance between city i and city j, the objective function to minimize is $\sum\limits_{i=1}^{n-1} d_{\pi(i)\pi(i+1)} + d_{\pi(n)\pi(1)}$. Here, we restrict our study to symmetric TSP (i.e., $d_{ij} = d_{ji}$ $\forall i,j \in \{1,2,..,n\}$).

Table 2. Diameter and average distances

Instance		Ham.	Adj.	Position Based	Lee	Swap	Interchange	Insertion	Ulam
Inst. 1	max	20	19	200	200	190	19	19	19
(size 20)	avg	19	17	133	100	95	16	13	14
Inst. 2	max	26	25	338	338	325	25	25	25
(size 26)	avg	25	23	225	169	162	22	18	19
Inst 3	max	100	99	5000	5000	4950	99	99	99
(size 100)	avg	99	97	3331	2500	2475	95	92	93

4 Search Landscape Analysis

In this section, we propose a first study of the search landscape corresponding to the previously selected permutation problems. Our purpose is to highlight how suitable but yet generic local search operators can be defined. We will first study the search landscape from a static point of view by using distance indicators on permutations that have been presented in Sect. 2.2. This will allow us to exhibit correlations between distances as well as correlations between problems and distances. Before starting our experiments, let us observe what are the typical topological characteristics of our problems, computing the maximal theoretical and the average distances between two permutations.

4.1 Search Space Diameters

The following results will help us to better interpret our further studies. Table 2 provides maximal theoretical (max) and average (avg) distances for three instances. Note that here the original problems have no influence since we consider only the search landscape, which only depends on the size of the permutation. We output here instances from size 20 to 100 in order to highlight the relative differences between distances. Results show similar properties when considering larger instances. Average results are obtained by computing the distance between 10^5 pairs of randomly generated permutations.

4.2 Search Space: Correlation Between Distances

The purpose of this first experiment is to compare the distance indicators. Table 3 shows correlations of distances obtained for 10^5 randomly generated pairs of permutations. Let us recall that the problem under consideration have no influence on the results. We considered two close sizes (20 and 26) and a larger one (100) in order to obtain different effects. Nevertheless, correlations do not depend on the size, as observed here. This study can be related to the work presented in [8] but with additional metrics. Moreover, the experimental process is slightly different even if similar conclusions are reported.

Table 3. Correlation distances - distances

Instance	Distance	Ham.	Adj.	Position Based	Lee	Swap	Inter-change	Insertion	Ulam
Size 20	Hamming	**1.000**	0.004	0.360	0.393	0.226	**0.672**	0.214	0.168
	Adjacency	0.004	**1.000**	−0.003	0.001	−0.005	0.026	0.087	−0.005
	PositionBased	0.360	−0.003	**1.000**	0.142	**0.939**	0.243	**0.605**	0.061
	Lee	0.393	0.001	0.142	**1.000**	0.090	0.261	0.088	0.308
	Swap	0.226	−0.005	**0.939**	0.090	**1.000**	0.148	**0.621**	0.040
	Interchange	**0.672**	0.026	0.243	0.261	0.148	**1.000**	0.124	0.098
	Insertion	0.214	0.087	**0.605**	0.088	**0.621**	0.124	**1.000**	0.038
	Ulam	0.168	−0.005	0.061	0.308	0.040	0.098	0.038	**1.000**
Size 26	Hamming	**1.000**	0.026	0.302	0.344	0.188	**0.640**	0.189	0.132
	Adjacency	0.026	**1.000**	0.005	0.005	0.000	0.021	0.089	0.008
	PositionBased	0.302	0.005	**1.000**	0.093	**0.941**	0.200	**0.589**	0.050
	Lee	0.344	0.005	0.093	**1.000**	0.050	0.213	0.061	0.286
	Swap	0.188	0.000	**0.941**	0.050	**1.000**	0.121	**0.594**	0.027
	Interchange	**0.640**	0.021	0.200	0.213	0.121	**1.000**	0.111	0.077
	Insertion	0.189	0.089	**0.589**	0.061	**0.594**	0.111	**1.000**	0.040
	Ulam	0.132	0.008	0.050	0.286	0.027	0.077	0.040	**1.000**
Size 100	Hamming	**1.000**	0.003	0.220	0.242	0.137	**0.575**	0.137	0.081
	Adjacency	0.003	**1.000**	0.001	0.003	0.000	0.009	0.059	0.001
	PositionBased	0.220	0.001	**1.000**	0.056	**0.944**	0.127	**0.557**	0.018
	Lee	0.242	0.003	0.056	**1.000**	0.037	0.137	0.035	0.228
	Swap	0.137	0.000	**0.944**	0.037	**1.000**	0.077	**0.532**	0.011
	Interchange	**0.575**	0.009	0.127	0.137	0.077	**1.000**	0.071	0.044
	Insertion	0.137	0.059	**0.557**	0.035	**0.532**	0.071	**1.000**	0.015
	Ulam	0.081	0.001	0.018	0.228	0.011	0.044	0.015	**1.000**

The correlation coefficient cf corresponds to the intensity of the connection between two sets of values and has a value ranging in $[-1, 1]$. Two sets are said to be strongly correlated if $|cf| > 0.5$.

$$cf(x, y) = \frac{\sum_{i=1}^{n}(x_i - \bar{x}) * (y_i - \bar{y})}{\sqrt{\sum_{i=1}^{n}(x_i - \bar{x})^2} * \sqrt{\sum_{i=1}^{n}(y_i - \bar{y})^2}}$$

where $\bar{x}$ and $\bar{y}$ are the mean values of x and y.

Note that the correlation between distances is measured only on the distances obtained between permutations. The objective function of the problem is not involved in the process (only the size of the studied instances influences the results rather than the type of problem). We can observe in Table 3 that distances can be grouped by sets of correlated distances:

$\{Hamming, Interchange\}, \{Adjacency\}, \{Lee\},$
$\{Swap, PositionBased, Insertion\}, \{Ulam\}.$

Table 4. Correlation problems - distances

Problem	Instance	Size	Ham.	Adj.	Position Based	Lee	Swap	Inter-change	Insertion	Ulam
QAP	nug20	20	0.105	0.091	0.031	0.006	−0.005	0.086	0.031	0.001
	lipa30_a	30	0.020	0.000	0.006	0.011	0.001	0.018	−0.007	0.006
	sko81	81	0.032	0.033	0.034	0.009	0.002	0.015	0.044	0.001
	bur26a	26	0.092	0.000	0.104	−0.009	0.012	0.059	0.089	0.032
	esc16a	16	−0.001	0.084	−0.001	0.004	−0.006	0.010	0.015	0.002
TSP	a280	280	0.001	0.166	0.007	0.002	0.001	0.001	0.021	0.006
	berlin52	52	0.000	0.392	0.006	−0.005	0.003	0.003	0.061	−0.002
	eil51	51	0.002	0.392	−0.010	0.002	0.000	−0.001	0.069	−0.005
	kroD100	100	0.003	0.268	0.003	0.001	0.001	0.000	0.039	0.002
	tsp225	225	0.002	0.192	−0.005	−0.001	0.000	−0.002	0.023	−0.001
FlowShop	20_5_01	20	0.089	0.049	0.440	0.020	0.474	0.062	0.278	0.017
	20_10_01	20	0.134	0.029	0.465	0.027	0.469	0.086	0.352	0.007
	20_20_01	20	0.110	0.018	0.331	0.076	0.315	0.075	0.253	0.031
	50_5_01	50	0.040	0.016	0.257	0.001	0.275	0.027	0.163	−0.004
	50_10_01	50	0.055	0.037	0.334	0.013	0.342	0.034	0.203	0.010

This first observation may help us to select distances for either building search operator or better controlling the search process with a specific heuristic. For instance, if a search algorithm using Hamming distance requires diversification, it seems intuitively appropriated to use another distance that is weakly correlated (for instance Lee distance).

4.3   Search Landscape: Correlation between Problems and Distances

The correlation between the objective function of a problem and the neighborhood relation used to define operators is obviously an important feature to ensure good performance for a LS algorithm. Ideally, if distances between configurations are proportional to their objective values difference, then it is easier to reach good solutions, since moves can be clearly guided by improvement strategies.

In Table 4 we examine the correlation between the distance indicators and the problems introduced previously, considering the search landscapes induced by their objectives functions. Instances whose known optimal solution are used to study this correlation. The distance between the optimal permutation and a randomly generated permutation is computed as well as the difference between their objective function values. This process is repeated 10^5 times in order to obtain a correlation value between distances and problems. Results are presented in Table 4.

In Table 4 no strong correlation can be observed. Considering QAP, due to its quadratic objective function, it is very difficult to define a metric that can be correlated to the fitness. For the flowshop problem, distances which induce less perturbations in the objective function values of the configurations, show better results as observed for Swap or Insertion indicators. Similar observation can be done for TSP using Adjacency. Nevertheless, the correlation depends on the size of the considered problems since random points can be far from the optimal solution for problems with large diameters.

5   Analysis of the Operational Landscape

We now turn to the operational point of view by considering search operators that can be used in a local search algorithm. Using the previous results and in order to avoid too combinatorial experiments, we consider only the following classic distances: Swap, Interchange and Insertion. The corresponding neighborhood are indeed often used to build operators for permutation problems. We also consider the three selectors: First, Best and Random. According to the definition of an operator provided in Sect. 3.1 we have thus nine possible operators.

The purpose of the following experiment is to assess the ability of an operator to reach an optimal solution using the shortest possible path (i.e., using the fewest number of permutations). Here, we aim at studying the short term convergence properties of operators for different problems, in order to identify good candidates for intensification. For different instances of each problem, a permutation is randomly generated at distance n, starting from the optimal known permutation, with the different neighborhoods associated to the operators. We observe then if the operator is able to come back to the optimal solution. Tests have been carried out at various distances from the optimal solution. Here, we are mainly interested by results obtained at a distance 5 from the optimal solution. For smaller distances it seems clear that all operators including the "best" selection mechanism are likely to return to the optimal solution. Oppositely, choosing too long distance for small instances leads to search the optimal solution from a totally random permutation. Table 5 shows results for a distance of 5. Values represent the probabilities of a path built by the operator to reach the optimal permutation.

The results show the efficiency of the operators with regards to intensification for the three problems. An operator that frequently reaches the optimal solution from a distance of 5 is indeed a pertinent operator for the intensification of the search. This experiment assesses that it is sufficient to reach a distance 5 from the optimal solution in order to easily reach it with this operator. Nevertheless, let us notice that maximal distances (i.e., diameters) related to operators are not of the same order of magnitude. For instance, the *insertion* and *interchange* have a smaller diameter than *swap*. We can remark that associations between operators and problems do not correspond the empirical intuitions. The low distance is certainly an explanation of this behavior.

6   Design of a Simple Hyperheuristic

In this section, the previous analysis are used in order to define a simple hyperheuristic approach for solving the three families of problems. The concept of hyperheuristic [5] has been initially introduced as "a heuristics to choose heuristics". Hyperheuristics manage indeed a set of heuristics and select or combine them in order to efficiently solve problems. Instead of manually designing a solving algorithm, an hyperheuristic is used in order to automate the process of selection, combination or generation of heuristics, aiming at solving different

Table 5. Ability to find the optimal solution starting at a distance of 5

Problem	Instances	Size	Swap		Interchange		Insertion	
			First Improve	Best Improve	First Improve	Best Improve	First Improve	Best Improve
FlowShop	20_5_01	20	0.7	0.84	0.05	0.13	0.14	0.3
	20_5_02	20	0.54	0.81	0.06	0.15	0.17	0.36
	20_10_01	20	0.22	0.48	0	0.2	0.03	0.27
	20_10_02	20	0.15	0.5	0.01	0.18	0.02	0.28
	50_5_01	50	0.81	0.99	0.44	0.63	0.75	0.87
QAP	nug12	12	0.25	0.45	0.13	0.27	0.03	0.09
	bur26a	26	0.57	0.77	0.11	0.78	0.01	0.03
	els19	19	0.41	0.71	0.1	0.7	0	0.07
	lipa40	40	0.58	0.8	0.84	1	0	0.01
	sko100a	100	0.86	0.99	0.01	1	0	0.07
	chr12a	12	0.12	0.31	0.02	0.25	0	0.02
	scr12	12	0.25	0.35	0.17	0.33	0	0.02
	lipa40a	40	0.56	0.77	0.9	0.99	0	0.02
	wil100	100	0.83	1	0.01	0.99	0	0.05
	tai80b	80	0.35	0.83	0	0.98	0	0.05
TSP	100	rd100	0.58	0.95	0.00	0.97	0.54	0.99
	berlin52	52	0.7	0.96	0.04	0.97	0.47	0.97
	eil51	51	0.49	0.92	0.03	0.95	0.29	0.97
	kroD100	100	0.62	0.96	0.00	0.97	0.40	0.98
	lin105	105	0.7	1	0	0.97	0.49	0.99
	tsp225	225	0.8	1	0	1	0.59	1
	st70	70	0.65	0.96	0.01	0.95	0.32	0.97

problems with a single generic solver. There is currently a very active community on hyperheursitics and a competition [15] has been launched to compare different approaches.

In the following, we show that the analysis of the search and operational landscape of the problems provided previously may help to design simple generic hyperheuristic for permutation-based problems. Our solving approach is rather simple and alternates two stages: an intensification phase and a diversification phase, which allows the search process to escape of local optima [6]. Using previous experiments, it is possible to characterize operators that promote intensification or diversification.

6.1 Algorithm

Our hyperheuristic algorithm takes as input a set of operators and an instance of a problem. Since the *swap* neighborhood has a larger diameter than the *insertion* or *interchange* neighborhoods, it is not considered in the experiments. Indeed, it is more difficult to use operators with different diameters if one wants to ensure fair comparisons. We consider here only two uncorrelated distances *interchange* and *insertion*. This is a restricted choice and other possible distances could have been considered. The set of possible operators is thus: *interchange*/first improve, *interchange*/best improve, *interchange*/random, *insertion*/first improve, *insertion*/best improve, *insertion*/random.

Using previous experiments, we consider the following methodology:

1. Select an operator for the the intensification stage: the algorithm starts with a study of paths on the instance in order to determine which operator should be considered for intensification. The algorithm selects thus the operator with the highest success rate as the intensification operator. The distances study of Sect. 4.2 is then used to select the diversification operator.
2. Select an operator for intensification: since the distances *interchange* and *insertion* are weakly correlated, we assume that if an operator using the *insertion* neighborhood is selected for intensification, then it may be interesting to select an operator that uses *interchange* neighborhood for diversification (and vice versa). The random selection mechanism will be considered as selector in order to ensure an efficient diversification.

The hyperheuristic is detailed in Algorithm 2.

---

input: Instance I, Set of operators
output: Value (according to an objective function f) of the best permutation found
Require:
    $(OpIntensification, OpDiversification) \leftarrow Select - operators(I)$ {Intensification operator and diversification obtained by experimentation}
    $p \leftarrow Random()$ {Randomly generated permutation}
    $best \leftarrow f(p)$ {Best fitness}
    while not stop condition **do**
        $p \leftarrow HillClimbing(OpIntensification, p)$ {HillClimbing process}
        if $f(p) < best$ **then**
            $best \leftarrow f(p)$
        end if
        $p \leftarrow diversification(p)$ {Diversification process}
    end while
    return $best$

Algorithm 2. Hyperheuristic algorithm

6.2   Results

Algorithm 2 has been evaluated on different instances of each problem. The results have been compared to the best known values and to an algorithm that selects uniformly an operator at each iteration among the possible ones. A basic Hill Climbing using a unique operator (combinations of *swap*, *interchange* or *insertion* and first, best or random selection) has been used, but obtains poor results in being stuck in local optima. Same experimental conditions are used to test the different algorithms: a maximum number of iterations is set for each

Table 6. Results

Instance	Size	Hyperheuristic			Best	Uniform		
		best	avg.	s.d.	known	best	avg.	s.d.
TSP								
Berlin52	52	**7542**	7912	175.6	**7542**	8282	8486.4	131.5
eil51	51	432	438.6	5.7	**426**	445	457.8	7.8
st70	70	690	711.4	13.9	**675**	772	821.8	25.5
kroD100	100	23627	25380.9	1332.73	**21294**	26333	32151.5	2292.3
lin105	105	15761	17785	1196.3	**14379**	17910	22387.5	1927.1
rd100	100	8748	9111.1	275.7	**7910**	10509	11483.3	635.6
tsp225	225	4824	5157.6	269.03	**3919**	8113	10038.68	2061.6
FlowShop								
20_5_01	20	**1278**	**1278**	0	**1278**	**1278**	1283.8	7.0
20_5_02	20	**1359**	1359.2	0.4	**1359**	1360	1360.4	1.2
20_10_01	20	1583	1584.9	2.9	**1582**	1600	1606.1	6.6
20_10_02	20	1660	1666.4	3.1	**1659**	1675	1689.9	9.2
50_5_01	50	**2724**	**2724**	0	**2724**	**2724**	**2724**	0
QAP								
Bur26a	26	**5426670**	5427870	1784	**5426670**	54322537	5435020	1460.8
tai50a	50	5067098	5074390	6361.6	**4938796**	5241678	5280400	27268.1
lipa40a	40	31645	31857.3	72.6	**31538**	32034	32052.7	11.3
sko100a	100	152560	153183	408.4	**152002**	155372	156912	1021.3
wil100a	100	274034	274553	365.7	**273038**	275750	278877	1335.3

instance, 20 runs are executed for each instance and the same initial permutations are used for the hyperheuristic and the random algorithms. Results are presented in Table 6, in which best values are indicated bold.

We can remark that whatever the instance, the hyperheuristic obtains significantly better results than those obtained by the algorithm with uniform selection, both in terms of best result and average. This difference shows the importance of carefully choosing operators for intensification and diversification. One can also note that as well as being generic, hyperheuristic obtained reasonable results with regards best-known ones – even for large instances, especially considering QAP and flowshop problems. Note that this algorithm is generic in comparison to problem-dedicated algorithms for these well-known problems. Our study can be extended to more operators and problems. Our purpose here was rather to highlight that studying the search space as well as the search landscape may be useful to devise generic hyperheuristics.

7   Conclusion

In this paper, an analysis of operators for permutation-based problems is proposed. Properties of distance measures, neighborhood and selection mechanisms were observed for different permutation-based problems, and provide a better understanding of the relative efficiency of operators. Known relationships between operators and problems have been confirmed. Moreover, collected informations can be used to automate the choice of operators for different permutation-based problems. We have proposed a simple hyperheuristic using these operators properties. Further studies will include more combinatorial experiments with more operators and problem instances.

References

1. Hamadi, Y., Monfroy, E., Saubion, F.: What is autonomous search? Technical report MSR-TR-2008-80, Microsoft Research (2008)
2. Hoos, H.H.: Automated algorithm configuration and parameter tuning. In: Hamadi, Y., Monfroy, E., Saubion, F. (eds.) Autonomous Search, pp. 37–71. Springer, Heidelberg (2012)
3. Maturana, J., Fialho, A., Saubion, F., Schoenauer, M., Lardeux, F., Sebag, M.: Adaptive operator selection and management in evolutionary algorithms. In: Hamadi, Y., Monfroy, E., Saubion, F. (eds.) Autonomous Search, pp. 161–190. Springer, Heidelberg (2012)
4. Burke, E.K., Kendall, G., Newall, J., Hart, E., Ross, P., Schulenburg, S.: Hyper-Heuristics: an emerging direction in modern search technology. In: Glover, F., Kochenberger, G.A. (eds.) Handbook of Meta-Heuristics, pp. 457–474. Springer, New York (2003)
5. Burke, E.K., Hyde, M., Kendall, G., Ochoa, G., Özcan, E., Woodward, J.: A classification of hyper-heuristic approaches. In: Gendreau, M., Potvin, J.-Y. (eds.) Handbook of Metaheuristics. International Series in Operations Research & Management Science, vol. 146, pp. 449–468. Springer, NewYork (2010)
6. Hoos, H., Stützle, T.: Stochastic Local Search: Foundations and Applications. Morgan Kaufmann Publishers Inc., San Francisco (2004)
7. Deza, M., Huang, T.: Metrics on permutations, a survey. J. Comb. Inf. Syst. Sci. **23**, 173–185 (1998)
8. Schiavinotto, T., Stützle, T.: A review of metrics on permutations for search landscape analysis. Comput. Oper. Res. **34**(10), 3143–3153 (2007)
9. Veerapen, N., Maturana, J., Saubion, F.: An Exploration-exploitation Compromise-based Adaptive Operator Selection For Local Search In: Genetic and Evolutionary Computation Conference, GECCO, pp. 1277–1284. ACM (2012)
10. Marmion, M.-E., Dhaenens, C., Jourdan, L.: A fitness landscape analysis for the permutation flowshop scheduling problem. In: International Conference on Metaheuristics and Nature Inspired Computing (META 2010), Djerba, Tunisie (2010). http://hal.inria.fr/inria-00523213
11. Praeger, C.: Finite primitive permutation groups: a survey. In: Kovács, L. (ed.) Groups-Canberra 1989. Lecture Notes in Mathematics, vol. 1456, pp. 63–84 (1990)
12. Koopmans, T.C., Beckmann, M.: Assignment problems and the location of economic activities. Econometrica **25**, 53–76 (1957)
13. Dudek, R.A., Panwalkar, S.S., Smith, M.L.: The lessons of flowshop scheduling research. Oper. Res. **40**(1), 7–13 (1992)
14. Reinelt, G.: TSPLIB–a traveling salesman problem library. ORSA J. Comput. **3**(4), 376–384 (1991)
15. Burke, E., Hyde, M., Ochoa, G.: Cross-domain heuristic search challenge (2011). http://www.asap.cs.nott.ac.uk/external/chesc2011/

Fitness Landscape of the Factoradic Representation on the Permutation Flowshop Scheduling Problem

Marie-Eléonore Marmion[1]($\boxtimes$) and Olivier Regnier-Coudert[2]

[1] Inria Lille Nord-Europe, LIFL, Université Lille 1, Lille, France
marie-eleonore.marmion@lifl.fr
[2] IDEAS Research Institute, Robert Gordon University, Aberdeen, UK
o.regnier-coudert@rgu.ac.uk

Abstract. Because permutation problems are particularly challenging to model and optimise, the possibility to represent solutions by means of factoradics has recently been investigated, allowing algorithms from other domains to be used. Initial results have shown that methods using factoradics can efficiently explore the search space, but also present difficulties to exploit the best areas. In the present paper, the fitness landscape of the factoradic representation and one of its simplest operator is studied on the Permutation Flowshop Scheduling Problem (PFSP). The analysis highlights the presence of many local optima and a high ruggedness, which confirms that the factoradic representations is not suited for local search. In addition, comparison with the classic permutation representation establishes that local moves on the factoradic representation are less able to lead to the global optima on the PFSP. The study ends by presenting directions for using and improving the factoradic representation.

Keywords: Permutation Flowshop Scheduling · Factoradics · Fitness landscape

1 Introduction

Many optimisation problems can be modelled by means of permutations. This is the case of many scheduling problems such as the Permutation Flowshop Scheduling Problem (PFSP) in which a set of jobs needs to be assigned to a set of machines in the same order on each machine. The space of permutations is known to be a challenging one to search due to the nature of the representation which is difficult to model in comparison with other domains. Efforts in the field of Estimation of Distribution Algorithms (EDAs) have led to the development of probabilistic models of the permutation space [1], but these models require adapted algorithms to make use of them.

The use of alternative genotypes has been proved useful [2] in many domains, including permutations [3]. By transferring a problem from one domain to another, alternative genotypes encourage re-use of state of the art methods across domains.

© Springer International Publishing Switzerland 2015
C. Dhaenens et al. (Eds.): LION 9 2015, LNCS 8994, pp. 151–164, 2015.
DOI: 10.1007/978-3-319-19084-6_14

In the context of permutations, the use of genotypes may allow common modelling frameworks to be adapted to the problem under consideration.

Recent work has shown that the factorial numbering system, also referred to as factoradics or Lehmer codes can be used to represent permutations [4–6]. The factoradic representation uses a string of integers, with different weights associated with each position. Consequently, optimising factoradics rather than permutations themselves allows the use of methods from the integer domain. Experiments have shown that factoradics can be successfully implemented within different types of algorithms. However, the fitness landscape associated with the factoradic representation and its move operators has not been studied to date. It is thus difficult to define precisely the characteristics of this representation and to understand what search principles are the most adapted to it.

The frequent use of the straightforward permutation representation in evolutionary computation to tackle permutation problems has not really been investigated in comparison to using other representations. This paper proposes a fitness landscape analysis of the factoradic representation and one of its simplest operator on the PFSP, one of the most widely used benchmark for permutation optimisation. The paper is organised as follows. First, the factoradic representation is described along with previous work on the topic. Section 3 highlights the core concepts of fitness landscape analysis, while Sect. 3.2 defines the metrics used for the analysis. The PFSP is explained in Sect. 4. Finally, experiments are presented and their results discussed in Sect. 5, providing recommendations on adapted search strategies for the factoradic representation.

2   The Factoradic Representation

2.1   Genotypes for Permutation Spaces

Introducing alternative genotypes may prove useful to overcome challenges encountered when handling permutations. In Evolutionary Algorithms (EAs), the term *genotype* is often used to describe the domain searched by the algorithms, that is the search space on which operators are applied. In order to assess solutions, a phenotype is required. The *phenotype* represents the domain in which a solution can be evaluated, or in other words, a domain that can be read by the fitness function. Not only does using alternative genotypes allow some problems to be modelled efficiently by EAs, but it may also map a problem to a domain which is more adapted to these algorithms. Consequently, it has been shown that using many representations within the same search procedure may yield improved results by balancing out between the biases introduced by each representation [7].

With respect to permutations, the random key (RK) genotype has widely been used [3]. Yet, it is known to display some features that may inhibit the search in some contexts [8]. The RK genotype represents the permutation phenotype as a real value encoding. To generate a new permutation using RK, a string of real values needs to be first generated, one per permutation index. This string is then sorted and the permutation index assigned to each value set at each position in the string. Figure 1 shows how a RK genotype is mapped into

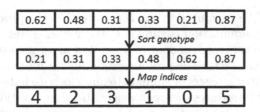

Fig. 1. Mapping from RK to permutation

a phenotype. Note that one of the properties of the RK representation is that many distinct real strings can represent the same solution.

The factoradic representation was used in [6] as an alternative to RK. It presents distinct characteristics and introduces a mutation operator that can be parametrised by the position of the gene to mutate, allowing specific algorithms such as the COMpeting Mutating Agents algorithm (COMMA) [9] to be considered.

2.2 Factoradics

The factoradic system is a numbering system of dimension n, which uniquely represents each number between 0 and $n! - 1$ as a string of factoradic digits. Each position i, $i \in [0, n-1]$ can be assigned a digit taking a value between 0 and i. The base of each position increases with i and so does its place value, i.e. the size of the factorial. Thus, the place value at position i is $i!$. The factoradic $a_{(!)}$ can be transformed into its decimal form $a_{(10)}$ as follows:

$$a_{(10)} = \sum_{i=0}^{n-1} a_{(!)_i} \times i! \tag{1}$$

where $a_{(!)_i}$ represents the i-th element of $a_{(!)}$. The potential of factoradics goes beyond the simple numbering system as it represents a way to easily represent permutations. For example, the factoradic *422100* denotes the permutation where the 4th, 2nd, 2nd, 1st, 0th and 0th items are drawn successively without replacement from the set of items. Figure 2 illustrates how this factoradic

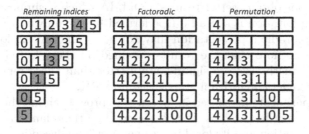

Fig. 2. Mapping from factoradic to permutation

number represents the permutation *423105*. Note that the procedure to decode a factoradic into a permutation is similar to the Fisher-Yates shuffle algorithm which is a means to generate random permutation in an unbiased way [10].

To summarize, the factoradic representation allows representation of a permutation by a string of integers of similar size, in which each digit is a number in $[0, i]$. In addition, it introduces different weights between positions. For instance, the first drawn item, i.e. at position $n - 1$ has a greater influence on the permutation than draws at lower positions.

2.3   Factoradic Algorithms

Most of the applications of factoradics in EAs have focused on Particle Swarm Optimisation (PSO) to turn permutations into a usable form for the algorithms [11,12]. Factoradics have also proved useful in allowing restriction of the search to sub spaces [5].

In [6], the factoradic representation is empirically assessed on four classic benchmark problems, PFSP, Travelling Salesman Problem (TSP), Quadratic Assignment Problem (QAP) and Linear Ordering Problem (LOP). In order to do so, three algorithms are adapted to handle the factoradic representation: a Genetic Algorithm (GA), a univariate EDA based on the Population-Based Incremental Learning algorithm (PBIL) [13] and COMMA.

The use of three distinct algorithms allowed to draw some conclusions on the potential of the factoradic representation. First the GA yielded poor results. GAs were originally developed to make use of building blocks in order to improve solutions. With respect to permutations, subsets of permutations can be seen as building blocks. However, with the factoradic representation, it is unclear how building blocks can be kept between successive generations. A GA using the factoradic representation is hence very prone to breaking building blocks during the search and may be unable to exploit any promising area of the search space. Despite competitive results on some problems, COMMA was not the most efficient at making use of factoradics. The nature of the algorithm which prevents some of the best solutions' genes to be mutated also prevented exploitation of the space around these solutions. Finally, the univariate EDA yielded the best performance, especially when used in conjunction with high learning rates, promoting exploration over exploitation. Overall, the GA is exclusively explorative and fails at reaching good solutions. By removing the notion of crossover, the two other methods obtain better results. COMMA which is supposed to allow a more refined exploitation is overall outperformed by the univariate EDA, showing that the factoradic representation may not allow efficient exploitation. It is also interesting to note that the results obtained on the selected benchmarks using factoradic methods were significantly better than those reported for the ordering messy genetic algorithm [14], which uses RK.

The sole experiments presented in [6] can only provide an insight on what can be achieved by using factoradics. In this paper, the fitness landscape associated with the representation and its local move operator is analysed in order to define the ways in which factoradics should be used.

3   Fitness Landscape Analysis

3.1   Fitness Landscape Definition

In optimization, the *search space* Ω is a set of feasible solutions and $f : \Omega \longrightarrow \mathbb{R}$ is a *fitness function* that assigns a quality to each solution $s \in \Omega$. A *neighborhood structure* can be defined as a mapping function $\mathcal{N} : \Omega \to 2^{\Omega}$ that assigns a set of solutions $\mathcal{N}(s) \subset \Omega$ to any admissible solution $s \in \Omega$. $\mathcal{N}(s)$ is called the *neighborhood* of s, and a solution $s' \in \mathcal{N}(s)$ is called a *neighbor* of s. A *fitness landscape* [15,16] can be defined by a triplet $(\Omega, \mathcal{N}, f)$.

The neighborhood structure links a solution with other solutions of the search space. This principle is used by local search algorithms that move from solution to neighboring solutions. Fitness landscape analysis is thus useful to understand and predict the behaviour of metaheuristic algorithms [17,18]. Different measures can be computed to do so.

3.2   Measures

In this paper, the fitness landscape will be analysed according to the following characteristics.

Autocorrelation of the Fitness. It measures the continuity of the fitness between neighboring solutions along a random walk. A random walk $W = (s_0, s_1, ..., s_m)$ from s to s' is a sequence of solutions belonging to Ω where $s = s_0$ and $s' = s_m$ such that $s_{i+1} \in \mathcal{N}(s_i)$ for all $i \in \{1, 2, ..., m-1\}$. The autocorrelation function $\rho(k)$ [19] is the correlation coefficient of the fitness between the solutions s_i and s_{i+k} of a random walk. The autocorrelation measures the correlation of the problem structure. The coefficient $\rho(1)$ represents the correlation between a solution and its successors in the random walk. By definition of a random walk, the solution's successor is one of its neighbors. If $\rho(1)$ is close to 1, the fitness variation between neighbors is low, and the search space may be considered as a structured graph where we can predict the variation of fitness between neighbors.

Distribution of Neighbors. It gives an insight about the problem structure and the quality of the neighbors. A neighbor of a solution can be considered as (i) improving, (ii) deteriorating or (iii) neutral. A *neutral neighbor* is a neighboring solution with the same fitness value than the current solution. In some case, neutral solutions can be useful to escape from a local optima [20,21]. In this paper, all solutions of the search space are evaluated and the average number of improving ($>$) and neutral ($=$) neighbors is computed for the global optima, the local optima and the other solutions.

Basins of Attraction. It measures the attraction strength of the optima of the search space. The basin of attraction of an optima s^* is the set of solutions $s_i \in \Omega$ where $s^* = climber(s_i)$ with *climber* being the basic local search where the current solution's best improving neighbor is chosen until a local optimum is found, that is a solution with no improving neighbor. The bigger the size of a basin of attraction, the more influential its optimum. In the present paper, the computation is done exhaustively for the whole search space. As a result, the number of local and global optima is known.

4   Permutation Flowshop Scheduling Problem

4.1   Problem Description

The Flowshop Scheduling Problem (FSP) is one of the most investigated scheduling problem from the literature. The problem consists in scheduling N jobs $\{J_1, J_2, \ldots, J_N\}$ on M machines $\{M_1, M_2, \ldots, M_M\}$. Machines are critical resources, *i.e.* two jobs cannot be assigned to the same machine at the same time. A job J_i is composed of M tasks $\{t_{i1}, t_{i2}, \ldots, t_{iM}\}$, where t_{ij} is the jth task of J_i, requiring machine M_j. A processing time p_{ij} is associated with each task t_{ij}. We here focus on a permutation FSP (PFSP), where the operating sequences of the jobs are identical and unidirectional for every machine. As consequence, a feasible solution can be represented by a permutation π_N of size N (the ordered sequence of scheduled jobs), and the size of the search space is then $|S| = N!$.

In this study, we will consider the minimisation of the makespan, *i.e.* the total completion time, as the objective function. Let C_{ij} be the completion date of task t_{ij}, the makespan (C_{max}) can be computed as follows:

$$C_{max} = \max_{i \in \{1, \ldots, N\}} \{C_{iM}\}$$

Minimizing the makespan has been proven to be NP-hard for three machines and more [22]. As a consequence, large-size problem instances can generally not be solved to optimality, and then metaheuristics may appear to be good candidates to obtain good quality solutions. Daolio et al. [17] showed that the difficulty of the PFSP increases with the number of jobs and the number of machines.

4.2   Neighborhood Operators for Permutations

The design of local search metaheuristics requires a proper definition of a neighborhood structure for the problem under consideration. A *neighborhood structure* is a mapping function $\mathcal{N} : S \to 2^S$ that assigns a set of solutions $\mathcal{N}(s) \subset S$ to any feasible solution $s \in S$. $\mathcal{N}(s)$ is called the *neighborhood* of s, and a solution $s' \in \mathcal{N}(s)$ is called a *neighbor* of s. A neighbor results from the application of a *move operator* performing a small perturbation to a solution s. The choice of the neighborhood operator is thus key to the local search efficiency.

For the PFSP, we will consider the *insertion* and the *exchange* operator. These operators are known to be classical for permutation problems [23]. The *insertion* operator can be defined as follows. A job located at position i is inserted at position $j \neq i$. The jobs located between positions i and j are shifted, as illustrated in Fig. 3. The number of neighbors per solution is $(N-1)^2$, where N stands for the size of the permutation (and corresponds to the number of jobs).

Insert operator   Exchange operator

Fig. 3. Illustration of the neighborhood operators for permutations.

The *exchange* operator can be defined as follows. Two jobs located at position i and j $(i \neq j)$ are swapped, as illustrated in Fig. 3. The number of neighbors per solution is $N * (N - 1)/2$.

In the literature, the *insertion* operator has shown better performance than the *exchange* operator when used in local search or evolutionary algorithms [24, 25]

4.3   Neighborhood Operators for Factoradics

The most straightforward operator that can be applied to factoradics is the *point mutation* (PM). The point mutation is similar to other point mutation from other domains. PM only affects one allele of the mutated solution. However, because of the characteristics of the factoradic representation, the position of the allele influences the amount of disruption brought to the solution. Hence, PM needs to be defined in conjunction with a mutation distance as defined in [9] for permutations. The mutation distance d denotes the position of the gene to mutate. Note that the gene at position zero can only take the zero value and is thus never considered during operations. An allele is mutated by sampling randomly its value from the range $[0, d]$. As is the case with the *exchange* operator for permutations, the number of neighbors per solution is $N * (N - 1)/2$. Table 1 shows all the neighbors of the solution represented by the factoradic sequence *422100* for different mutation distances. It also shows the resulting permutations, illustrating how an increase in d generally leads to an increase in the number of positions being altered in the permutation.

Although alternative mutation operators were introduced in [6], they are not considered in the present study. The focus is instead set on the simple PM operator. It is however interesting to point out that the two alternative mutation operators named *multi-point mutation* and *random multi-point mutation* were used in order to bring more disruption to a solution.

5   Experiments

In this section, an exhaustive landscape analysis of different problem sizes of the PFSP is conducted in order to compare two different representations with their own move operators.

5.1   Experimental Setup

Landscapes. The factoradic and the permutation representations are considered. The PM move operator is used for the factoradics as it is the most natural for

Table 1. Full neighborhood of the factoradic *422100* using point mutation

	Mutated factoradic	Resulting permutations
d = 0	**1** 4 2 2 1 0 0	4 2 3 1 5 0 4 2 3 1 0 5
d = 1	**2** 4 2 2 1 0 0 **0**	4 2 3 5 0 1 4 2 3 1 0 5 4 2 3 0 1 5
d = 2	**3** 4 2 2 1 0 0 **1** **0**	4 2 5 1 0 3 4 2 3 1 0 5 4 2 1 3 0 5 4 2 0 3 1 5
d = 3	**4** **3** 4 2 2 1 0 0 **1** **0**	4 5 2 1 0 3 4 3 2 1 0 5 4 2 3 1 0 5 4 1 3 2 0 5 4 0 3 2 1 5
d = 4	**5** 4 2 2 1 0 0 **3** **2** **1** **0**	5 2 3 1 0 4 4 2 3 1 0 5 3 2 4 1 0 5 2 3 4 1 0 5 1 3 4 2 0 5 0 3 4 2 1 5

this representation of solutions. In the previous section, an insight have been given about the disruption it brings to the solution when the mutation of all the point is allowed. Thus, it seems interesting to compare this PM operator with the *partial* PM operator, which considers only p points from the end of the factoradic, that is the index at which the value can only be 0 or 1. Note that p is not directly set in the experiments presented here but computed by a percentage P of the whole neighborhood given by the *complete* PM operator. We investigate the PM operator with the percentage $P \in \{30, 50, 70, 100\}$ of the points mutation allowed. For example, on a problem of size $n = 10$, setting $P = 50$, will restrict the genes that can be mutated to the last 5 indices. Note that $P = 100$ gives the *complete* PM neighborhood. The resulting PM operators will be named as PM_P.

The insertion (IN) and the exchange (EX) move operators are used for the permutations as they are classical for this type of representation. Although previous landscape analysis shown that IN operator is the best adapted for the PFSP with a permutation representation [17], in the following, we will perform the analysis of the landscape corresponding to the EX operator. Indeed, the EX operator leads to the same number of neighbors as the *complete* PM operator, and it allows to present an interesting comparison between two search space where solutions are exactly connected to the same number of neighbors.

Benchmark Instances. Taillard proposed a set of benchmark instances [26] where the processing time p_{ij} of job $i \in N$ and machine $j \in M$ is generated randomly, according to a uniform distribution $\mathcal{U}([0; 99])$. These instances are widely used in the literature, but the number of jobs is too large and gives a search space with at least 20! solutions, thus impossible to evaluate exhaustively. An exhaustive analysis of the fitness landscape of the instances being needed, we generated our own instances in the same way as the Taillard instances, investigating different values for the number of jobs $N \in \{5, 6, 7, 8, 9\}$ and for the number of machines $M \in \{3, 5, 10\}$. For each problem size $(N \times M)$, ten instances are generated.

Measures. For all instances, a random walk of 100 solutions is performed and the autocorrelation value $\rho(k)$ is computed.

As the analysis is made exhaustively for all solutions of the search space, the distribution of the solutions is known. The exact numbers of global optima (GO), local optima (LO) or of other solutions $(\overline{LO})$ of the search space are computed.

The basins of attraction of every global and local optima are computed. The sum of solutions belonging to each type of optima $sum(\textbf{basin}_{LO})$ and $sum(\textbf{basin}_{GO})$ is computed.

From the distribution of the neighborhood of each solution, three different values are computed: $avg(\overline{LO}_>)$, $avg(\overline{LO}_=)$ and $avg(LO_=)$, which stands for the average number of strictly better $(>)$ and neutral $(=)$ neighbors of a solution $\overline{LO}$ and the average number of neutral neighbors of a local optimum LO respectively.

Note that the results presented in the following section corresponds to the mean obtained over the 10 instances of each problem size.

Table 2. $\rho(1)$: First value of the autocorrelation function for each move operators.

$\rho(1)$	Permutations		Factoradics			
Instance	IN	EX	PM_{100}	PM_{70}	PM_{50}	PM_{30}
5×3	0.46	0.41	0.34	0.14	-0.15	-0.13
5×5	0.44	0.38	0.33	0.15	-0.13	0.04
5×10	0.45	0.39	0.3	0.06	-0.19	0.02
6×3	0.59	0.53	0.43	0.28	0.14	-0.1
6×5	0.58	0.48	0.41	0.31	0.17	-0.12
6×10	0.52	0.45	0.43	0.28	0.07	-0.24
7×3	0.7	0.57	0.42	0.34	0.34	0.16
7×5	0.67	0.53	0.41	0.28	0.27	0.2
7×10	0.61	0.53	0.42	0.27	0.33	0.23
8×3	0.77	0.62	0.52	0.38	0.22	0.16
8×5	0.72	0.59	0.49	0.32	0.21	0.08
8×10	0.68	0.62	0.52	0.3	0.24	0.09
9×3	0.73	0.64	0.6	0.43	0.44	0.31
9×5	0.72	0.67	0.6	0.41	0.37	0.17
9×10	0.73	0.63	0.52	0.37	0.27	0.3

Table 3. Distribution of the search space Ω. $|\Omega|$ is the size of the search space. $|GO|$ is the number of global optima in Ω and %GO its proportion. For each move operators, the column gives the percentage of local optima (LO).

%LO				Permutations		Factoradics			
Instance	$\|\Omega\|$	$\|GO\|$	%GO	IN	EX	PM_{100}	PM_{70}	PM_{50}	PM_{30}
5×3	120	3.1	2.58	3	5.25	7.75	19	36	36
5×5	120	2.6	2.16	3.92	6.75	9.83	24.67	45	45
5×10	120	1	0.83	1.58	4.08	5.17	11.67	25.08	25.08
6×3	720	9.3	1.29	1.39	3.75	4.35	10.33	20.33	34.89
6×5	720	10.9	1.51	1.76	3.14	4.57	12.57	25.97	43.57
6×10	720	1.1	0.15	0.57	1.4	2.64	5.85	13.03	28.67
7×3	5040	26.8	0.53	0.81	1.6	3.44	15.7	15.7	26.79
7×5	5040	10	0.2	0.39	1.04	2.25	14.77	14.77	25.74
7×10	5040	2.2	0.04	0.17	0.7	1.55	10.82	10.82	21.89
8×3	40320	138.3	0.34	0.52	1.59	3.1	21.12	32.41	45.2
8×5	40320	72.9	0.18	0.31	1.14	2.01	19.95	31.47	44.15
8×10	40320	2.9	0.01	0.15	0.36	0.98	4.66	9.37	18.11
9×3	362880	3491.2	0.96	1.03	2.2	3.98	36.78	50.64	62.78
9×5	362880	727.6	0.2	0.6	1.24	2.31	21.67	33.67	45.98
9×10	362880	3.4	$\ll 10^{-2}$	0.03	0.14	0.5	2.29	5.02	10.36

5.2   Experimental Results

Experiments have been conducted for 6 different landscapes denoted by: IN, EX, PM_{100}, PM_{70}, PM_{50}, PM_{30}.

Table 2 shows the first value $\rho(1)$ of the autocorrelation function of the fitness computed along a random walk. For problems of very small size ($N = \{5, 6\}$), $\rho(1)$ is meaningless, the fitness of a solution being not correlated with the fitness of its neighbor. Moreover, the landscapes induced by PM operators ($\mathcal{P}_{PM}$) seem to be random since $\rho(1)$ is always under 0.6. The landscapes are locally highly rugged. On the landscape $\mathcal{P}_{IN}$ and $\mathcal{P}_{EX}$ induced by IN and EX operators respectively, $\rho(1)$ shows that the fitness between neighbors is more *continuous*. These landscapes are locally smoother than the landscapes $\mathcal{P}_{PM}$ that could help learning strategies in local search algorithms.

Table 3 shows the distribution of solution in the search space Ω. This computation is feasible since the search space is exhaustively enumerated. The proportion of global optima in the search space decreases when the number of jobs and the number of machines increases. Therefore, the difficulty of the problem is proportional to the problem size. The proportions of local optima show that the factoradic representation induces more local optima than the permutation representation. Furthermore, the landscape $\mathcal{P}_{PM}$ is more and more rugged, i.e. with more and more local optima, when the percentage of PM decreases. This high number of local optima represents an issue when dealing with exploitation methods.

Table 4. Size of the basin of attraction of the search space. For each move operators, the two columns give the proportion of solutions in the search space belonging to a basin of a local optimum (LO) and a global optimum (GO) respectively.

$[sum(\text{basin}_{LO}), sum(\text{basin}_{GO})]$

| Instance | Permutations | | | | Factoradics | | | | | | | |
	IN		EX		PM_{100}		PM_{70}		PM_{50}		PM_{30}	
5×3	7.18	92.82	15.29	84.71	33.48	66.52	74.3	25.7	87.49	12.51	87.49	12.51
5×5	23.15	76.84	42.92	57.08	64.22	35.78	86.05	13.95	93.11	6.89	93.11	6.89
5×10	20.35	79.65	48.04	51.96	63.56	36.44	88.35	11.65	95.22	4.78	95.22	4.78
6×3	5.72	94.28	40.78	59.22	60.57	39.43	80.25	19.75	89.69	10.31	94.18	5.82
6×5	22.37	77.63	50.99	49.01	64.64	35.36	90.2	9.8	95.55	4.46	97.56	2.44
6×10	40.55	59.45	64.73	35.27	83.78	16.22	94.53	5.47	97.77	2.23	99.16	0.84
7×3	23.26	76.74	32.29	67.72	71.62	28.38	93.42	6.58	93.42	6.58	96.2	3.8
7×5	34.17	65.83	56.75	43.25	82.39	17.61	97.37	2.63	97.36	2.64	98.58	1.42
7×10	37.3	62.7	66.18	33.82	86.96	13.04	98.62	1.38	98.62	1.38	99.37	0.63
8×3	17.3	82.7	50.74	49.26	77.38	22.62	96.88	3.12	98.14	1.86	98.78	1.21
8×5	37.57	62.43	59.5	40.5	78.23	21.77	98.11	1.89	98.95	1.05	99.22	0.78
8×10	60.27	39.73	80.52	19.48	94.63	5.37	99.43	0.57	99.84	0.16	99.94	0.06
9×3	7.01	92.99	30.58	69.42	56.18	43.82	94.67	5.33	96.01	3.99	96.4	3.6
9×5	45.35	54.65	59.7	40.3	79.5	20.5	97.83	2.17	98.56	1.44	98.96	1.04
9×10	75.69	24.31	90.46	9.54	97.53	2.47	99.82	0.18	99.93	0.07	99.97	0.03

Table 4 shows the average size of the basin of attraction for global and local optima. The less the percentage of neighbors, the more difficult it is to reach global optima. As expected, reaching global optima becomes more diffi cult as the problem size increases, as it generally correlates with an increase in $sum(\text{basin}_{LO})$, which represents a trick where it is difficult to escape from for *naive* local search algorithms. These results reinforce the conclusion that the landscape $\mathcal{P}_{IN}$ is more adapted for local search algorithms and that the landscape $\mathcal{P}_{PM}$ is not adapted for local search algorithms.

Table 5 shows the proportion of better or equal neighbors of a solution and the proportion of equal neighbors of local optima. Neutrality arises when neighbors have the same fitness values. This leads to plateaus in the landscape. The landscape $\mathcal{P}_{IN}$ is known to be neutral and this neutrality can be exploited efficiently to solve it [21]. The landscape $\mathcal{P}_{EX}$ and $\mathcal{P}_{PM}$ are also neutral. Besides, $\mathcal{P}_{PM_P}$ is more neutral when P decreases. This neutrality might be useful to escape from the numerous local optima and to move to another basin of attraction.

These experimental results on these small instances gives more insight on the use of the factoradic representation for the PFSP. Indeed, the landscape $\mathcal{P}_{PM}$ is locally rugged and local optima are numerous. The first idea that consists in taking into account less neighbors by selecting a percentage of the PM neigh borhood is not adapted to the PFSP. In fact, we can think that the landscape is a partition of independent networks of solutions. A study about local optima networks should be undertaken to confirm this intuition.

Table 5. Distribution of the neighborhood. For each move operators, the three columns give the percentage of the average number of better ($>$) and neutral ($=$) neighbors of $\overline{LO}$ and the average number of neutral ($=$) neighbors of LO respectively.

$\%[avg(\overline{LO}_>), avg(\overline{LO}_=), avg(LO_=)]$

Instance	Permutations					
	IN			EX		
5×3	44.84	12.99	7.29	44.35	15.32	28.06
5×5	45.74	11.81	14.81	45.95	13.71	21.45
5×10	49.63	2.34	1.56	50.75	2.53	6.34
6×3	41.49	18.01	10	41.2	20.23	28.18
6×5	45.08	11.09	9.33	44.57	13.22	18.27
6×10	48.48	3.6	3.33	48.35	4.64	6.38
7×3	40.49	19.61	18.48	39.49	22.13	28.41
7×5	43.66	13	15.19	42.84	15.13	21.42
7×10	46.89	6.38	7.9	46.56	7.49	11.52
8×3	38.17	24	30.09	38.13	24.73	32.94
8×5	41.19	17.85	22.12	40.75	19.23	27.71
8×10	47.31	5.48	10.03	46.91	6.47	12.1
9×3	33.7	33.15	36.21	33.34	34.45	47.33
9×5	39.76	20.85	31.07	39.37	21.99	33.75
9×10	47.54	4.94	6	47.13	5.87	6.92

Instance	Factoradics											
	PM_{100}			PM_{70}			PM_{50}			PM_{30}		
5×3	47.61	11.56	18.71	51.8	12.46	31.95	62.39	9.37	38.78	62.39	9.37	38.78
5×5	48.32	11.95	18.6	53.2	13.18	34.52	64.17	9.89	50.27	64.17	9.89	50.27
5×10	51.52	2.26	4	54.87	2.78	5.4	63.99	2.48	8.73	63.99	2.48	8.73
6×3	45.81	12.01	18.76	47.76	13.16	24.84	52.73	11.87	31.72	63.76	7.34	33.83
6×5	46.68	10.33	15.76	48.65	11.79	27.8	53.78	11.48	37.77	64.85	8.29	44.73
6×10	49.62	3.34	4.73	50.7	4.29	7.57	53.85	5.07	13.74	63.56	4.24	20.84
7×3	45.17	12.63	15.95	48.37	15.56	32.86	48.37	15.56	32.86	53.58	14.6	39.69
7×5	45.96	9.98	15.37	48.76	12.47	34.21	48.76	12.47	34.21	53.39	11.82	40.27
7×10	47.71	6	9.27	49.79	8.63	27.95	49.79	8.63	27.95	53.95	8.55	37.99
8×3	42	18.08	26.75	44.95	20.7	53.89	47.94	20.45	61.06	53.02	18.82	66.11
8×5	42.84	15.74	23.36	45.79	18.04	53.16	49.04	17.54	60.05	54.36	15.62	64.63
8×10	48.29	4.3	8.65	48.8	6.42	16.11	50.29	7.41	21.42	54.11	7.65	26.91
9×3	36.44	29.36	44.34	42.81	29.38	73.83	46.38	27.87	79.97	50.73	25.44	84.3
9×5	41.42	18.69	30.62	44.23	21.61	59.63	46.82	21.45	67.51	50.38	20.27	73.27
9×10	48.31	3.86	4.96	48.19	5.65	11.08	48.78	6.54	18.67	50.52	7.03	26.05

6   Conclusions

The fitness landscape associated with the factoradics was analysed using the PFSP. This has shown that the use of the representation along with its simplest operator, PM, leads to a rugged landscape. The size of the basins of attraction of the global optima is also small in comparison with those of local optima, making the former very difficult to reach. Naive local search methods are thus not adapted for this representation and are better used in conjunction with IN and EX. This observation is in line with the findings from [6], where the suite of algorithms chosen for optimising factoradics yielded poor exploitation behaviors.

The fact that the landscape is more rugged with PM_P when P decreases also proves that parameterising this operator in order to manage the amount of disruption brought to a solution is not a viable strategy on the PFSP. Modification in the lower rank of a factoradic solution, as is the case when P is small, results in the last elements in the permutation to be altered. On the PFSP, this will affect the last jobs being processed. Changes on these is more likely to bring changes in terms of fitness because the idle times they may introduce are unlikely to be compensated by the processing of successive jobs. Because other permutation problems have different characteristics, this comment on the potential benefit of the parameterisation of PM_P may not hold on other problems.

The results presented in this paper open avenues for future research. First, fitness landscape analysis should be performed including the study of local optima networks and on other types of permutation problems. Including alternative genotypes such as RK in the landscape study would further benefit the analysis. Experiments suggest that more efforts be spent on improving ways to refine the locality for the factoradic representation. Altering the representation in order to have a landscape exhibiting a lesser degree of ruggedness could be achieved by investigating a possible application of the gray code principle to the factoradic representation [27].

References

1. Ceberio, J., Irurozki, E., Mendiburu, A., Lozano, J.A.: A distance-based ranking model estimation of distribution algorithm for the flowshop scheduling problem. IEEE Trans. Evol. Comput. **18**(2), 286–300 (2014)
2. Rothlauf, F.: Representations for Genetic and Evolutionary Algorithms. Springer, Heidelberg (2006)
3. Bean, J.C.: Genetic algorithms and random keys for sequencing and optimization. ORSA J. Comput. **6**(2), 154 160 (1994)
4. Kromer, P., Platos, J., Snasel, V.: Modeling permutations for genetic algorithms. In: International Conference of Soft Computing and Pattern Recognition, pp. 100–105. IEEE (2009)
5. Mehdi, M.: Parallel hybrid optimization methods for permutation based problems. Ph.D. thesis, Université des Sciences et Technologie de Lille (2011)
6. Regnier-Coudert, O., McCall, J.: Factoradic representation for permutation optimisation. In: Bartz-Beielstein, T., Branke, J., Filipič, B., Smith, J. (eds.) PPSN 2014. LNCS, vol. 8672, pp. 332–341. Springer, Heidelberg (2014)
7. Schnier, T., Yao, X.: Using multiple representations in evolutionary algorithms. In: Proceedings of the 2000 Congress on Evolutionary Computation, vol. 1, pp. 479–486. IEEE (2000)
8. Ashlock, D.: Evolutionary Computation for Modeling and Optimization. Springer, New York (2006)
9. Regnier-Coudert, O., McCall, J.: Competing mutating agents for bayesian network structure learning. In: Coello, C.A.C., Cutello, V., Deb, K., Forrest, S., Nicosia, G., Pavone, M. (eds.) PPSN 2012, Part I. LNCS, vol. 7491, pp. 216–225. Springer, Heidelberg (2012)
10. Durstenfeld, R.: Algorithm 235: random permutation. Commun. ACM **7**(7), 420 (1964)
11. Samarghandi, H., ElMekkawy, T.Y.: A meta-heuristic approach for solving the no-wait flow-shop problem. Int. J. Prod. Res. **50**(24), 7313–7326 (2012)

12. Hosseini-Nasab, H., Emami, L.: A hybrid particle swarm optimisation for dynamic facility layout problem. Int. J. Prod. Res. **51**(14), 4325–4335 (2013)
13. Baluja, S.: Population-based incremental learning. a method for integrating genetic search based function optimization and competitive learning. Technical report, Carnegie Mellon University (1994)
14. Knjazew, D., Goldberg, D.E.: Omega-ordering messy ga: solving permutation problems with the fast messy genetic algorithm and random keys. In: Proceedings of Genetic and Evolutionary Computation Conference, pp. 181–188 (2000)
15. Wright, S.: The roles of mutation, inbreeding, crossbreeding and selection in evolution. In: Proceedings of the Sixth International Congress on Genetics, vol. 1 (1932)
16. Stadler, P.F.: Landscapes and their correlation functions. J. Math. Chem. **20**, 1–45 (1996)
17. Daolio, F., Verel, S., Ochoa, G., Tomassini, M.: Local optima networks of the permutation flow-shop problem. In: Legrand, P., Corsini, M.-M., Hao, J.-K., Monmarché, N., Lutton, E., Schoenauer, M. (eds.) EA 2013. LNCS, vol. 8752, pp. 41–52. Springer, Heidelberg (2014)
18. Marmion, M.-E., Jourdan, L., Dhaenens, C.: Fitness landscape analysis and metaheuristics efficiency. J. Math. Model. Algorithms **12**(1), 3–26 (2011)
19. Weinberger, E.: Correlated and uncorrelated fitness landscapes and how to tell the difference. Biol. Cybern. **63**, 325–336 (1990)
20. Verel, S., Collard, P., Clergue, M.: Scuba Search : when selection meets innovation. In: Proceedings of the 2004 Congress on Evolutionary Computation, CEC 2004, pp. 924–931. IEEE Press (2004)
21. Marmion, M.-E., Dhaenens, C., Jourdan, L., Liefooghe, A., Verel, S.: NILS: a neutrality-based iterated local search and its application to flowshop scheduling. In: Merz, P., Hao, J.-K. (eds.) EvoCOP 2011. LNCS, vol. 6622, pp. 191–202. Springer, Heidelberg (2011)
22. Lenstra, J.K., Rinnooy Kan, A.H.G., Brucker, P.: Complexity of machine scheduling problems. Ann. Discret. Math. **1**, 343–362 (1977)
23. Schiavinotto, T., Stützle, T.: A review of metrics on permutations for search landscape analysis. Comput. Oper. Res. **34**, 3143–3153 (2007)
24. Reeves, C.R.: A genetic algorithm for flowshop sequencing. Comput. Oper. Res. **22**, 5–13 (1995)
25. Ruiz, R., Maroto, C.: A comprehensive review and evaluation of permutation flowshop heuristics. Eur. J. Oper. Res. **165**(2), 479–494 (2005)
26. Taillard, E.D.: Benchmarks for basic scheduling problems. Eur. J. Oper. Res. **64**, 278–285 (1993)
27. Johnson, S.M.: Generation of permutations by adjacent transposition. Math. Comput. **17**(83), 282–285 (1963)

Exploring Non-neutral Landscapes with Neutrality-Based Local Search

Matthieu Basseur, Adrien Goëffon$^{(\boxtimes)}$, and Hugo Traverson

LERIA, Université D'Angers, Angers, France
{matthieu.basseur,adrien.goeffon,hugo.traverson}@univ-angers.fr

Abstract. In this paper, we present a generic local search algorithm which artificially adds neutrality in search landscapes by discretizing the evaluation function. Some experiments on NK landscapes show that an adaptive discretization is useful to reach high local optima and to launch diversifications automatically. We believe that a hill-climbing using such an adaptive evaluation function could be more appropriated than a classical iterated local search mechanism.

1 Context

In combinatorial optimization, fitness landscapes study abstracts problem specificities and aims at evaluating the pertinence of generic metaheuristics. More formally, a fitness landscape is a triplet $(\mathcal{X}, \mathcal{N}, f)$, where $\mathcal{X}$ is a discrete set of solutions, $\mathcal{N} : \mathcal{X} \rightarrow 2^{\mathcal{X}}$ a neighborhood function, and $f : \mathcal{X} \rightarrow (0,1)$ a fitness (evaluation) function. In [1], we compared the efficiency of several hill-climbing variants (denoted as *climbers*) to determine neighborhood based moving strategies that are likely to reach *high* solutions (with high fitness values). In particular, we focused on the ways to handle neutrality [4]. To achieve this, we introduced rounded landscapes, by setting a discretization level of the fitness function, referring to its codomain size r. The rounded function f_r is then defined from an original fitness function f as follows:

$$f_r(x) = \frac{\lfloor r.f(x) \rfloor}{r}$$

Let us notice that f_r gives a partial order which is compliant with the order relation induced by f. It means that, $\forall x, y \in \mathcal{X}, f(x) < f(y) \Rightarrow f_r(x) \leqslant f_r(y)$. A second property of f_r functions ($f_r(x) \leqslant f(x) < f_r(x) + \frac{1}{r}$) makes possible to compare fitnesses reached on original and rounded corresponding landscapes. In [1], we observed on rounded landscapes that some r values allow *stochastic climbers* (first improvement which accepts indifferently neutral and improving moves) to reach higher solutions than while considering original fitness functions.

In this paper, we propose to extend the principle of rounding fitness function to help neutrality-based local searches to reach high local optima. We propose a generic local search algorithm based on an adaptive evaluation function $LS_{\bar{f}}$

© Springer International Publishing Switzerland 2015
C. Dhaenens et al. (Eds.): LION 9 2015, LNCS 8994, pp. 165–169, 2015.
DOI: 10.1007/978-3-319-19084-6_15

which incorporates an artificial rate of neutrality chosen with respect to information collected during the search. It simulates deteriorating moves during intensification phases, as well as automatic perturbations when local improvements become rare.

Here, the efficiency of the proposed mechanism is evaluated on NK landscapes [2], where size and ruggedness are tunable by means of parameters N and K.

2   Determining Appropriate Neutrality Rates of Landscapes

Intuitively, adding neutrality to landscapes necessarily induces to decrease their ruggedness, which make local searches more efficient as long as they exploit neutral moves. Nevertheless, a too large neutrality level creates flat areas which can drastically increase the number of moves needed to reach high solutions. In an extreme case, on totally flat landscapes, stochastic hill-climbings behave like random walk processes. Here, we propose to control the neutrality rate by means of the fitness function f_r. Setting r adequately consists in reducing the ruggedness while keeping a moderate rate of neutrality.

We have extended our previous study on NK and NKr landscapes in order to determine the most appropriate rounding values with respect to landscapes under consideration (with $N \in \{128, 256, 512, 1024\}$ and $K \in \{1, 2, 4, 8\}$). To estimate the neutrality of these most appropriate rounded landscapes, we define a search neutrality indicator $\tilde{\nu}$ which aims to estimate the average rate of neutrality encountered during the search:

Definition 1 (Search Neutrality). *Let $\mathcal{P}$ be a fitness landscape, N the neighborhood size and C the history of a climber execution (given by a sequence of evaluated and selected solutions). Let p the number of strictly improving moves in C. l_i and n_i ($i \in \{1, \ldots, p\}$) refer respectively to the number of evaluations and the number of neutral moves realized between the $(i-1)^{th}$ and i^{th} improving moves. The search neutrality depends on C and P, and is defined by:*

$$\tilde{\nu}(C, \mathcal{P}) = \frac{\sum_{i=1}^{p} \frac{n_i}{l_i} \log_N l_i}{\sum_{i=1}^{p} \log_N l_i}$$

Table 1 reports the ranges of rounding values r from which climbers executed on corresponding NKr landscapes are not statistically outperformed by climbers executed on derived landscapes with other values of r, as well as their associated search neutrality. Non-dominated ranges have been determined using a dichotomic sampling of r values. Statistical analysis were performed using a binomial test based on 100 runs per r value. We observe that optimal r values depend on landscape properties. The most noticeable information is that the $\tilde{\nu}$ values which make a neutrality-based climbing more efficient, are similar on all (N, K) parameterizations.

Table 1. Climber comparison on NK and non-dominated NKr landscapes. We also indicate ranges of optimal values of r and their associated $\tilde{\nu}$ (in %).

N, K	NK f mean	NKr (NK using f_r)			N, K	NK f mean	NKr (NK using f_r)		
		r range	$\tilde{\nu}$ range				r range	$\tilde{\nu}$ range	
128_1	.7021	.7227	[206, 285]	[20, 26]	512_1	.6897	.7069	[731, 824]	[24, 27]
128_2	.7021	.7395	[94, 148]	[23, 32]	512_2	.7135	.7484	[454, 523]	[25, 28]
128_4	.7254	.7837	[54, 81]	[23, 31]	512_4	.7200	.7790	[234, 304]	[24, 28]
128_8	.7142	.7685	[40, 64]	[19, 25]	512_8	.7206	.7810	[173, 206]	[19, 22]
256_1	.7021	.7206	[321, 350]	[26, 28]	1024_1	.6969	.7157	[1371, 1562]	[25, 28]
256_2	.7066	7410	[212, 264]	[24, 28]	1024_2	.7146	.7507	[959, 1178]	[23, 27]
256_4	.7235	.7843	[117, 152]	[24, 28]	1024_4	.7246	.7844	[449, 570]	[24, 28]
256_8	.7166	.7755	[84, 112]	[19, 22]	1024_8	.7216	.7836	[331, 403]	[19, 21]

Input : a fitness landscape $(\mathcal{X}, \mathcal{N}, f)$, a set of n rounding values
$\mathcal{R} = \{r_1, \ldots, r_n\}$ (with $r_i > r_{i+1}$), parameters $d, \nu_{\text{ref}}, \theta, \epsilon$.
Output: the best solution found x_{opt}
Randomly select $x \in \mathcal{X}$;
$F \leftarrow f(x)$;
$t \leftarrow 0$;
$F_{\text{opt}} \leftarrow F$;
$x_{\text{opt}} \leftarrow x$;
repeat
    Randomly select $x' \in \mathcal{N}(x)$;
    $t \leftarrow t + 1$;
    $\alpha \leftarrow 1 - \frac{1}{\min(d,t)}$;
    for all $r_i \in \mathcal{R}$ **do**
        $\nu_{\text{est}}[r_i] \leftarrow \alpha \times \nu_{\text{est}}[r_i]$;
        $\mu_{\text{est}}[r_i] \leftarrow \alpha \times \mu_{\text{est}}[r_i]$;
        if $f_{r_i}(x') = f_{r_i}(x)$ **then** $\nu_{\text{est}}[r_i] \leftarrow \nu_{\text{est}}[r_i] + 1 - \alpha$ **else if**
        $f_{r_i}(x') > f_{r_i}(x)$ **then** $\mu_{\text{est}}[r_i] \leftarrow \mu_{\text{est}}[r_i] + 1 - \alpha$
    $\mathcal{R}' \leftarrow \{r_i \in \mathcal{R}, \nu_{\text{est}}[r_i] > 0\}$;
    if $t > d$ **then** $\mathcal{R}' \leftarrow \{r_i \in \mathcal{R}', \mu_{\text{est}}[r_i] \geqslant \epsilon\}$ **if** $\mathcal{R}' = \emptyset$ **then** $t \leftarrow 0$
    {*Diversification*} **else**
        $R \leftarrow \text{argmin}_{r_i \in \mathcal{R}'} \frac{\max(\nu_{\text{est}}[r_i], \nu_{\text{ref}})}{\min(\nu_{\text{est}}[r_i], \nu_{\text{ref}})}$;
        $\mathcal{R}' \leftarrow \{r_i \in \mathcal{R}', \nu_{\text{est}}[R] - \theta \leqslant \nu_{\text{est}}[r_i] \leqslant \nu_{\text{est}}[R]\}$;
        $R \leftarrow \text{argmin}_{r_i \in \mathcal{R}'} \nu_{\text{est}}[r_i]$;
        if $f_R(x') \geqslant f_R(x)$ **then**
            $x \leftarrow x'$;
            if $f(x) > f(x_{\text{opt}})$ **then**
                $x_{\text{opt}} \leftarrow x$;
until *Stopping criterion*;

Algorithm 1: $LS_{\tilde{f}}$

3   Local Search with Adaptive Evaluation Function ($LS_{\overline{f}}$)

Previous results showed that there exists an appropriate range of neutrality rate which leads to reach high solutions by hill-climbing. About 25 % of neutrality seems appropriate for climbing efficiently NK landscapes. However, additional experiments emphasized that the local neutrality in NKr landscapes is negatively correlated with the height (fitness value) of solutions. Then, we propose to set r dynamically thanks to an adaptive mechanism which aims at preserving a reference neutrality rate ν_{ref}.

The local search algorithm we introduce here, $LS_{\overline{f}}$, is a stochastic climber which selects at each iteration a rounding value r_i among a set of candidate roundings $\{r_1, \ldots, r_n\}$. To each rounding value r_i is associated an estimated neutrality $\nu_{\text{est}}[r_i]$, which is dynamically updated at each iteration as follows:

$$\nu_{\text{est}}[r_i] \leftarrow \begin{cases} \alpha \times \nu_{\text{est}}[r_i] + 1 - \alpha, & \text{if } f_{r_i}(x') = f_{r_i}(x) \\ \alpha \times \nu_{\text{est}}[r_i], & \text{otherwise} \end{cases}$$

Additionally, at each iteration, we select the rounding value which is the closest to ν_{ref} in terms of ratio (more precisely $\text{argmin}_{r_i} \frac{\max(\nu_{\text{est}}(r_i), \nu_{\text{ref}})}{\min(\nu_{\text{est}}(r_i), \nu_{\text{ref}})}$).

Maintaining a certain level of neutrality can prevent to reach local optima (in the sense of the original fitness function) and also requires to use a predefined number of iterations as a stopping criterion. Therefore, we propose to associate an improving move rate estimation μ_{est} to each rounding value r_i. $\mu_{\text{est}}[r_i]$ is estimated similarly to $\nu_{\text{est}}[r_i]$, by considering strictly improving moves. This allows the detection of search stagnation with respect to each rounding value. Then we refine the r_i selection mechanism by forbidding *stagnant* r_i values to be selected (r_i such that $\mu_{\text{est}}[r_i]$ is lower than a threshold ϵ). As a consequence, the search will be naturally driven to a local optimum.

To simulate a perturbation mechanism, we just need to reset the neutrality estimations. This can be done when every μ_{est} value, which estimate improving move rates, is smaller than a threshold ϵ. Such mechanism partially randomizes the search during several steps by considering flat landscapes. Algorithm 1 provides a detailed description of $LS_{\overline{f}}$.

To assess the relevance of the proposed climbing technique, $LS_{\overline{f}}$ has been compared with a classical Iterated Local Search (ILS) process, where diversification has been parameterized as follows:

- random restart;
- random walk from the last local optimum found (5 variants: 5, 10, 15, 20, 30 moves);
- random walk from the best local optimum found (5 variants: 5, 10, 15, 20, 30 moves).

On each instance, $LS_{\overline{f}}$ is compared to the 11 ILS parameterizations (with 200 perturbations for each).

Table 2 compares average fitnesses reached by $LS_{\overline{f}}$ (with and without perturbations), with local searches (with — LS — and without perturbation — ILS) using original evaluation functions, requiring an equivalent computational effort. These results emphasize the relevance of adapting the shape of a landscape according to its local properties. Moreover, $LS_{\bar{f}}$ allows the simulation of deteriorating moves during intensification and diversification phases without explicitly dealing with them. It is obvious that the behavior of such a mechanism can be linked with Simulated Annealing (SA) [3]. In future work, it should be interesting to provide a deep analysis of these two ways to simulate diversification during an intensification process.

Table 2. Average efficiency of $LS_{\bar{f}}$ on NK landscapes, without (0) or after 200 perturbations. We also report comparison with local search (LS for hill-climbing, ILS for iterated local search) on original NK landscapes. The ILS column contains the best average results obtained among the 11 tested parameterizations.

Instance	LS	$LS_{\bar{f}\ (0)}$	$LS_{\bar{f}\ (200)}$	ILS	Instance	LS	$LS_{\bar{f}\ (0)}$	$LS_{\bar{f}\ (200)}$	ILS
128_1	.7021	**.7245**	**.7245**	**.7245**	512_1	.6897	.7066	**.7088**	**.7088**
128_2	.7021	.7422	**.7424**	.7414	512_2	.7135	.7500	**.7520**	.7502
128_4	.7254	.7947	**.7959**	.7921	512_4	.7200	.7819	**.7888**	.7769
128_8	.7142	.7919	**.8050**	.7807	512_8	.7206	.7852	**.7920**	.7837
256_1	.7021	.7217	**.7221**	**.7221**	1024_1	.6969	.7160	**.7174**	**.7174**
256_2	.7066	.7441	**.7448**	.7443	1024_2	.7146	.7509	**.7532**	.7518
256_4	.7235	.7910	**.7940**	.7882	1024_4	.7246	.7840	**.7910**	.7778
256_8	.7166	.7813	**.7087**	.7815	1024_8	.7210	.7835	**.7893**	.7817

Acknowledgment. This work was partially supported by the Fondation mathématique Jacques Hadamard within the Gaspard Monge Program for Optimization and operations research.

References

1. Basseur, M., Goëffon, A.: Hill-climbing strategies on various landscapes: an empirical comparison. In: Proceeding of the Fifteenth Annual Conference on Genetic and Evolutionary Computation Conference (GECCO), pp. 479–486. ACM (2013)
2. Kauffman, S.A., Weinberger, E.D.: The NK model of rugged fitness landscapes and its application to maturation of the immune response. J. Theor. Biol. **141**(2), 211–245 (1989)
3. Kirkpatrick, S., Gelatt, C.D., Vecchi, M.P.: Optimization by simulated annealing. Science **220**(4598), 671–680 (1983)
4. Smith, T., Husbands, P., Layzell, P.J., O'Shea, M.: Fitness landscapes and evolvability. Evol. Comput. **10**, 1–34 (2002)

A Selector Operator-Based Adaptive Large Neighborhood Search for the Covering Tour Problem

Leticia Vargas[1,2]([⊠]), Nicolas Jozefowiez[1,2], and Sandra Ulrich Ngueveu[1,3]

[1] CNRS, LAAS, 7 Avenue du Colonel Roche, 31400 Toulouse, France
[2] INSA, LAAS, Univ de Toulouse, 31400 Toulouse, France
lgvargas@laas.fr
[3] INP, LAAS, Univ de Toulouse, 31400 Toulouse, France

Abstract. The Covering Tour Problem finds application in distribution network design. It includes two types of vertices: the covering ones and the ones to be covered. This problem is about identifying a lowest-cost Hamiltonian cycle over a subset of the covering vertices in such a way that every element not of this type is covered. In this case, a vertex is considered covered when it is located within a given distance from a vertex in the tour. This paper presents a solution procedure based on a *Selector* operator that allows to convert a giant tour into an optimal CTP solution. This operator is embedded in an adaptive large neighborhood search. The method is competitive as shown by the quality of results evaluated using the output of a state-of-the-art exact algorithm.

Keywords: Covering tour problem · *Split* procedure · ALNS algorithm

1 Introduction

This study aims at solving a tour location problem (TLP), namely the *Covering Tour Problem* (CTP) through a new splitting operator, *Selector*, embedded into an *Adaptive Large Neighborhood Search* (ALNS) metaheuristic. The overall goal of TLPs is to construct an optimal tour through a *subset* of the vertices of a network, subject to a set of constraints. They differ from classical vehicle routing problems since the assumption shared by problems of the TSP and VRP families is that *all* vertices of the network should be served, something which is not valid in many real applications. In TLPs, the visits are optional. The new *Selector* operator finds a minimum-cost tour (subject to a given sequence) which passes through a subset of vertices and meets side constraints. In the case of the CTP, every vertex not in the elementary cycle must lie within a prespecified radius from at least one vertex in the cycle.

The CTP is a generalization of the *Traveling Salesman Problem* (TSP) and it can be formally described as follows. Let $G = (N, E)$ be an undirected graph, where $N = V \cup W$ represents the vertex set and $E = \{(v_i, v_j)|v_i, v_j \in N, i < j\}$ is the edge set. V is the subset of n vertices that *can* be visited at most once,

© Springer International Publishing Switzerland 2015
C. Dhaenens et al. (Eds.): LION 9 2015, LNCS 8994, pp. 170–185, 2015.
DOI: 10.1007/978-3-319-19084-6_16

$T \subseteq V$ is the subset of vertices that *must* be visited exactly once, while W is the subset of vertices that must be *covered*. Vertex $v_0 \in T$ is the depot. Let d_{ij} be the distance associated with edge $(i, j) \in E$, and $D = (d_{ij})$ the distance matrix that satisfies the triangle inequality. The solution of the CTP is to find a minimum-length elementary cycle in V such that each vertex $w_i \in W$ is covered by the cycle, and also all vertices in T are found in the cycle. A vertex $w_i \in W$ is covered if there exists at least one vertex $v_j \in V$ in the cycle for which $d_{ij} \leqslant c$, where c is known as the *covering distance*. Figure 1 shows a feasible CTP tour for an instance where $|V| = 8$, $|T| = 2$ and $|W| = 17$, and exemplifies how vertex $v_A \in V$ covers vertices $\{w_1, w_2, w_3\}$.

A very closely related problem is the *Covering Salesman Problem* (CSP) where the aim is to identify a minimum-length tour visiting a subset of the vertices in N and covering all the vertices not on the tour. When the subset of vertices that must be on the tour is empty, $T = \emptyset$, the CTP reduces to a CSP, and when T consists of the entire vertex subset, $T = V$, the CTP reduces to the TSP. Therefore, it is NP-hard. The CTP can be formulated, with suitable definitions, as a *Generalized Traveling Salesman Problem* (GTSP) where vertices are clustered and the aim is to identify a minimum-length cycle which visits at least a vertex of each cluster, as explained by Fischetti et al. (1997).

Despite its practical importance, the CTP has not been widely studied. It is introduced and formulated in Current and Schilling (1989) as the *Covering Salesman Problem* (CSP), where they also describe some real world routing problems that can be modelled by the CTP, such as the design of bimodal distribution systems. For instance, cities in the tour are served by air, while the ones not in the tour are served by trucks which originate at their nearest city on the air route. Another application explained is the routing of rural health care delivery teams in developing countries where medical services are only delivered to a subset of villages, but individuals living in villages not in the route are able to reach the medical team at the nearest stop. In the formerly cited work, and in Current and Schilling (1994) the problem is treated as a bicriterion routing problem: the length of the tour and the number of vertices included in it.

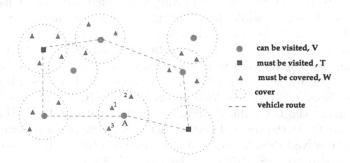

Fig. 1. Example of CTP tour

In the literature, only one exact method, a branch-and-cut algorithm by Gendreau et al. (1997), has been presented so far to solve the CTP. Hodgson et al. (1998) successfully applied this exact method to the routing of a mobile medical facility in Ghana. Jozefowiez et al. (2007) also propose a bi-objective treatment of the problem: minimization of the tour length and minimization of the covering distance, and develop a two-phase cooperative strategy that combines a multi-objective evolutionary algorithm with the branch-and-cut algorithm of Gendreau et al. (1997). Salari and Naji-Azimi (2012) combine heuristic search and integer linear programming techniques to solve the CSP.

Other heuristic algorithms have also been studied. The ones proposed by Current and Schilling (1989) and Gendreau et al. (1997) are based upon solution procedures for the *Set Covering Problem* (SCP) and the TSP. Motta et al. (2001) have proposed a GRASP metaheuristic to solve a generalized version of the CTP where the tour may also include vertices of set W, while Baldacci et al. (2005) have presented three scatter-search heuristic algorithms for the CTP. In addition, Golden et al. (2012) developed a generalized version of the CSP, which they named the *Generalized Covering Salesman Problem* (GCSP), and defined three variants of it for which they proposed two local search heuristics.

The contribution of this study is the development of a new operator that optimally splits a giant tour into visited and not-visited vertices. In other words, it selects the vertices of a given giant tour that comprise the optimal solution for the CTP. A second contribution is the proposal of a state-of-the-art meta-heuristic for solving the CTP.

The remainder of this paper is structured as follows. Section 2 presents the Selector operator, while in Sect. 3 we describe our implementation of an ALNS-based metaheuristic that incorporates the Selector operator. Computational results are presented in Sect. 4, and conclusions are reported in Sect. 5.

2   Selector Operator

Our solution method is based on the *route first–cluster second* approach proposed by Beasley (1983). The first phase, routing also called ordering, is handled by the ALNS metaheuristic, while the second phase, clustering also called splitting, is handled by our new operator. When solving the CTP, the *Selector* operator splits a giant tour (GT), which is a permutation of all n vertices that can be in the tour, into subsequences of visited and not-visited vertices in a similar way as the *Split* operator segments a GT into feasible vehicle routes when applied to solve the *Capacitated Vehicle Routing Problem* (CVRP) as proposed by Prins (2004). Splitting the GT entails solving a shortest-path problem. However, in the case of the CTP, the covered vertices act as a constraining resource and the problem to be solved then becomes an *Elementary Shortest Path Problem with Resource Constraints* (ESPPRC) (see Feillet et al. 2004).

In Sect. 2.1, Beasley's approach is explained. Section 2.2 presents the method used by our *Selector* operator to solve the ESPPRC in general terms, whereas Sect. 2.3 demonstrates its specific algorithmic implementation.

2.1   Split Method

The *split* method was firstly presented by Beasley (1983) as the second phase of a route first–cluster second heuristic to solve the CVRP. Relaxing vehicle capacity and maximum route length, the first phase solves a TSP to form a GT that determines the order in which the customers are to be visited. The second phase constructs an auxiliary cost network and then applies a shortest-path algorithm to obtain an optimal partition of the GT into least-cost, capacity-feasible, vehicle routes. This shortest path can be computed using Bellman-Ford's algorithm for directed, acyclic graphs. Beasley provided no computational results for his proposal, and the method neither outperformed more traditional CVRP heuristics nor was it given adequate recognition (see Laporte and Semet 2002). However, when Beasley's seminal method was efficiently implemented within a genetic algorithm (GA) (Prins 2004), it proved to be the first GA able to compete with the best methods available at that time for the solution of the CVRP, i.e. tabu search heuristics. It is since known as the basic *Split* procedure, and other versions of it have been developed to tackle additional constraints as presented in Prins et al. (2009). In the last decade, the route first–cluster second approach has led to successful constructive heuristics and metaheuristics for routing problems as explained in Prins et al. (2014) where a more general name, order–first split–second, is given to the methodology, and an analysis of 70 articles involving splitting procedures is made.

2.2   Resource-Constrained Shortest Path Problem

The ESPPRC requires the computation of an elementary shortest path in a network such that the overall resource usage does not exceed the limits. Resources are used when visiting vertices or traversing arcs, hence, record of used resources should be kept. Such problems are NP-hard (Feillet et al. 2004). The standard approach to solve an ESPPRC is dynamic programming (DP) and has pseudopolynomial complexity.

The ESPPRC for the CTP can be solved quickly enough in practice by adapting Desrochers' algorithm (1988), a multi-label version taking resource constraints into consideration of the Bellman-Ford algorithm, which is a label-correcting approach where labels on a vertex are repeatedly extended to its successors. The basic principle of Desrochers' algorithm is to associate with each partial path a label indicating the cost of the path and its consumption of resources, and to eliminate unnecessary labels as the search progresses. Throughout the search, then, every vertex receives labels, and these labels are iteratively extended toward every possible successor vertex until no new labels are created.

Every label representing a feasible path can be understood as a vector $V = (\zeta | r_1, r_2, \ldots, r_k)$ that memorizes the path cost ζ and the resource consumptions r_i that enable to know if a partial path can still be extended. The effectiveness of the DP algorithm outlined relies upon the feasibility of pruning labels that cannot lead to an optimal solution. For this purpose, suitable dominance tests are always performed when labels are extended, so that the algorithm records only non-dominated labels.

2.3   Selector Algorithm Proposed

Auxiliary Graph. *Selector* works on a directed, acyclic graph $M = (V', A)$, where V' represents the position in the giant tour of the n vertices that can be visited in the original graph, i.e., the representation of $GT = (GT_0, GT_1, \ldots, GT_{n-1})$ in V' implies $v_i = GT_i$. Thus, in the following, when refering to vertex v_i we imply the vertex located in position i in GT. Figure 2 shows an example for $n = 6$ and depicts only some of the possible arcs. An arc $(i,j) \in A \mid j \geqslant i+2$ models a subpath that visits only vertices v_i and v_j. Such arc exists only if it is feasible to skip the points located between vertices v_i and v_j. For instance, arc (v_1, v_4) in Fig. 2 indicates a subpath that visits vertex v_1, skips v_2 and v_3, and ends at vertex v_4. This arc will be kept if no $w_i \in W$ remains uncovered despite skipping vertices v_2 and v_3. This means that the subset $\{v_1, v_4, v_5\}$ covers all $w_i \in W$. The algorithm creates a vector of size n to represent M, and this vector maintains the labels (subpaths) generated to reach each vertex $v_i, i \in \{1, 2, \ldots, n-1\}$. The weight of an arc is equal to the Euclidean distance, d_{ij}, between the vertex located in position i and the one in position j in GT.

An optimal solution for a given GT indicates a minimum-cost path σ from 0 to $n-1$ in M. This result can be seen as the splitting of the giant tour into visited and not-visited vertices. Finding σ has pseudopolynomial complexity.

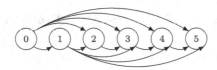

Fig. 2. Auxiliary graph M representing optional visits for vertices 0 and 1.

Labels and Their Control. Starting from v_0, a vertex in M may be reached through different permutations of visited and not-visited predecessors and each with a different cost and a different coverage of the vertices in W. In practice, this information is stored in labels. As a result, several labels might exist at each vertex. A label λ_j stored in vertex $v_i \in M \backslash \{v_0\}$ represents a path that starts at the depot and ends at vertex v_i. It contains five fields $\lambda_j^i = (\zeta, \omega[k], v_i, \pi, \nu)$ which are useful for the decision making at different stages of the algorithm. ζ memorizes the cost (sum of the arc weights) of the path represented, $\omega[k]$ is a vector in which location i either stores the number of vertices in V that can still cover vertex $w_i \in W$ or stores a flag indicating vertex w_i is covered, v_i keeps the last visited vertex (site where label is stored), π stores the path from the depot to vertex v_i, and ν keeps the number of vertices already covered.

The dominance rule applied to control label proliferation is as follows. Let $\Lambda^i = \{\lambda_1^i, \lambda_2^i, \ldots, \lambda_k^i\}$ be the set of labels associated with vertex $v_i \in M \backslash \{v_0\}$, and Ω_j the set of vertices covered by label λ_j. A label $\lambda_1 \in \Lambda^i$ dominates $\lambda_2 \in \Lambda^i$, with $\lambda_1 \neq \lambda_2$, if $\zeta_1 \leq \zeta_2$ and $\Omega_2 \subseteq \Omega_1$.

A look ahead mechanism also allows to reduce the number of labels created. If when extending a label it is found that a vertex $v_i \in V\backslash T$ must be visited in the future because it is the only one that can cover a set $\Gamma \subset W$, then mark all the vertices $w_i \in \Gamma$ as covered. The result of this look ahead is that it is known then there is no need to visit any vertex $v_j \in V\backslash T$ that only usefully covers vertices in $\Gamma \subset W$, and as a consequence, less labels are produced.

On the other hand, the following feasibility rule is used. Let $\bar{\Omega}_j$ denote the subset of vertices of W that are not covered by the subpath represented by label λ_j. A vertex $v_i \in V\backslash T$ can only be skipped if for each $w_i \in \bar{\Omega}_j$, there still remain vertices ahead that can cover it. The number of such vertices is kept through field $\omega[k]$. Thus, when the decision to skip a node is evaluated, for each $w_i \in \bar{\Omega}_j$, feasible labels yield $w[i] > 0$. Such value indicates that no vertex is left uncovered, so it is feasible to skip vertex $v_i \in V\backslash T$. For computational efficiency, a matrix relating the coverage of the vertices $w_i \in W$ by the vertices $v_i \in V$ is precomputed and kept at hand.

Other way to control label proliferation in this algorithm is the computation and updating of an upper bound as it will be explained in the ensuing section.

Algorithm 1. Selector

Input: giant tour GT, distance matrix D, set T
Output: optimal tour of visited vertices, S, and cost value of tour, $c(S)$
1: $L^u \leftarrow$ search upper bound {See Algorithm 2}
   {build an initial set of labels}
2: **while** ($\exists$ arc(v_0, v_i)) **do**
3:     $\Lambda^i \leftarrow \Lambda^i \cup \{L\}$ {L is the label being treated}
4:     Extend Horizontally(L) {see Algorithm 3}
5:     $i \leftarrow i + 1$
6: **end while**
   {extend labels created}
7: **while** ($\exists$ an Λ^i) **do**
8:     $L \leftarrow \min_{i \in N}\{\lambda_j\}$ {find label of lowest cost}
9:     Extend Skipping(L) {see Algorithm 4}
10: **end while**

Finding the Shortest Path. Algorithm 1 illustrates the core procedure of *Selector*. It executes three main steps: (i) search for an initial feasible solution or upper bound (UB), (ii) build an initial set of labels, and (iii) extend the created labels. In the explanations that follow the term *horizontal extension* means to iteratively visit in M the adjacent successor vertex until a complete feasible solution is built or any other of the stopping criteria is met (see Algorithm 3). The worst-case time complexity of the horizontal extension process is $O(n)$.

As mentioned, besides the feasibility and dominance rules, the upper bound is helpful to limit the creation of labels, and it is computed as follows (refer to Algorithm 2). For every vertex $v_i \in M$ for which arc (v_0, v_i) exists, it builds arc (v_i, v_{i+1}) and from this point continues the construction with a horizontal extension. Next, it constructs the arc to the next successor, arc (v_i, v_{i+2}), and proceeds with the same horizontal extension, and so on. The process stops when arc (v_i, v_{i+k}) can no longer be constructed, and restarts with the creation for the next vertex of arc (v_0, v_{i+1}). In this first step, every built path is compared and the best one is kept, no labels are stored in order to execute it fast. The worst-case time complexity of the search is $O(n^3)$.

The second step, the generation of an initial set of labels, iteratively constructs arc (v_0, v_i) followed by a horizontal extension. However, at each step (at every vertex) a non-dominated label documenting the subpath is stored. The process repeats as long as it is possible to construct arc (v_0, v_i). The worst-case time complexity of this step is $O(n^2)$.

Finally, the created labels are extended (see Algorithm 4). This means that from the last visited vertex stored in the label, v_{last}, it tries to reach successor v_{last+2} and from this point does a horizontal extension storing non-dominated labels at every step. The process repeats as long as arc (v_{last}, v_{last+k}) exists. The label chosen for extension is always the one that documents the shortest path and the execution of Algorithm 4 continues until there are no labels to extend. The worst-case time complexity of this extension is $O(n^2)$. Nevertheless, it might be executed for several thousands of labels in large instances.

As can be observed in Algorithm 3, at every step of the label extension the following conditions are verified: (a) the vertex to be included is not redundant, (b) $\zeta_{subpath} < \zeta_{best}$, (c) $UB_{current} < UB_{best}$, and (d) label is not dominated. Any vertex that turns out to be redundant is simply skipped and the construction continues, no labels are kept for not-visited vertices. If the cost of the path being built is worse than the cost of the best known solution, the search in that trajectory is abandoned. The initial upper bound built at the onset is updated throughout the search to improve the limits for the creation of labels.

A distinctive and important characteristic of our operator is that aside from the constraints mentioned in the definition of the problem, it does not impose any further restrictions on the selected vertices of $V \backslash T$ such as adjacency, for example. This operator is capable of discarding any vertex $v_i \in V \backslash T$ at any point in the tour.

Algorithm 2. Search Upper Bound

Input: giant tour GT, distance matrix D, set T
Output: feasible tour of visited vertices, S, and cost value of tour, $c(S)$
1: $i \leftarrow 1$
2: **while** ($\exists$ arc(v_0, v_i)) **do**
3:     $k \leftarrow i$
4:     **while** ($\exists$ arc (v_i, v_{k+1})) **do**
5:         Extend Horizontally(L) {see Algorithm 3}
6:         $k \leftarrow k + 1$
7:     **end while**
8:     $i \leftarrow i + 1$
9: **end while**

Algorithm 3. Extend Horizontally(L)

Input: label to be extended, L
Output: labels derived from L
    {only nondominated labels that can be extended are kept}
1: **for** ($j = lastVisitedNode + 1$ to $j \leqslant n$) **do**
2:    $clientsCovered \leftarrow clientsCovered + coverage\ of\ j$
3:    **if** (node is not redundant) **then**
4:        $cost\ of\ L \leftarrow cost\ of\ L + cost\ of\ visiting\ j$
5:        **if** ($cost\ of\ L < cost\ of\ L^*$) **then**
6:            **if** ($clientsCovered \neq clients$) **then**
7:                **if** (L not dominated) **then**
8:                    $\Lambda^i \leftarrow \Lambda^i \cup \{L\}$
9:                **end if**
10:            **else**
11:                $L^* \leftarrow L$
12:                **return**
13:            **end if**
14:        **else**
15:            **return**
16:        **end if**
17:    **end if**
18: **end for**

Algorithm 4. Extend Skipping(L)

Input: label to be extended, L
Output: labels derived from L
1: $i \leftarrow lastVisitedNode$
2: $k \leftarrow 2$
3: **while** ($\exists$ arc (v_i, v_{i+k})) **do**
4:    Extend Horizontally(L)
5:    $k \leftarrow k + 1$
6: **end while**

---

3   Adaptive Large Neighborhood Search (ALNS)

In our methodology, the overall task of the ALNS metaheuristic is to build suboptimal giant tours from which efficient CTP solutions are extracted. This is, according to Beasley's method (1983), the ordering phase, and the splitting phase is performed by the *Selector* operator embedded into this heuristic. ALNS, a local search framework which uses several competing destroy and repair methods and chooses amongst them using statistics gathered during the search, competes strongly with genetic algorithms (GA) in vehicle routing. However, the efficiency of GAs relies on sophisticated local search methods and population management techniques, while in ALNS neighborhoods are searched by simple and fast heuristics. ALNS has provided good solutions for a wide variety of VRPs as shown in Pisinger and Ropke (2010) and Ribeiro and Laporte (2011).

The three backbones of our implementation are (i) removal operators, (ii) insertion operators and (iii) metaheuristic that defines the criteria to accept a new solution. Three removal and three insertion heuristics were implemented. In the following, the word facility indicates a vertex that belongs to the giant tour, and lower-case Greek letters indicate user-controlled parameters.

3.1   Removal Operators

Shaw Removal Heuristic (SRH). Originally proposed by Shaw (1997), its general idea is to remove facilities that exhibit similitude, characteristic computed by a *relatedness measure* $R(i, j)$. For this implementation, the similarity between two facilities is measured by $R(i, j) = d_{ij}$, where d_{ij} is the Euclidian distance between facilities i and j. This relatedness measure is used to remove facilities in the same way as described by Shaw (1998). In order to avoid the sorting of facilities required at each iteration, a nearest facility matrix is precomputed and kept at hand. The worst-case time complexity of the SRH is $O(n^2)$.

Worst Removal Heuristic (WRH). Ropke and Pisinger (2006) propose a heuristic that randomly removes facilities with a high cost in the current solution X and tries to insert them in better positions. It iterates recalculating the costs until it has removed the indicated number of facilities. The removal, though random, is user-controlled by parameter ρ. The worst-case time complexity of the WRH is $O(n^2)$.

Random Removal Heuristic (RRH). This procedure simply selects γ facilities at random and removes them from the current solution X. Though it tends to generate a poor set of removed members, it is useful to diversify the search. The worst-case time complexity of the RRH is $O(n)$.

How Many to Remove. The number of facilities removed, γ, from the current solution X is key to the ALNS performance. When few elements are removed, the heuristic has a higher probability of being trapped in one suboptimal area of the search space. On the other hand, when too many are removed, it is almost like starting from scratch and the insertion heuristics cannot build a good solution from such situation. In addition, the larger the number removed, the larger the execution time of both insertion and removing heuristics. We choose γ randomly between a lower and upper limit. The lower limit is fixed at a value given according to the number of vertices in set V, 20 % to 25 % of its size, while the upper limit is fine-tuned with parameter ϵ. This parameter indicates the maximum percentage of elements removed from the complete solution size.

3.2   Insertion Operators

Best Greedy Heuristic (BGH). This simple construction heuristic performs at most γ iterations as it inserts one facility into solution X in each iteration.

The minimum cost position value is computed for all facilities waiting insertion, set F, and the one with the minimum global cost position is chosen. This process is repeated until $F = \emptyset$. The worst-case time complexity of the BGH is $O(n^2)$.

First Greedy Heuristic (FGH). This heuristic works similarly to the previous one. However, instead of inserting the facility having the minimum global cost position, it inserts the one sitting in the first position. That is to say, it respects the order of the facilities in F. After the first facility has been inserted, the minimum cost position for each is recalculated and the process repeats until all facilities in set F have been inserted.

Ropke and Pisinger(2006) add a noise term to the objective function during the insertion phase of the BGII and regret-k heuristics in order to randomize them and avoid always making the move that seems best locally. In our implementation, the FGH is used mainly to introduce this noise into the insertion process as done by Ribeiro and Laporte(2011). This heuristic obviously runs faster than the BGH.

Regret-k Heuristic (RKH). This heuristic tries to improve the myopic behaviour of the greedy heuristics by incorporating a kind of look ahead information when selecting the facility to insert, as done by Ropke and Pisinger(2006) and Pisinger and Ropke(2007). Let Δf_i^1 denote the change in objective value incurred by inserting facility i at its minimum cost position, and Δf_i^2 denote the change by inserting it at its second best position. The regret value is defined in terms of the former values as $c_i^* = \Delta f_i^2 - \Delta f_i^1$. In each iteration, the regret heuristic chooses to insert the facility i that maximizes $max_{i \in F}\{c_i^*\}$, and such facility is inserted at its minimum cost position. Ties are broken by selecting the facility with lowest cost insertion. This is a time-consuming operator but unnecessary computations were avoided when computing Δf_i^n. The worst-case time complexity of the RKH is $O(n^3)$.

Choosing a Removal and an Insertion Heuristic. In order to select a heuristic, weights are assigned to them and a *roulette wheel selection principle* is applied. The removal heuristic is selected independently of the insertion heuristic and vice versa. Initially, all heuristics are equally likely.

Adaptive Weight Adjustment. The probability of selecting a heuristic changes based on its performance. To enable this change, a score is kept for each and it is updated at each iteration. Our implementation keeps track of visited solutions using a hash table. A hash key is assigned to every solution and this key is stored in the table. We followed the scheme of scores and updating procedure of probability weights proposed by Ropke and Pisinger (2006).

3.3   General Framework with Simulated Annealing

Algorithm 5 depicts the ALNS process implemented with simulated annealing (SA) as the outer metaheuristic that guides the search. We followed the SA

Table 1. Values of the ALNS parameters after experimental tuning.

Parameter	Meaning	Value
γ	Number of facilities removed at each ALNS iteration (instance size dependent)	[5, 30]
ς	Segment size for updating probabilities in number of ALNS iterations	50
τ	Reaction factor that controls the rate of change of the weight adjustment	0.3
δ	Avoids determinism in the SRH	6
ρ	Avoids determinism in the WRH	2
σ_1	Score for finding a new global best solution	50
σ_2	Score for finding a new solution that is better than the current one	20
σ_3	Score for finding a new non-improving solution that is accepted	5
β	Cooling factor used by simulated annealing	0.99999
ϵ	Fixes the upper limit of facilities removed at each iteration	0.3

scheme suggested by Pisinger and Ropke (2007). The algorithm works on two items: the giant tour, GT, constructed by the ALNS heuristics, and S, the solution covering tour computed by *Selector* when applied to GT. The cost of GT is labelled l, while the cost value for solution S is identified as c. The variable $GT_{current}$ indicates the solution obtained at the beginning of an iteration, and the variable GT_{new} is the temporary solution obtained during the iteration. For the sake of clarity, the updating processes for scores, hash keys and probabilities are not shown.

The 2-*opt* procedure is used to rapidly improve the length of the starting GT and avoid a long, random, initial walk. In addition, the acceptance of GT_{new} is controlled by the cost value of solution S. The latter is because experimentation showed that improvements on the length of the giant tour do not necessarily lead to improvements on the length of the tour computed by *Selector*. However, in the long run, the length of the CTP tour benefits from improvements on the length of the GT. An important consequence of this finding is that execution of *Selector* to optimality at each ALNS iteration is of low benefit. The upper bound computed by Algorithm 2 can serve as a probe to determine if the complete process is worth executing. This derives in important time savings.

4   Computational Results

The results obtained by our metaheuristic are compared against the optimal solutions computed by the branch-and-cut algorithm of Gendreau et al. (1997). The exact algorithm is written in Python 2.7 and uses 5.6 Gurobi callbacks. Library Python-Igraph 0.7.0 helps to solve graph problems occurring in the valid

---

Algorithm 5. The General Framework of the ALNS with Simulated Annealing

---

Input: Giant tour GT, distance matrix D
Output: S_{best} and $c(S_{\text{best}})$
1: 2-opt(GT_0)
2: compute $l_0(GT_0)$
3: initialize, to the same value, probability P_r^t for each removal operator $r \in R$, and
   likewise probability P_i^t for each insertion operator $i \in I$.
4: $t \leftarrow l_0$, {set initial temperature, variable used in probability function}
5: $l_{\text{current}} \leftarrow l_0$
6: $GT_{\text{current}} \leftarrow GT_0$
7: $UB_{\text{best}} \leftarrow$ Search Upper Bound(GT_0) {see Algorithm 2}
8: $c(S_{\text{best}}) \leftarrow c(S_{\text{current}}) \leftarrow$ Selector(GT_0) {see Algorithm 1}
9: $i \leftarrow 1$ {iteration counter}
10: **repeat**
11:    select a removal operator $r \in R$ with probability P_r^t {roulette wheel}
12:    obtain GT_{new}^- by applying r to GT_{current}
13:    select an insertion operator $i \in I$ with probability P_i^t
14:    obtain GT_{new} by applying i to GT_{new}^-
15:    $UB_{\text{current}} \leftarrow$ Search Upper Bound(GT_{new})
16:    **if** ($UB_{\text{current}} < UB_{\text{best}}$) **then**
17:      $c(S_{\text{new}}) \leftarrow$ Selector(GT_{new})
18:      $UB_{\text{best}} \leftarrow UB_{\text{current}}$
19:    **else**
20:      $c(S_{\text{new}}) \leftarrow UB_{\text{current}}$
21:    **end if**
   {decide acceptance of new solution}
22:    **if** ($c(S_{\text{new}}) < c(S_{\text{current}})$) **then**
23:      $c(S_{\text{current}}) \leftarrow c(S_{\text{new}})$
24:      $GT_{\text{current}} \leftarrow GT_{\text{new}}$
25:    **else**
26:      $p \leftarrow e^{-\frac{c(S_{\text{new}}) - c(S_{\text{current}})}{t}}$
27:      generate a random number $n \in [0,1]$
     {new solution might be accepted with a computed probability even it is worse}

28:      **if** ($n < p$) **then**
29:        $c(S_{\text{current}}) \leftarrow c(S_{\text{new}})$
30:        $GT_{\text{current}} \leftarrow GT_{\text{new}}$
31:      **end if**
32:    **end if**
33:    **if** ($c(S_{\text{new}}) < c(S_{\text{best}})$) **then**
34:      $c(S_{\text{best}}) \leftarrow c(S_{\text{new}})$
35:      $S_{\text{best}} \leftarrow S_{\text{new}}$
36:    **end if**
37:    $t \leftarrow \beta \cdot t$ {cooling rate set to be very slow}
38:    **if** (segment size $= \varsigma$) **then**
39:      update probabilities using the adaptive weight adjustment procedure
40:    **end if**
41:    $i \leftarrow i + 1$
42: **until** (defined number of iterations is met)

---

cut separation. The heuristic algorithms are coded in C++ and the benchmark was done on a computer with 8 GiB of memory, processor Intel Core i7-4770 CPU@3.40 GHz, and Linux OS type 64 bits.

Since test problems for the CTP are not found in the literature, we created data sets based on 9 Euclidean TSPLIB instances (Reinelt 1991) whose sizes range from 100 to 200 vertices. Sets of vertices of $|V \cup W| \in \{100, 150, 200\}$ were created using kroX100 (X $\in$ {A,B, ... ,E}), kroX150 and kroX200 (X $\in$ {A,B}) respectively. T and V are defined by taking the first $|T|$ and $|V| - |T|$ points, respectively, while W is defined by the remaining points. Tests were run for $|V| \in \{25, 50, 75, 100\}$. $|T| = 1$, only the depot is compulsory, which is the worst case regarding the number of labels created.

The costs $\{\zeta_{ij}\}$ are treated as integer values equal to $\lfloor d_{ij} + .5 \rfloor$, where d_{ij} is the Euclidean distance between points i and j (Reinelt 1991). The value of c is resolved using $c = \max \left(\max_{v_k \in V \setminus T} \min_{w_l \in W} \{\zeta_{l,k}\}, \max_{w_l \in W} \{\zeta_{l,k(l)}\} \right)$, where $k(l)$ indicates the second nearest vertex $v_k \in V \setminus T$. Computing this value in such way ensures that each vertex $v_i \in V \setminus T$ covers at least one vertex $w_i \in W$, and each vertex $w_i \in W$ is covered by at least two different vertices $v_i \in V \setminus T$ as explained in Gendreau et al. (1997).

Several independent executions were done to test our randomized heuristic. Each instance was run 30 times with a different seed each time and for 30,000 iterations. To define an efficient parameter set, we used a *ceteris paribus* approach based on sets of three or four values for each parameter. The resulting set is listed in Table 1. On the other hand, both the optimal values and the quality of the solutions computed by the heuristic can be observed in Table 2 where the first three columns document the instance information, the next three report the findings of the exact method, and the last five those of the approximate approach. Columns UB and Opt show the time in seconds needed to reach an upper bound and the optimal value respectively. Column $\bar{\theta}$ indicates the deviation of the heuristic solution from the optimum value in percentage, and $\bar{t}$ corresponds to the total run time in seconds. These two figures are average values over the 30 runs. Column Found indicates how many times the heuristic found the optimum value in the set of runs. Column Best Gap shows how close (in percentage) the heuristic came to the optimum value, and the last one, labeled S_{N-1}, exhibits the corrected sample standard deviation.

On the whole, the heuristic is very accurate and its performance is highly satisfactory, since for 96 % of the instances it was capable of finding the optimum value rapidly. In the few cases where the optimum was not reached, the minimum value computed was less than 1 % away from the optimal solution value. In addition, the average deviation is typically within 1 % of optimality. Also, it repeatedly found the optimum value for 63 % of the instances. In general, given an instance, this number worsens as $|V|$ increases. Results are reported for 30,000 iterations. However, observing the evolution of the search, we could see that optimal solutions were identified for approximately 75 % of the instances as early as in the first 1,000 iterations. Furthermore, in general, the spread around the optimum of the values computed is very moderate. We can, thus, state that it is a heuristic capable of identifying very good solutions quite quickly.

Table 2. Performance of heuristic compared to the branch-and-cut algorithm.

Instance Based on	$|V|$	$|W|$	Optimum	UB(s)	Opt(s)	$\bar{\theta}(\%)$	$\bar{t}(s)$	Found	Best Gap(%)	S_{N-1}
kroA100	25	75	7985	0.17	0.17	2.36	0.42	14	0	272.51
kroA100	50	50	8608	22.20	44.95	0.21	0.95	11	0	23.06
kroB100	25	75	6449	0.21	0.27	0.16	0.50	24	0	22.79
kroB100	50	50	8043	1.18	21.54	0.70	1.25	1	0	60.20
kroC100	25	75	6161	0.01	0.01	0	0.81	30	0	0
kroC100	50	50	7942	0.81	0.81	0	2.27	30	0	0
kroD100	25	75	6651	0.24	0.38	0	0.31	30	0	0
kroD100	50	50	8411	3.75	4.33	0.02	1.13	27	0	4.64
kroE100	25	75	7417	0.26	0.27	0.02	0.42	29	0	8.71
kroE100	50	50	8493	1.10	1.11	0	1.00	30	0	0
kroA150	25	125	8050	0.13	0.13	1.43	0.54	3	0	131.72
kroA150	50	100	9623	118.80	121.58	0.37	1.16	2	0	38.56
kroA150	75	75	9971	1569.38	2884.34	0.60	2.93	0	0.59	59.81
kroB150	25	125	6165	0.01	0.01	0	1.50	30	0	0
kroB150	50	100	7818	1.16	1.16	0.02	2.32	29	0	7.23
kroB150	75	75	7434	13.34	38.24	0.01	4.32	26	0	2.16
kroA200	25	175	6165	0.01	0.01	0	1.63	30	0	0
kroA200	50	150	8273	0.46	0.49	0	5.09	30	0	0
kroA200	75	125	8499	141.97	266.19	0	6.31	30	0	0
kroA200	100	100	8355	1110.87	1709.22	0	14.21	30	0	0
kroB200	25	175	6450	0.15	0.15	0.18	1.17	23	0	24.62
kroB200	50	150	8171	2.69	3.46	0.78	2.6	3	0	77.29
kroB200	75	125	10007	0.65	0.65	1.42	4.5	5	0	166.52
kroB200	100	100	9988	17.20	17.68	1.73	11.60	5	0	202.63

5   Conclusions

This paper presents a study on a novel resolution method for a difficult combinatorial optimization problem which finds application in network design and vehicle routing. Its key feature is the *Selector* operator that optimally splits an initial sequence of facilities into subsequences of visited and not-visited vertices. We have proposed an approximate method capable of obtaining very high quality solutions in very short periods of time. It is a simple, easy to implement heuristic and its core, the *Selector* operator, is new and creative in its own right. Given the practical relevance of the CTP, we will look into other heuristic mechanisms to solve large-scale instances. We believe the approach developed in this work can be translated or adapted to solve related TLPs like the *Orienteering Problem*, also known as the *Selective TSP*, and the *Prize Collecting TSP*.

Acknowledgement. This work has been partially supported by the French National Research Agency through the ATHENA project under the grant ANR-13-BS02-0006. Leticia Vargas also benefits from an Erasmus Mundus Project LAMENITEC fellowship, agreement 372362-1-2012-1-ES, and from partial support provided by CONACyT Mexico for her doctoral studies. This support is gratefully acknowledged. Thanks are also due to the referees for their through review and valuable comments.

References

Baldacci, R., Boschetti, M.A., Maniezzo, V., Zamboni, M.: Scatter search methods for the covering tour problem. In: Rego, C., Alidaee, B. (eds.) Metaheuristic Optimization Via Memory and Evolution, pp. 59–91. Kluwer, Boston (2005)

Beasley, J.: Route first-cluster second methods for vehicle routing. OMEGA Int. J. Manag. Sci. **11**, 403–408 (1983)

Current, J.R., Schilling, D.A.: The covering salesman problem. Transp. Sci. **23**, 208–213 (1989)

Current, J.R., Schilling, D.A.: The median tour and maximal covering problems. European J. Oper. Res. **73**, 114–126 (1994)

Desrochers, M.: An algorithm for the shortest path problem with resource constraints. Technical report G-88-27, GERAD. Montréal (1988)

Feillet, D., Dejax, P., Gendreau, M., Gueguen, C.: An exact algorithm for the elementary shortest path problem with resource constraints: application to some vehicle routing problems. Networks **44**, 216–229 (2004)

Fischetti, M., Salazar, J.J., Toth, P.: A branch-and-cut algorithm for the symmetric generalized traveling salesman problem. Oper. Res. **45**, 378–394 (1997)

Gendreau, M., Laporte, G., Semet, F.: The covering tour problem. Oper. Res. **45**, 568–576 (1997)

Golden, B., Naji-Azimi, Z., Raghavan, S., Salari, M., Toth, P.: The generalized covering salesman problem. INFORMS J. Comput. **24**, 534–553 (2012)

Hodgson, J., Laporte, G., Semet, F.: A covering tour model for planning mobile health care facilities in Suhum district. Ghana. Jl Reg. Sci. **38**, 621–638 (1998)

Jozefowiez, N., Semet, F., Talbi, E.G.: The bi-objective covering tour problem. Comput. Oper. Res. **34**, 1929–1942 (2007)

Laporte, G., Semet, F.: Classical and new heuristics for the vehicle routing problem. In: Toth, P., Vigo, D. (eds.) The Vehicle Routing Problem, pp. 109–128. SIAM, Philadelphia (2002)

Motta, L., Ochi, L.S., Martinhon, C.: GRASP metaheuristics to the generalized covering tour problem. In: MIC 2001-4th Metaheuristics International Conference, Porto, Portugal, pp. 387–391 (2001)

Pisinger, D., Ropke, S.: A general heuristic for vehicle routing problems. Comput. Oper. Res. **34**, 2403–2435 (2007)

Pisinger, D., Ropke, S.: Large neighborhood search. In: Gendreau, M., Potvin, J.-Y. (eds.) Handbook of Metaheuristics, pp. 399–419. Springer, New York (2010)

Prins, C.: A simple and effective evolutionary algorithm for the vehicle routing problem. Comput. Oper. Res. **31**, 1985–2002 (2004)

Prins, C., Labadi, N., Reghioui, M.: Tour splitting algorithms for vehicle routing problems. Int. J. Prod. Res. **47**, 507–535 (2009)

Prins, C., Lacomme, P., Prodhon, C.: Order-first split-second methods for vehicle routing problems: a review. Transp. Res. **40**, 179–200 (2014)

Reinelt, G.: TSPLIB - a traveling salesman problem library. ORSA J. Comput. **3**, 376–384 (1991)

Ribeiro, G., Laporte, G.: An adaptive large neighborhood search heuristic for the cumulative capacitated vehicle routing problem. Comput. Oper. Res. **39**, 728–735 (2012)

Ropke, S., Pisinger, D.: An adaptive large neighborhood search heuristic for the pickup and delivery problem with time windows. Transport. Sci. **40**, 455–472 (2006)

Salari, M., Naji-Azimi, Z.: An integer-programming-based local search for the covering salesman problem. Comput. Oper. Res. **39**, 2594–2602 (2012)

Shaw, P.: A new local search algorithm providing high quality solutions to vehicle routing problems. Technical report. University of Strathclyde, Scotland (1997)

Shaw, P.: Using constraint programming and local search methods to solve vehicle routing problems. In: Maher, M., Puget, J.-F. (eds.) CP 1998. LNCS, vol. 1520, pp. 417 431. Springer, Heidelberg (1998)

Metaheuristics for the Two-Dimensional Container Pre-Marshalling Problem

Alan Tus[1,2](✉), Andrea Rendl[3], and Günther R. Raidl[4]

[1] Dynamic Transportation Systems, AIT – Austrian Institute of Technology,
Mobility Department, Giefinggasse 2, 1180 Vienna, Austria
altus@microsoft.com
[2] MDCN - Microsoft Development Center Norway, Torggata 2, 0181 Oslo, Norway
[3] Optimization Research Group, NICTA and Monash University,
School of IT, 878 Dandenong Road, Caulfield East 3145, Australia
andrea.rendl@nicta.com.au
[4] Institute of Computer Graphics and Algorithms, Vienna University of Technology,
Favoritenstraße 9/1861, Vienna, Austria
raidl@ads.tuwien.ac.at

Abstract. We introduce a new problem arising in small and medium-sized container terminals: the *Two-Dimensional Pre-Marshalling Problem* (2D-PMP). It is an extension of the well-studied Pre-Marshalling Problem (PMP) that is crucial in container storage. The 2D-PMP is particularly challenging due to its complex side constraints that are challenging to express and difficult to consider with standard techniques for the PMP. We present three different heuristic approaches for the 2D-PMP. First, we adapt an existing construction heuristic that was designed for the classical PMP. We then apply this heuristic within two metaheuristics: a Pilot method and a Max-Min Ant System that incorporates a special pheromone model. In our empirical evaluation we observe that the Max-Min Ant System outperforms the other approaches by yielding better solutions in almost all cases.

Keywords: Ant colony optimization · Construction heuristics · Container terminal · Pilot-Method · Pre-Marshalling problem

1 Introduction

Containers are an essential component of today's shipping industry. They are standardized to facilitate shipment and storage. Container shipment typically proceeds in chains of different transport modes, such as trains, ships or trucks. *Container terminals* are crucial to the overall shipment process since they act as hubs between the different transport modes. They deal with container *exchange* between vessels, and container *storage* until the appropriate vessels arrive.

Container storage planning is a key factor in container terminal organization and affects all other operations of the terminal. It is concerned with storing containers in such a way that they can be quickly retrieved when they are due for further shipment. Containers are stored in so-called *container bays* which are rows

C. Dhaenens et al. (Eds.): LION 9 2015, LNCS 8994, pp. 186–201, 2015.
DOI: 10.1007/978-3-319-19084-6_17

of adjacent container stacks. When a container is due for shipment it is typically removed from its bay by a gantry crane that can access the topmost containers of each stack. Due containers may sometimes be blocked by other containers that are stacked upon them. In this case, the blocking containers have to be relocated to access the due containers, which increases the loading time. This can cause severe delays for vessel loading and vessel departures, resulting in displeased terminal clients and ultimately additional costs. Therefore, container terminals rearrange (*pre-marshal*) container bays to assure that each container is reachable when it is due. The *Pre-Marshalling Problem* (PMP) is concerned with finding a sequence of container relocations of minimal length, such that in the resulting bay no container is blocked when removed according to its due-time.

In this paper we introduce a new variant of the PMP, the *Two-Dimensional Pre-Marshalling Problem* (2D-PMP), which occurs in smaller and medium-sized container terminals. Those terminals do not use gantry cranes to load vessels but so-called *reach stackers*. Reach stackers are powerful forklifts that can carry fully loaded containers, however, most conventional reach stackers can only access the leftmost and rightmost stacks of a container bay. This means that due containers can be blocked by containers stacked *next* to them, in addition to containers stacked *upon* them. Thus, a new 2-dimensional restriction on how containers may be stacked such that they can be removed according to their due-times without having to relocate blocking containers needs to be considered. Our use-case is based on the Ennshafen container terminal[1] in Enns, Austria. Another example of a smaller to medium-sized container terminal is the Frihamnen container terminal in Stockholm, Sweden. Space is an issue for them since they are located in the city center, so agile vehicles like the reach stacker are required to perform most of the work.

This work is based on a master thesis [1] which can be referred to for detailed algorithm explanations, further implementation details and more elaborate results.

Related Work

The classical PMP is known to be NP-hard [2] and is well-studied [3]. The 2D-PMP is a novel problem for which, to the best of our knowledge, no solution approach has yet been published. In the following, we therefore give a chronological overview of the most prominent methods for the PMP.

Lee and Hsu [4] present a MIP model in form of a multi-commodity flow formulation in which nodes represent slots in the container bay, and arcs connect slots. Thus, container moves are represented with flows, and constraints assure that the moves are valid. The number of moves is restricted by an upper bound, and the objective is to find a flow that yields a desirable bay in a minimal number of moves. This formulation grows substantially in the number of moves and therefore has difficulties to scale.

---

[1] http://www.ennshafen.at/en/container_terminal/technical_data.

Caserta and Voß [5] propose a corridor-method based approach for solving the PMP. Given an initial solution (constructed by a method similar to the GRASP heuristic), the approach builds a corridor around the incumbent solution and performs low-level decisions using a combination of greedy heuristics and a roulette-wheel approach. In their experiments, the authors obtained new upper bounds for existing benchmarks with their approach.

Lee and Chao [6] introduce a neighbourhood search method that improves an initial feasible solution by applying different local search heuristics and an integer program that possibly reduces the length of the move sequence in the incumbent solution, yielding the same final bay configuration.

Exposito et al. [7] propose a "lowest priority first" construction heuristic, see Sect. 3, as well as an instance generator for the PMP. The heuristic aims at placing containers at positions that seem most suitable, starting with the container that will be removed last.

Bortfeldt and Forster [8] present a tree-search heuristic for the PMP, where the tree depth is restricted by an upper bound. The tree search incorporates a classification and ordering of the possible moves at each configuration, which drives the search towards promising directions. This method is competitive with the approaches from [5,6].

Prandtstetter [9] proposes a dynamic programming approach for the PMP where states represent the container bay and are extended by container moves. This method is quite successful since symmetric and dominated states (with a weaker lower bound) are easily detected and discarded.

Rendl and Prandtstetter [10] introduce a constraint programming model for the PMP where decision variables represent the container bay state and moves are set exclusively by the search strategy, which applies the heuristic from [8].

Tierney et al. [11] present a novel solution technique for solving pre-marshalling problems to optimality using the A* and IDA* algorithms. Both algorithms perform a path-based search guided by a cost estimation heuristic. Additionally, the search is directed by combining branching rules, symmetry breaking and strong lower bounds. Branching rules are used to prevent move reversal, unrelated and transitive moves, and empty stack symmetry.

2   The Two-Dimensional Pre-Marshalling Problem

Pre-marshalling problems arise in container terminals where containers are stored in *stacks* that are arranged in container *bays*. A container bay consists of a number of adjacent stacks that may not exceed a maximum height; see Fig. 1 for a sample bay with four stacks. Each container is assigned a *priority value* that indicates when the container is due to be removed from the bay. The smaller the priority value, the sooner the container will be removed from the container bay.

The priority of a container is often not known at its arrival in the bay, therefore containers are frequently stacked more or less arbitrarily. As a consequence, due containers would often be blocked by other containers that are due at a later time. In the container bay in the left of Fig. 1, for example, the shaded containers are blocked since containers with higher priority values are placed upon them.

Fig. 1. A container bay with four stacks where the numbers represent the priority values of the containers. Shaded containers are blocked for the gantry crane. The left figure shows the bay before pre-marshalling, the right thereafter.

Fig. 2. Left: The classically pre-marshalled container bay from Fig. 1, now serviced by reach stackers. The shaded container is blocked horizontally. Right: A configuration where all containers can directly be removed by reach stackers.

The classical Pre-marshalling Problem (PMP) is concerned with relocating containers in a bay most economically in such a way that thereafter all containers can be immediately retrieved from the bay in the order given by the priority values. More specifically, the PMP seeks a minimal sequence of container moves that yields a container bay without blocked containers.

In large container terminals all container movements are performed exclusively by gantry cranes that can access the topmost container of each stack. In small and medium-sized terminals gantry cranes are only used to move containers within a bay, while reach stackers remove containers to load vessels.

Reach stackers are forklifts that can carry one container at a time, but their access is restricted to the top containers of the leftmost and rightmost stacks of a bay. This is a significant restriction, since containers can be blocked *vertically* (by containers in the same stack) but also *horizontally* (by containers in neighbouring stacks). The classical PMP only considers vertical blocking, and therefore does not sufficiently improve the container bay configuration in the scenario with reach stackers. For instance, Fig. 2(left) illustrates that the pre-marshalled configuration from Fig. 1(right) still has a blocked container when considering reach stackers for removal.

Therefore, we introduce the *Two-Dimensional Pre-Marshalling Problem* (2D-PMP) that considers blocking from two dimensions, i.e., vertically and horizontally.

2.1 Formal Definition of the 2D-PMP

The 2D-PMP considers a container bay with S stacks holding a set of containers C. Stacks may not exceed a maximum height H. We denote the set of all stacks as S and the set of all tiers as T. Each container $c \in C$ is referred to by its priority value, i.e., $c \in \mathbb{N}$, which we assume without loss of generality to be unique; i.e., a complete ordering is specified for the containers.

We are given an initial bay configuration $\mathcal{B}$, where $\mathcal{B}_{s,t}$ is the container in the t-th tier in the s-th stack, with $(s,t) \in \mathcal{S} \times \mathcal{T}$ and write, $\mathcal{B}_{s,t} = 0$ if slot $\mathcal{B}_{s,t}$ is empty. The bay can be altered by performing moves: the topmost container from a (non-empty) stack can be moved to the top of another stack whose height is less than H. We denote such a container move from the top of stack i to the top of stack j by $m = (i,j)$, where $i, j \in \{1, \ldots, S\}$.

A container c that is located at $\mathcal{B}_{s,t}$ is called *vertically* blocked, if there exists another container c' at $\mathcal{B}_{s,t'}$ with $t' > t$ (container c' is placed above c) and $c' > c$. A stack s is called *v-perfect*, if no container in stack s is vertically blocked. A container c that is located at $\mathcal{B}_{s,t}$ is called *horizontally* blocked, if there exist two containers c' and c'' where c' is located at $\mathcal{B}_{s',t'}$ and c'' at $\mathcal{B}_{s'',t''}$ with $s' < s$ (stack s' is left of stack s) and $s'' > s$ (stack s'' is right of stack s) and $c < c'$ and $c < c''$ (container c has the smallest priority value of the three containers). A stack s in which no container is horizontally blocked, is called *h-perfect*. The aim of the 2D-PMP is to obtain with minimum effort a bay configuration in which all stacks are v-perfect as well has h-perfect.

A solution is thus a sequence of container moves $\sigma = \{m_1, m_2, \ldots, m_k\}$ that makes the initial bay configuration v- and h-perfect, and the 2D-PMP seeks an optimal solution having minimum length k.

3   Lowest Priority First Heuristic (LPFH)

The *Lowest Priority First Heuristic* (LPFH) has been proposed in [7] The general idea is to distinguish between *well-located* containers that may *remain* at their current position and *non-located* containers that are *moved* to obtain an unblocked configuration.

After classifying each container as well-located or non-located, the LPFH iteratively performs the following three steps until all containers are well-located:

1. select the non-located container c with highest priority value,
2. select a destination stack s for c so that c becomes well-located,
3. and compute feasible moves to relocate c to s and perform them.

3.1   The 2D Lowest Priority First Heuristic (2D-LPFH)

We extend the LPFH for the 2D-PMP into a new heuristic, 2D-LPFH, by applying two main changes. First, we adapt the notion of well-located and non-located containers, and second, we adapt the destination stack selection. The complete 2D-LPFH algorithm for the 2D-PMP is outlined in Algorithm 1. In the following, we highlight the differences to the classical LPFH.

Container Location. A container is called *well-located* in the 2D-PMP, if it is neither vertically nor horizontally blocked, and if it is located in one of the stacks foreseen for its priority (see below). A container is called *non-located* in the 2D-PMP, if at least one of the well-located criteria is violated.

Destination Stack Selection. The destination stack selection consists of three main steps: First, adequate stacks are identified that are as second step ordered according to the number of moves that are necessary to move the selected container to the respective stack. Third, the destination stack is randomly selected upon the best λ_2 adequate stacks, where λ_2 is an appropriately chosen strategy parameter. In the 2D-LPFH, containers with higher priority value are usually best placed in a central stack, while containers with low priority values are usually best placed in one of the outermost stacks. This way it is less likely that they are horizontally blocking or blocked by other containers. We incorporate this intuition in the stack selection, where we denote the set of adequate stacks $\mathcal{A}_c$ for container c: First, we calculate the average number of containers per stack as $cps = |\mathcal{C}|/S$. Second, we determine the first preferred stack as $s_1 = \lfloor c/(2 \cdot cps) \rfloor$. Third, we add the "mirror" stack $s_2 = S - s_1$ as the second preferred stack $(if\, s_2 \neq s_1)$, i.e., if the first preferred stack s_1 is closer to the left edge of the bay, then the second preferred stack s_2 will be closer to the right edge of the bay and have an equal number of stacks between itself and the right edge as the first preferred stack s_1 and the left edge. In case our configuration has an odd number of stacks, the middle stack does not have a second preferred stack. Fourth, in case s_1 and s_2 are not the outermost stacks, we allow some flexibility by adding the neighbouring stacks that are closer to the outer border to the set of adequate stack as $s_{1'} = s_1 - 1$ and $s_{2'} = s_2 + 1$. Finally, our set of *adequate stacks* is $\mathcal{R}_c = \{s_{1'}, s_1, s_2, s_{2'}\}$. This calculation assumes unique priority values. When moving the selected container to the destination stack we have to enable the move by removing all containers placed on top of the selected container and all containers placed on top of the designated slot in the destination stack. These containers are called *interfering containers* and they are moved to *temporary stacks*; any stack with free slots except the source and destination stacks.

In case the source and destination stack are the only stacks with free slots and there is at least one interfering container, the algorithm is blocked and can not continue. In such cases it is recommended to restart the algorithm because it might perform a different sequence of moves due to it's stochastic nature.

Compound Moves. In each iteration, the 2D-LPFH returns a *compound move* $\mathcal{K} = \{m_1, \ldots, m_k\}, k \geq 1$, i.e., a sequence of moves, corresponding to all the individual moves necessary to relocate a non-located container to a better location. This especially needs to be taken into account when applying the heuristic within a metaheuristic. Applying only a subsequence $\hat{\mathcal{K}} \in \mathcal{K}$ would in general result in a completely different and usually unintended and worse behaviour. For instance, $\hat{\mathcal{K}}$ might easily lead to a state having a higher objective value than when applying the complete $\mathcal{K}$. Therefore, when using the 2D-LPFH within a metaheuristic, we always consider only complete compound moves $\mathcal{K}$. However, this does not affect the behaviour of our metaheuristic algorithms as each regular or compound move can be viewed as a changeset applied to our current bay; the exact actions or their count are irrelevant to the metaheuristic.

Algorithm 1. Lowest Priority First Heuristic for the 2D-PMP

Input: $\mathcal{B}$: initial state; λ_2, λ_3: heuristic parameters
1: $\mathcal{N} \Leftarrow$ non-located containers in $\mathcal{B}$
2: $\mathcal{W} \Leftarrow$ well-located containers in $\mathcal{B}$
3: $\sigma \Leftarrow$ empty list
4: **while** $\mathcal{N} \neq \emptyset$ **do**
5:    $c \Leftarrow$ container with highest priority value in $\mathcal{N}$ located at stack s
6:    $\mathcal{A}_c \Leftarrow$ set of adequate stacks for c
7:    Sort $\mathcal{A}_c$ ascending by the lowest number of interfering containers in each stack
8:    $s' \Leftarrow$ select random stack from $\mathcal{A}_c$ among the top λ_2 stacks
9:    $\mathcal{G} \Leftarrow$ set of all interfering containers in c's stack and s'
10:    **for each** $g \in \mathcal{G}$ that is positioned at stack s_g **do**
11:      $\mathcal{V} \Leftarrow$ set of available temporary stacks for g
12:      sort $\mathcal{V}$ ascending by the highest priority valued non-located container in each stack
13:      $s'' \Leftarrow$ select random stack from $\mathcal{V}$ among the top λ_3 stacks
14:      perform move (s_g, s'') on $\mathcal{B}$, obtaining new bay configuration $\mathcal{B}'$
15:      append move (s_g, s'') to σ
16:      $\mathcal{B} \Leftarrow \mathcal{B}'$
17:    **end for**
18:    perform move (s, s'), obtaining new bay configuration $\mathcal{B}'$
19:    append move (s, s') to σ
20:    $\mathcal{N} \Leftarrow$ non-located containers in $\mathcal{B}'$
21:    $\mathcal{W} \Leftarrow$ well-located containers in $\mathcal{B}'$
22:    $\mathcal{B} \Leftarrow \mathcal{B}'$
23: **end while**
24: **return** sequence of moves σ

---

4   Pilot Method

The Pilot method [12] is a meta-heuristic that applies a simpler construction heuristic H as a lookahead subheuristic to guide a master construction process towards a more promising solution. Starting from an initially "empty" solution θ, the subheuristic H constructs n so-called *pilot solutions* θ_i, $\forall i = 1, \ldots, n$, which are partial solutions built from up to k greedy construction steps. Each pilot solution θ_i is evaluated by a quality estimation function f that is able to evaluate partial solutions. The idea is that these quality estimates provide a better guidance for the master construction process than a simple greedy criterion. A most promising partial solution θ_i with the best quality estimation is then chosen and its first construction step adopted by the master procedure, i.e., θ is extended by the corresponding step. Then, the same construction process with the help of the subheuristic is repeated from the next step onward until a complete solution is obtained.

We use the Pilot method to solve the 2D-PMP by applying different construction methods operating on partial solutions. Recall that a solution to the 2D-PMP is a sequence of (compound) moves, therefore, a pilot solution θ_i is a sequence of (compound) moves $\theta_i = \{m_1, \ldots, m_j\}$ with $j \geq 1 \leq k$.

Evaluation Function. Given a bay configuration $\mathcal{B}_{\theta_i}$ that has been obtained by applying the pilot solution θ_i on the initial configuration $\mathcal{B}$, the evaluation function $f(\mathcal{B}_{\theta_i})$ returns an estimate of the number of (compound) moves that are necessary to get an unblocked bay configuration. We define f as

$$f(\mathcal{B}_{\theta_i}) = b_v(\mathcal{B}_{\theta_i}) + \frac{1}{2}b_h(\mathcal{B}_{\theta_i}) + b_t(\mathcal{B}_{\theta_i}) \tag{1}$$

where $b_v(\mathcal{B}_{\theta_i})$ and $b_h(\mathcal{B}_{\theta_i})$ represent the number of containers in $\mathcal{B}_{\theta_i}$ that are blocked vertically and horizontally, respectively, and $b_t(\mathcal{B}_{\theta_i})$ is the cardinality of the set of containers placed upon all vertically blocked containers (excluding blocked containers).

Subheuristic. The subheuristic H is a heuristic that extends a current partial master solution θ, by a (compound) move $m \in \mathcal{M}_{\mathcal{B}_\theta}$ where $\mathcal{M}_{\mathcal{B}_\theta}$ is the set of all legal (compound) moves for the current container bay $\mathcal{B}_\theta$. Therefore, applying k construction steps corresponds to appending k legal (compound) moves to θ. We apply 2D-LPFH as a subheuristic.

5   An Ant Colony Optimization Approach

We developed a $\mathcal{MAX}\text{-}\mathcal{MIN}$ Ant System (MMAS) [13] to tackle the 2D-PMP. MMAS is a well known variant of Ant Colony Optimization (ACO) that strongly favours so-far best solutions. It is characterized by four features: First, only the best ant may update pheromone trails. Second, pheromone values have strict upper and lower bounds $\tau_{\min}$ and $\tau_{\max}$, respectively. Third, all pheromone values are initiated with $\tau_{\max}$ and the evaporation rate ρ is kept low. Fourth, the algorithm performs restarts after finding no improvement for a certain number of iterations, to tackle stagnations.

For solving the 2D-PMP with MMAS, we consider the problem a path-construction problem, where *nodes* represent container bay states, and *edges* represent container movements. Thus, we search for a (shortest) path from the initial node to a node that is a valid final state.

We outline the main steps of the MMAS in Algorithm 2. In each iteration the algorithm constructs n ant solutions (line 3) that are subsequently improved using a simple local search procedure that is discussed in more detail in Sect. 5.3. The best constructed solution σ is selected in line 11 and the global best solution σ^* possibly updated in line 13. In case the number of consecutive iterations without improvement exceeds the threshold $i_{\max}$ (line 17), the pheromone values are reinitialized (line 18). Finally the pheromone values are updated in lines 22 to 26. The decision whether to use the iteration best or global best solution is based on the probability $p_{\sigma*}$. These steps are repeated until either a maximum number of iterations or a time limit has been reached.

In the following sections we give a more detailed description of the different components of our approach. Section 5.1 outlines the pheromone model, Sect. 5.2 discusses the ant construction algorithm and Sect. 5.3 the local search procedure.

Algorithm 2. Max-Min Ant System for 2D-PMP

Input: $\mathcal{B}$: initial state; n: ant count; $i_{\max}$: maximum number of consecutive iterations without improvement; $p_{\sigma*}$: probability for using the global best solution for
pheromone update; f: evaluation function
1: $i \Leftarrow 0$
2: **while** stopping criteria not satisfied **do**
3:     $\{\sigma_1 \ldots, \sigma_n\} \Leftarrow$ construct set of n ant solutions from $\mathcal{B}$
4:     **if** $\{\sigma_1 \ldots, \sigma_n\} = \emptyset$ **then**
5:         continue and register failed attempt
6:     **end if**
7:     $\{\sigma_1 \ldots, \sigma_n\} \Leftarrow$ apply local search to all solutions in $\{\sigma_1, \ldots, \sigma_n\}$
8:     **if** σ^* not assigned **then**
9:         $\sigma^* \Leftarrow$ set σ_1 as new global best solution
10:     **end if**
11:     $\sigma \Leftarrow$ find best solution in $\{\sigma_1 \ldots, \sigma_n\}$
12:     **if** $f(\sigma) < f(\sigma^*)$ **then**
13:         $\sigma^* \Leftarrow$ set σ as new global best solution
14:         $i \Leftarrow 0$
15:     **else**
16:         $i \Leftarrow i + 1$
17:         **if** $i < i_{\max}$ **then**
18:             reinitialize pheromone values
19:             $i \Leftarrow 0$
20:         **end if**
21:     **end if**
22:     **if** random value $\in [0, 1) < p_{\sigma*}$ **then**
23:         update pheromones with global best solution σ^*
24:     **else**
25:         update pheromones with iteration best solution σ
26:     **end if**
27: **end while**
28: **return** σ^*

5.1   Pheromone Model

We studied different pheromone models in the design of the MMAS. First, we
considered a *state-based* pheromone model where each container bay state $\mathcal{B}$ is
associated with a pheromone value $\tau_{\mathcal{B}}$. However, since a state can be reached by
different (compound) moves that are not taken into account by the pheromone
model, this model gives little guidance to the agents/ants. We assume that
this is the reason why the state-based model performed poorly in our initial
experiments.

We thus extended that model to the *move-based* pheromone model. In it, we
associate a pheromone value $\tau_{(\mathcal{B},m)}$ to each state-move pair $(\mathcal{B}, m)$. This way
the pheromone values direct the ants in a more traditional way into following
promising (compound) moves, given the current container bay state. A crucial aspect hereby was to dynamically create pheromone values on the fly and

---

Algorithm 3. Ant construction algorithm

Input: $\mathcal{B}$: initial state, f: evaluation function, h: heuristic function
 1: $f_e \Leftarrow f(\mathcal{B})$
 2: $\sigma \Leftarrow$ empty list
 3: **while** $f_e > 0 \wedge$ stopping criterion **do**
 4:     $\mathcal{M} \Leftarrow$ query h for set of all possible (compound) moves in state $\mathcal{B}$
 5:     **if** $\mathcal{M}$ contains a (compound) move m leading to a final state **then**
 6:         append m to σ
 7:         **return** σ
 8:     **end if**
 9:     $P \Leftarrow$ calculate probabilities for all $m \in \mathcal{M}$
10:     $p_{sum} \Leftarrow 0$
11:     $r \Leftarrow$ random number between 0 and 1
12:     **for each** $m \subset \mathcal{M}$ **do**
13:         $p_{sum} \Leftarrow p_{sum} + P_m$
14:         $m' \Leftarrow m$
15:         **if** $p_{sum} \geq r$ **then**
16:             break
17:         **end if**
18:     **end for**
19:     append m' to σ
20:     apply (compound) move m' to $\mathcal{B}$ yielding new state $\mathcal{B}$
21:     $f_e \Leftarrow f(\mathcal{B})$
22: **end while**
23: **return** σ

---

efficiently maintain them in a hash table, as obviously we cannot store them all in a simple statically allocated matrix due to the state-space's exponential size.

5.2   Ant Construction Algorithm

The ant construction algorithm is given in Algorithm 3. It takes three arguments: an initial state, an evaluation function and a heuristic function. In each iteration, it first determines all possible (compound) moves for the current state $\mathcal{B}$ (line 4) by querying the given heuristic algorithm h (e.g. 2D-LPFH). The heuristic algorithm returns a set of all possible (compound) moves $\mathcal{M}$ it can execute for the given state $\mathcal{B}$; i.e. the heuristic algorithm explores the whole search space for one step (one (compound) move) from the current state and returns all possibilities, allowing the ant construction heuristic to choose the next (compound) move. Then, for each possible (compound) move m, the probability $p_{\mathcal{B},m}$ of the ant applying (compound) move m to $\mathcal{B}$ is computed (line 9) by

$$p_{\mathcal{B},m} = \frac{[\tau_{\mathcal{B},m}]^\alpha [\eta_{\mathcal{B},m}]^\beta}{\sum_{l \in \mathcal{M}_\mathcal{B}} [\tau_{\mathcal{B},l}]^\alpha [\eta_{\mathcal{B},l}]^\beta}, \text{if } m \in \mathcal{M}_\mathcal{B}, \tag{2}$$

where $\tau_{\mathcal{B},m}$ refers to the pheromone value of (compound) move m from state $\mathcal{B}$, α refers to the priority given to pheromone values, $\eta_{\mathcal{B},m}$ refers to the heuristic value of (compound) move m for state $\mathcal{B}$, and β refers to the priority given to heuristic values. The heuristic value $\eta_{\mathcal{B},m}$ is acquired by evaluating the state $\mathcal{B}_m$ that is reached by applying (compound) move m to the current state $\mathcal{B}$. We then select one of the (compound) moves according to their probability and a random factor (line 11-19). This procedure is repeated until a complete solution has been constructed or a maximum number of moves has been reached (line 3).

In addition, we check if the final state can be directly reached by a (compound) move and in this case immediately select it without calculating probabilities (line 5). All ant solutions can be constructed in parallel, since the only shared resource are the read-only pheromone values.

5.3 Improvement Heuristic

We use an improvement heuristic that, given a solution σ, tries to find "shortcuts" in the solution. More specifically, the algorithm tries to detect non-consecutive states $\mathcal{B}_1$ and $\mathcal{B}_2$ that are connected with moves $\{m_1, \ldots, m_k\} \in \sigma, k \geq 2$, which can be connected by a single move m'. In this case, the sequence $\{m_1, \ldots, m_k\}, k \geq 2$ can be replaced by m' in the solution and thus shortens the solution σ. Figure 3 illustrates such a shortcut between states $\mathcal{B}_2$ and $\mathcal{B}_5$, eliminating moves m_2, m_3 and m_4 and thus shortening the solution by two moves.

The heuristic works as follows. For each move $m \in \sigma$ yielding bay state $\mathcal{B}_m$, two sets are computed: the set of states that are reachable (within one move) from $\mathcal{B}_m$, $\mathcal{M}_\mathcal{B}$, and the set of states that are visited in the solution after the successor of $\mathcal{B}_m$, denoted by $\mathcal{M}_\sigma$. If $\mathcal{M}_\mathcal{B} \cap \mathcal{M}_\sigma \neq \emptyset$, then a shortcut has been found, since the state(s) in $\mathcal{M}_\mathcal{B} \cap \mathcal{M}_\sigma$ can be reached by one move from $\mathcal{B}_m$. Therefore, we chose the state $\mathcal{B}' \in \mathcal{M}_\mathcal{B} \cap \mathcal{M}_\sigma$ with the largest number of moves from $\mathcal{B}_m$ and replace the moves between $\mathcal{B}_m$ and $\mathcal{B}'$ with move m'.

6   Experimental Evaluation

We perform experiments with the LPFH, the Pilot method and the MMAS. The experiments are all executed on the same machine in a sequential manner and each experiment was run on the same set of instances.

Problem Instances. We obtained problem instances from the PMP instance generator provided in [7]. Our instances have $s = \{4, 6, 8, 10, 12, 14\}$ stacks with height $H = 4$ and occupancy rates $q = \{50\%, 75\%\}$ (i.e., fill level of the container bay). This yields 11 instance categories, because we leave out instances

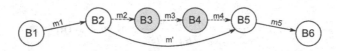

Fig. 3. Finding a shortcut move m' from $\mathcal{B}_2$ to $\mathcal{B}_5$ saving two moves.

with $s = 4$ and $q = 75\%$ because they are too difficult to solve. Containers have unique priorities and are randomly positioned within the given bay. The instances are available online[2].

Algorithm Parameter Settings. The 2D-LPFH uses $\lambda_{2,3} = 2$ (see [7]), after having applied parameter tuning tests with values $\lambda_2, \lambda_3 \in \{1, 2, 3, 5, 10\}$: restricting the search space ($\lambda_{2,3} = 1$) does not allow the algorithm enough flexibility to find good solutions, but loosening the search space too much ($\lambda_{2,3} \in \{5, 10\}$) does not yield good results within a short time. The semi-greedy approach ($\lambda_{2,3} = 2$) performs best. However, if given more time (5 min), the $\lambda_{2,3} \in \{5, 10\}$ results improved significantly, but still remained inferior to the semi-greedy approach. The Pilot method experiments were run with 2D-LPFH as the subheuristic and $k \in \{2, 3, 4, 5, 6, 7\}$ construction steps for each heuristic. Test have shown that $k \in \{5, 6, 7\}$ always performs significantly better than $k \in \{2, 3, 4\}$ with only slight differences among $k \in \{5, 6, 7\}$. After careful review of all results, $k = 7$ is chosen as the representing result since it yielded slightly better results than $k = 5$, but did not affect run time significantly. Finally, the MMAS uses 2D-LPFH to calculate the heuristic values during construction and the following parameters: $n = 8$, $\alpha = 1$, $\beta = 2$, $\rho = 0.02$, $i_{max} = 75$ and $p_{\sigma*} = 0.1$. All experiments use the same evaluation function stated in Eq. (1), and after each algorithm has finished, the improvement heuristic (see Sect. 5.3) is applied. All of the mentioned parameter values were determined in preliminary tests or are based on recommended values from [7] and [14]. For more details on the parameter selection, see [1].

Experimental Setup. The experiments are carried out on a machine with four Intel Xeon E5645 processors, each with six cores at 2.40 GHz, along with 200 GB of RAM. The underlying system is Ubuntu 13.04. with Java 1.7. Our implementation of MMAS runs the ant construction algorithms in parallel.

All experiments have a time limit of 5 min (wall clock time, the machine not otherwise utilized) or 500 moves. The ant construction algorithm has a different move limit of 250 moves since we do not want bad solutions appearing in the pheromone trails. As you will notice in the results section, 500 moves is a very generous move limit since we expect solutions with less than 100 moves for the biggest instances. All our experiments are of stochastic nature so we repeat them until the final solution is not improved a consecutive number of runs. The 2D-LPFH, Pilot method and MMAS had a no-improvement iteration number of 25, 25 and 5, respectively.

We perform experiments with the 2D-LPFH, Pilot method and MMAS. Due to space limitations we only present each approach with its best setup:

1. **2D-LPFH:** a single-run of the 2D-LPFH with $\lambda_{2,3} = 2$
2. **Pilot:** the Pilot method with 2D-LPFH as construction-heuristic and $k = 7$ lookahead moves

---

[2] http://www.ads.tuwien.ac.at/w/Research/Problem_Instances.

Table 1. Objective values over 50 instances for each category, where "s" denotes the number of stacks and q the occupancy rate; all stacks have maximum height 4. "avg" denotes the average objective value and "std" the standard deviation over all solved instances in the category. "sol" is the number of instances solved per category.

s	q	2D-LPFH			Pilot			2D-LPFH 5 min			MMAS		
		avg	std	sol	avg	std	sol	avg	std	sol	avg	std	sol
4	50%	5.92	±1.861	50	6.440	±2.168	50	5.86	±1.784	50	**5.68**	±1.634	50
6	50%	11.20	±2.955	50	11.04	±2.523	50	10.54	±2.367	50	**10.36**	±2.371	50
8	50%	16.88	±2.512	50	16.94	±2.645	50	15.18	±2.067	50	**14.86**	±2.232	50
10	50%	23.30	±3.046	50	22.94	±3.040	50	20.38	±2.221	50	**20.08**	±2.311	50
12	50%	29.46	±3.309	50	29.20	±3.676	50	**25.36**	±2.562	50	25.54	±2.667	50
14	50%	36.24	±2.911	50	37.26	±3.573	50	**31.74**	±2.465	50	**31.74**	±2.732	50
6	75%	29.18	±4.839	28	27.28	±5.163	32	24.76	±4.513	34	**24.65**	±4.478	34
8	75%	43.19	±5.497	36	39.21	±5.292	38	34.17	±3.946	46	**32.36**	±3.275	42
10	75%	59.81	±5.849	37	53.79	±5.910	43	47.47	±5.358	45	**44.57**	±4.217	44
12	75%	72.18	±8.314	34	68.23	±8.026	43	58.77	±5.704	47	**55.93**	±4.763	46
14	75%	85.10	±8.837	31	79.16	±6.109	38	71.07	±4.729	42	**68.57**	±5.735	42

3. **2D-LPFH 5-min:** the 2D-LPFH run sequentially for the same time as the MMAS, returning the best found result; same configuration as the 2D-LPFH
4. **MMAS:** the Max-Min Ant System with $n = 8, \alpha = 1, \beta = 2$ and the 2D-LPFH as the heuristic function.

6.1   Results

Table 1 shows the detailed results for all four approaches: the average objective value (number of moves) for each instance category, the standard deviation, and the number of solved instances. We first see that the MMAS mostly provides the best results, closely followed by the 2D-LPFH with a 5-minute runtime. The third best results come from the Pilot method, followed by the single-run 2D-LPFH. This is clearly illustrated in Fig. 4, where the average objective is shown for each solving approach for both occupancy rates, clearly demonstrating that the MMAS provides the best results.

Furthermore, we observe in Table 1 that all approaches managed to solve all $q = 50\%$ instances. For the $q = 75\%$ categories, we notice that the MMAS and extended run time 2D-LPFH solved almost the same number of instances with the extended run time 2D-LPFH in a slight lead. They are followed by the Pilot method and 2D-LPFH.

Note that the Pilot method, as well as the single-run 2D-LPFH only take several milliseconds, while the MMAS and 2D-LPFH take 5 min to run, so we do expect the latter approaches to provide better results. Using only the average

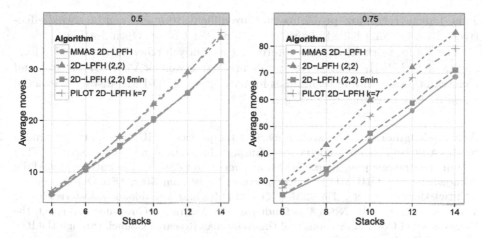

Fig. 4. Average number of moves per category for best algorithm configurations split by occupancy rate. Values for $q = 0.50$ are shown on the left and $q = 0.75$ on the right.

number of moves and number of solved instances, the MMAS and extended run time 2D-LPFH are the overall best performing test cases. We confirmed this conclusion using the Wilcoxon Paired Rank Sum test.

Due to space constraints only the most interesting results are included in this section. For further experimental results and analysis please refer to [1].

7   Conclusions

In this work, we presented a novel challenging problem: the two-dimensional pre-marshalling problem (2D-PMP). This problem is an extension of the well studied NP-hard pre-marshalling problem (PMP). The 2D-PMP occurs in small to medium-sized container terminals and is characterized by complex side constraints imposed by the reach stackers used for moving containers to ships. These side constraints make it difficult to apply existing approaches for the PMP to the 2D-PMP.

We first extended an existing PMP construction heuristic, the LPFH, to consider the additional constraints and came up with 2D-LPFH. For obtaining better solutions, we further integrated it in two metaheuristics: a Pilot method and a novel Max-Min Ant System (MMAS) approach. The MMAS approach yielded in our tests mostly the best solutions. Also, we observed that by using 2D-LPFH as a heuristic, the MMAS and Pilot method are able to solve instances that the 2D-LPFH is not capable of solving itself.

Applying a MMAS approach to this kind of problem is rather unconventional. In fact, for the classical PMP, no ACO-based approaches have been published so far. However, for the 2D-PMP that incorporates complex side constraints that are cumbersome to express, MMAS could be shown to be an amenable solving approach, since the specialized pheromone model can guide the ants effectively.

This demonstrates how problems that are difficult to model can be solved effectively by a learning-based algorithm such as Ant Colony Optimisation.

For future work it appears interesting to study further variants and refinements of 2D-LPFH,' alternative metaheuristics, as well as A* or IDA* search and constraint programming techniques for solving small 2D-PMP instances exactly. Possibly look into adapting work from [11].

Acknowledgments. This work is part of the project TRIUMPH, partially funded by the Austrian Federal Ministry for Transport, Innovation and Technology (BMVIT) within the strategic programme I2VSplus under grant 831736. The authors thankfully acknowledge the TRIUMPH project partners Logistikum Steyr (FH OÖ Forschungs &Entwicklungs GmbH), Ennshafen OÖ GmbH, and via donau – Österreichische Wasserstrassen-GmbH. NICTA is funded by the Australian Government through the Department of Communications and the Australian Research Council through the ICT Centre of Excellence Program.

References

1. Tus, A.: Heuristic solution approaches for the two dimensional pre-marshalling problem. Master's thesis, Vienna University of Technology, Vienna, Austria (2014). https://www.ads.tuwien.ac.at/publications/bib/pdf/tus_14.pdf
2. Caserta, M., Schwarze, S., Voß, S.: Container rehandling at maritime container terminals. In: Böse, J.W. (ed.) Handbook of Terminal Planning. Operations Research/Computer Science Interfaces Series, vol. 49, pp. 247–269. Springer, New York (2011)
3. Carlo, H.J., Vis, I.F., Roodbergen, K.J.: Storage yard operations in container terminals: literature overview, trends, and research directions. Eur. J. Oper. Res. **235**(2), 412–430 (2014). Maritime Logistics
4. Lee, Y., Hsu, N.Y.: An optimization model for the container pre-marshalling problem. Compu. Oper. Res. **34**(11), 3295–3313 (2007)
5. Caserta, M., Voß, S.: A corridor method-based algorithm for the pre-marshalling problem. In: Giacobini, M., Brabazon, A., Cagnoni, S., Di Caro, G.A., Ekárt, A., Esparcia-Alcázar, A.I., Farooq, M., Fink, A., Machado, P. (eds.) EvoWorkshops 2009. LNCS, vol. 5484, pp. 788–797. Springer, Heidelberg (2009)
6. Lee, Y., Chao, S.L.: A neighborhood search heuristic for pre-marshalling export containers. Eur. J. Oper. Res. **196**(2), 468–475 (2009)
7. Expósito-Izquierdo, C., Melián-Batista, B., Moreno-Vega, M.: Pre-marshalling problem: heuristic solution method and instances generator. Expert Syst. Appl. **39**(9), 8337–8349 (2012)
8. Bortfeldt, A., Forster, F.: A tree search procedure for the container pre-marshalling problem. Eur. J. Oper. Res. **217**(3), 531–540 (2012)
9. Prandtstetter, M.: A dynamic programming based branch-and-bound algorithm for the container pre-marshalling problem. Technical report, AIT Austrian Institute of Technology (2013) submitted to European Journal of Operational Research. http://matthias.prandtstetter.at/papers/pre2013.pdf
10. Rendl, A., Prandtstetter, M.: Constraint models for the container pre-marshalling problem. In: Proceedings of the Twelfth International Workshop on Constraint Modelling and Reformulation. ModRef 2013, PP. 44–56 (2013)

11. Tierney, K., Pacino, D., Voß, S.: Solving the pre-marshalling problem to optimality with a* and ida*. In: Conference of the International Federation of Operational Research Societies (2014)
12. Duin, C., Voß, S.: The pilot method: a strategy for heuristic repetition with application to the steiner problem in graphs. Networks **34**(3), 181–191 (1999)
13. Stützle, T., Hoos, H.H.: Max-min ant system. Future Gener. Comput. Syst. **16**(9), 889–914 (2000)
14. Dorigo, M., Stützle, T.: Ant Colony Optimization. Bradford Company, Scituate (2004)

Improving the State of the Art in Inexact TSP Solving Using Per-Instance Algorithm Selection

Lars Kotthoff[1]([✉]), Pascal Kerschke[2], Holger H. Hoos[3], and Heike Trautmann[2]

[1] Insight Centre for Data Analytics, Cork, Ireland
lars.kotthoff@insight-centre.org
[2] University of Münster, Münster, Germany
{kerschke,trautmann}@uni-muenster.de
[3] University of British Columbia, Vancouver, Canada
hoos@cs.ubc.ca

Abstract. We investigate per-instance algorithm selection techniques for solving the Travelling Salesman Problem (TSP), based on the two state-of-the-art inexact TSP solvers, LKH and EAX. Our comprehensive experiments demonstrate that the solvers exhibit complementary performance across a diverse set of instances, and the potential for improving the state of the art by selecting between them is significant. Using TSP features from the literature as well as a set of novel features, we show that we can capitalise on this potential by building an efficient selector that achieves significant performance improvements in practice. Our selectors represent a significant improvement in the state-of-the-art in inexact TSP solving, and hence in the ability to find optimal solutions (without proof of optimality) for challenging TSP instances in practice.

1 Introduction

The travelling salesman problem (TSP) is arguably the most prominent NP-hard combinatorial optimisation problem. Given a set of n locations – which, by convention, are called *cities* – and pairwise distances between those cities, the objective in the TSP is to find the shortest round-trip or *tour* through all cities, i.e., a sequence in which every city is visited exactly once, except for the last city, which is the same as the first, and the sum of the distances between successively visited cities along the tour is minimal. Here, we consider the *2D Euclidean TSP* in which the cities correspond to points in the Euclidean plane and the distances between them are simply the Euclidean distances between those points. This is the most commonly studied special case of the TSP, and, like the general TSP, it is known to be NP-hard. The Euclidean TSP has important applications (e.g., in the fabrication of printed circuit boards) and also arises in the context of various transportation and logistics applications.

There are two types of TSP algorithms: exact algorithms, which are guaranteed to find an optimal solution to any TSP instance and, when run to completion, produce a proof of optimality; and inexact algorithms, which cannot guarantee or prove the optimality of the solutions found. Intriguingly, the state

C. Dhaenens et al. (Eds.): LION 9 2015, LNCS 8994, pp. 202–217, 2015.
DOI: 10.1007/978-3-319-19084-6_18

of the art for both types of algorithms has been defined by a single solver each for many years: the exact solver Concorde [1] and the inexact solver LKH [4]. Furthermore, LKH typically finds high-quality and even optimal solutions much more quickly than Concorde, and therefore, for the purpose of finding such solutions, per-instance algorithm selection techniques (see, e.g., [8]) were inapplicable to the TSP.

Recently, however, an improvement in the state of the art in inexact TSP solving in the form of a new evolutionary algorithm, EAX, has been reported [13], and from the performance comparison against LKH, it appeared possible that per-instance selection between those two solvers might yield further improvements.

In this work, we pursue this possibility and show, for the first time, that per-instance algorithm selection techniques can be used to improve the state of the art in inexact TSP solving. After providing some preliminary information about the TSP solvers, benchmark instances and algorithm selection techniques we use in our study in Sect. 2, we report performance results for LKH and EAX that clearly indicate the potential benefit of per-instance algorithm selection (Sect. 3). Next, we report the performance that can be obtained from actual algorithm selectors, using broad sets of instance features from the literature [6,12,15,19] (Sect. 4). Finally, we demonstrate how an effective selector can be constructed based on a small number of efficiently computable probing features extracted from the initial phase of EAX runs (Sect. 5), before concluding with some general observations and directions for future work.

2   Background and Experimental Setup

TSP Solvers. We consider two state-of-the art inexact TSP solvers in this work: LKH [4] and EAX [13].

LKH is a stochastic local search algorithm based on the Lin-Kernighan procedure. It uses an improved variant of the Lin-Kernighan algorithm, based on 5-exchange moves in combination with a construction procedure loosely related to the nearest neighbour heuristic. LKH has defined the state of the art in inexact TSP solving since it was first introduced in 2000.

Besides the reference implementation of LKH, we used a modification of version 1.3, developed in the context of a study of LKH's scaling behaviour [9].[1] This modification adds a simple dynamic restart mechanism to the original LKH algorithm, based on the observation that the performance of the former suffered frequently from stagnation of the underlying stochastic search process. We dub this variant LKH+restart.

EAX is a recently introduced evolutionary algorithm for inexact TSP solving. Its key ingredient is a new edge assembly crossover procedure, which obtains

---

[1] A similar modification can in principle be applied to the current version 2.0.3 of LKH, but as we will see, the performance of version 1.3, for which the modification was made available to us, is sufficient to obtain better performance than EAX in many cases.

high-quality tours by combining edges from two parent tours with a small number of new, short edges. EAX uses 2-opt local search to determine the initial population, as well as a specific tabu search procedure for generating offspring from very high-quality parent solutions. Furthermore, an entropy-based mechanism is used to preserve diversity in the population of candidate solutions. A rather complex combination of termination criteria is used to determine when a run of EAX is ended, at which point the best tour encountered during the run is returned. Nagata and Kobayashi [13] provide empirical evidence that EAX often, but not always, outperforms LKH on several sets of commonly studied Euclidean TSP instances in terms of the solution qualities reached within similar or shorter running times.

We modified the official implementation of EAX to permit setting the random seed (which had previously been fixed to one value) and to terminate when a given solution quality or bound in running time is reached (or exceeded). These modifications were necessary to facilitate our comparative performance analysis and did not compromise performance. During initial experiments, we noticed that EAX often terminates prematurely. We therefore created two variants, which we studied in the following. The first, simply dubbed EAX, disables the original termination criterion and ends a run *only* when a given solution quality or bound in running time is reached (or exceeded). We verified that single runs of this variant performed no worse than the original version of EAX. Our second variant uses the original termination criterion to trigger a restart, by initialising another run; this is done until a given solution quality or bound in running time is reached (or exceeded). We dub this variant EAX+restart.

Benchmark Instances. Consistent with other work in this area, we use four types of benchmark instances.

Random uniform Euclidean (RUE) instances are obtained by placing n points uniformly at random in a square, with integer coordinates between 1 and 1 000 000; each point corresponds to a city to be visited. Distances between these cities are defined as Euclidean distances between the respective points, rounded to the nearest integer. We generated instances with 1 000, 1 500, and 2 000 cities, 1 000 each. After filtering the instances that no solver could solve within 1 CPU hour on our reference machine and instances for which features could not be computed because the computation ran out of memory, we were left with 999 instances with 1 000 cities, 1 000 with 1 500 cities, and 998 with 2 000 cities. The RUE instances used in our experiments were generated using the `portgen` generator from the 8th DIMACS Implementation Challenge. Optimal solution qualities for all RUE instances were obtained using Concorde [1].

TSPLIB is a widely used collection of TSP instances with different characteristics, including instances from various applications of the TSP. In our experiments, we used 74 instances with edge types EUC 2D, CEIL 2D and ATT and sizes between 48 and 11 849. Again we excluded instances that no solver was able to solve within 1 CPU hour on our reference machine and instances for which we were unable to compute features.

Finally, we used two sets of instances from the TSP webpage at http://www.math.uwaterloo.ca/tsp/index.html. The *National* instances are based on the locations of cities within different countries, and we used 8 National instances with 734 to 9 882 cities. The *VLSI* instances stem from an application in VLSI circuit design, and we used 27 VLSI instances with 662 to 2 924 cities. These instances are known to be particularly hard for many TSP solvers, including Concorde and EAX.

We limited our study to instances for which the optimal solution is known, since we were interested in the ability of our solvers to find optimal solutions and in the time required for doing so. This is the most ambitious goal for any TSP solver, and even though inexact solvers, such as the ones we consider here, cannot prove optimality, they are typically able to find solutions whose optimality is later proven using other methods much more effectively than the best exact solvers.

Automated Algorithm Selection. The per-instance algorithm selection problem [16] involves selecting from a set of candidate algorithms the one expected to perform best on a given problem instance. It is relevant where algorithm portfolios [3,5] are employed – instead of tackling a set of problem instances with just a single solver, a set of them is used with the best being selected for each instance.

Algorithm selection systems build performance models of the algorithms or the portfolio they are contained in to forecast which algorithm to use in a particular context. Usually, these models are induced using machine learning. Using the model predictions, one or more algorithms from the portfolio are selected to be run sequentially or in parallel.

Here, we consider the case where exactly one algorithm is selected for solving the problem. One of the most prominent and successful systems that employs this approach is SATzilla [20], which defined the state of the art in SAT solving for a number of years. Since then, additional algorithm selection systems have been developed and proved their worth in the annual SAT competition (e.g. CSHC [10], which has also been applied to MaxSAT). Other successful application areas have been constraint solving [14], continuous black-box optimization [2,11], mixed integer programming [21], and AI planning [18].

The interested reader is referred to a recent survey [8] for additional information on algorithm selection.

Construction and Evaluation of Algorithm Selectors. In the following, we use the LLAMA algorithm selection toolkit [7], version 0.7.2, to build algorithm selectors for the TSP and consider a range of different approaches to algorithm selection used in the literature. We build models that treat algorithm selection as a classification problem and predict the algorithm to use. We furthermore build models that use regression to predict the performance of the individual algorithms in the portfolio separately and choose the algorithm with the best predicted performance. Finally, we consider models that, for each pair of algorithms, use regression to predict the performance difference between them. The solver with the largest performance improvement over all other algorithms is chosen.

In addition to a range of algorithm selection models, we also consider a range of different machine learning techniques. For classification, we use C4.5 decision trees (J48), random forests (RF), and recursive partitioning trees (RPART). For regression, we consider random forests (RF), support vector machines (KSVM), and multivariate adaptive regression spline (MARS) models. All machine learning models were used with their default parameters.

We generally consider the portfolio that contains all four solvers – LKH and EAX as well as their respective restart variants. From our original set of instances, we selected all that at least one of these solvers was able to find the optimal solution for within the specified cutoff time of one hour. We also filter instances for which we were unable to compute feature values because the computation ran out of memory or unsupported constructs in the input. This leaves us with a total of 3 106 instances.

We use 10-fold cross-validation to determine the performance of the algorithm selection models. The entire set of instances was randomly partitioned into 10 subsets of approximately equal size. Of the 10 subsets, 9 were combined to form the training set for the algorithm selection models, which were evaluated on the remaining subset. This process was repeated 10 times for all possible combinations of training and test sets. At the end of this process, each problem instance in the original set was used exactly once to evaluate the performance of the algorithm selection models.

Execution Environment and Performance Measurement. All experiments were run on 24-core 2.5 GHz Intel XEON machines with 64 GB of RAM running CentOS 6.4 64 Bit. We measured execution times using the `time` command and limited the CPU time of solvers with the `runsolver` tool [17] where necessary. We set the cutoff time to 3 600 CPU seconds. We ran each solver 10 times on an instance with different random seeds and took the median of the results.

The mean PAR10 score over all instances is 2 062.15 for LKH, 422.48 for LKH+restart, 11 462.98 for EAX, and 104.01 for EAX+restart. The PAR10 score is the penalized average runtime. That is, if the solver chosen for the respective instance was able to solve it within the cutoff time of one hour, the actual runtime is the score. Otherwise, we penalise the solver by multiplying the cutoff time by a factor of 10.

3   Potential for Portfolios

Figure 1 shows scatter plots of the CPU times on our benchmark sets of TSP instances for the four inexact TSP solvers we considered in our study. It is obvious that there is substantial potential for algorithm selection – the solvers show very different behaviour on different sets of instances. There are many instances with large performance differences; in particular, many instances are easily handled by one solver, while the other times out after an hour.

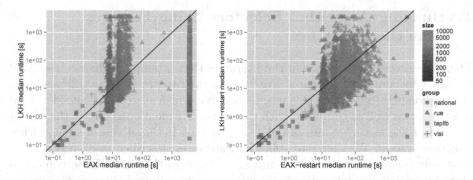

Fig. 1. Performance differences for EAX and LKH (left) and the respective restart variants (right). Each point represents a problem instance. The axes show the CPU time consumed by the respective solver as the median over 10 runs on a log scale. Both solvers exhibit the same performance for points on the diagonal line. The points at the top and right of the plots represent instances on which one of the solvers timed out, the instances in the top right corner could not be solved by either of the solvers.

The RUE instances (triangles), which comprise the vast majority of our instance set, are clustered in the centers of the plots – most of them can be solved by all solvers, and often there are only small performance differences. Still, there are a few instances that at least one of the solvers cannot solve within the time limit of one CPU hour. The TSPLib instances are more varied. While most of them are easily solvable within a few seconds by all solvers, a few are very hard for one solver, but easily solvable by another. The VLSI and National instances are in between very easy and very hard.

The left hand side of Fig. 1 shows that there is a large set of instances that EAX is unable to solve within the time limit. However, the right hand side, which compares the restart variants of the solvers, shows that EAX+restart is able to solve the vast majority of these instances within the time limit. This suggests that EAX+restart effectively improves over plain EAX; further analysis of the performance correlation between the two variants indicates potential for automated selection between those. Similar observations apply to LKH vs. LKH+restart.

While in general solving times tend to increase with instances size, the solver behaviour is not completely consistent with the size of the problem instances. For example, there is a large number of relatively small instances on which EAX times out after an hour. Similarly, there are small instances where LKH exhibits the same behaviour. This suggests, consistent with earlier work on performance modelling of TSP solver performance (e.g. [6]) that more information is required to forecast solver behaviour.

We note that, as can be clearly seen from Fig. 1 and from the performance of the single and virtual best solvers shown in Table 1, by simply running the algorithms we consider (and in particular: LKH+restart and EAX+restart) in parallel, an improvement can be achieved over the single best solver (EAX+restart), and hence over the current state of the art in incomplete TSP solving.

4 Building Algorithm Selectors Using Features from the Literature

There are several approaches in the literature that attempt to characterise TSP instances by computing features. We focus on the two presented in [12][2] and [6][3], as they comprise a large set of syntactic and dynamic features, and consider them in isolation as well as combined with each other. As mentioned above, the cost of computing the feature values can play a major part in the success of an algorithm selection system. We therefore split the feature set described in [6] further into relatively cheap features and the full set of features that in addition comprises more expensive characteristics and ones that are computed through probing.

We denote the feature set described in [12] **TSPmeta** and the one from [6] **UBC**. Based on these, we use the following four sets of features in our experiments.

UBC (cheap) The feature set from [6] without the more expensive features, in particular, the local search, branch and cut, and clustering distance features (13 features). The mean time of computing this set of features was 0.98 s per instance, with the median at 0.97 s (standard deviation 0.42).

UBC The full feature set from [6] (50 features). The mean time of computing this set of features was 20.71 s per instance, with the median at 16.47 s (standard deviation 46.36).

TSPmeta The full feature set from [12] (64 features). The mean time of computing this set of features was 33.61 s per instance, with the median at 28.51 s (standard deviation 39.47).

UBC ∪ TSPmeta The union of **UBC** and **TSPmeta** (114 features). Some of the features in the constituent sets contain the same information.

An additional set of features based on k-nearest neighbour analysis has been introduced very recently in [15]. These features will be included in future studies.

4.1 Results

The results we achieve with the feature sets described above are detailed in Table 1 (we report PAR10 scores over the union of our four benchmark sets).

We are able to improve upon running the single best solver (EAX+restart) only in two cases overall. All other selectors are (sometimes much) worse than simply choosing the single best solver statically. In particular, the classification-based models exhibit very bad performance. The regression-based models perform much better, in particular, the random forest and MARS models.

To what extent these results are caused by the cost of computing the features becomes clear when examining the results that ignore this cost, presented in Table 2. While the differences for the classification-based models are relatively

---

[2] http://cran.r-project.org/web/packages/tspmeta/index.html.
[3] http://www.cs.ubc.ca/labs/beta/Projects/EPMs/TSP_features_UBC2012.tar.gz.

Table 1. Summary of algorithm selector results using sets of features from the litera-ture. The numbers represent mean PAR10 scores, *including the cost of feature compu-tation*, over the entire set of instances and rounded to two digits. We show the scores for the virtual best and single best solver for comparison. The scores for the models that are better than the single best algorithm are shown in **bold face.**

		UBC (cheap)	UBC	TSPmeta	UBC ∪ TSPmeta
Virtual best		18.52			
Single best		104.01			
Classification	J48	3077.42	3725.07	3773.81	3542.36
	RF	2676.62	2176.89	2312.16	2252.69
	RPART	1931.51	1580.83	1628.55	1612.98
Regression	RF	119.96	126.40	151.20	158.14
	MARS	**95.88**	223.23	204.23	204.97
	KSVM	295.76	911.49	3906.04	2140.11
Regression pairs	RF	144.48	139.35	151.50	170.33
	MARS	**95.08**	138.87	208.21	205.86
	KSVM	345.48	850.06	1733.45	1948.49

small, there are major changes for the random forest and MARS regression models.

The cost of computing the probing features can be substantial; this can be seen, e.g., when comparing the performance of the random forest regression model with the TSPmeta feature set without costs (118.57) with the performance including the overhead (151.20). The average cost of computing this feature set is almost twice as large as the average PAR10 score of the virtual best solver.

Figure 2 (right) shows the performance of the best overall model, MARS regression on pairs of solvers trained using the UBC (cheap) feature set, com-pared to the single best solver. There is a large number of instances where the solver the selector chooses is better than the single best (points below the diag-onal); in particular, there are 3 instances where the single best solver times out (right margin of plot), while the selector chooses a solver that does not. There are, however, a significant number of instances where the choice made by the selector is incorrect, and EAX+restart exhibits better performance than the chosen solver. In particular, there are two instances that are easy for the single best solver, while the solver chosen by the selector times out (top margin of plot). Unsurprisingly, as can be seen when comparing the left and right plots in Fig. 2, the cost of feature computation mainly affects selector performance on easy instances. Detailed inspection of our results indicates that on struc-tured TSP instances, the selector tends to achieve more substantial performance improvements than on RUE instances.

Additionally, we performed forward feature selection, where we start with an empty set and repeatedly add the feature that gives most additional information,

Table 2. Summary of algorithm selector results using features from the literature. The numbers represent PAR10 scores over the entire set of instances *without taking the cost for feature computation* into account and rounded to two digits. We show the scores for the virtual best and single best solver for comparison. The scores for the models that are better than the single best algorithm are shown in **bold face**.

		UBC (cheap)	UBC	TSPmeta	UBC ∪ TSPmeta
Virtual best		18.52			
Single best		104.01			
Classification	J48	3076.55	3696.60	3734.90	3495.63
	RF	2675.73	2148.12	2271.91	2194.63
	RPART	1930.59	1551.95	1597.61	1553.29
Regression	RF	119.00	106.40	118.57	106.00
	MARS	**94.91**	192.36	171.74	152.90
	KSVM	294.80	892.43	3867.60	2069.93
Regression pairs	RF	143.52	119.86	118.87	118.19
	MARS	**94.10**	107.93	175.72	154.00
	KSVM	344.52	831.01	1702.98	1877.95

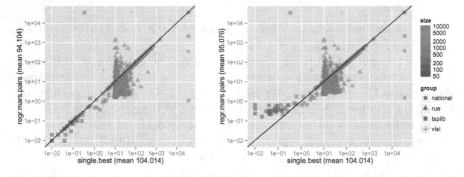

Fig. 2. Algorithm selector performance for the best model trained with features from the literature without and with taking feature costs into account (left and right plot, respectively) – in both cases, the best model was MARS regression on pairs of solvers with the UBC (cheap) feature set. The *x*-axis shows the log PAR10 score of the single best solver, the *y*-axis the log PAR10 score of the selector. Each point represents a TSP instance. Points on the diagonal indicate that the selector chose the single best solver, below the diagonal that the selector chose a better solver than the single best.

based on entropy and correlation on the full feature set UBC ∪ TSPmeta to determine the features that are most important for determining the solver to run. No cost-sensitive feature selection strategy was applied (we plan to improve on this approach in future work). The resulting set included eight features from [6] (the mean and standard deviation cluster distances, the average tour cost from the construction heuristic, the skew of the probability of edges in local minima, the time required for the local search probing feature computation, the maximum depth, the median and standard deviation of the distances of the minimum spanning tree) and one from [12] (the fraction of nodes on the convex hull).

We also performed feature selection on the features used by the best overall model, MARS on pairs of solvers with the UBC (cheap) feature set. Just a single feature was chosen, the average length of the minimum spanning tree.

While feature selection was able to improve the performance slightly in some cases, selectors trained on the reduced feature set showed worse performance in other cases. There is significant overlap in the type of features computed in the UBC and TSPmeta feature sets, which may explain the inconsistent results we achieved with feature selection. All results reported in this paper are without feature selection, as feature selection does not significantly and consistently improve the results and increases the conceptual complexity of selector construction.

5 Building Algorithm Selectors Using EAX Probing Features

In the previous sections, we have shown that there are significant complementarities in performance between the four solvers we consider and therefore significant potential for algorithm selection to improve the current state of the art in TSP solving. Using features described in the literature, we can already achieve a significant performance improvement over the single best solver on our set of instances. In this section, we investigate whether we can improve on this by using a different, novel set of features.

As explained earlier, there is a trade-off between the cost of computing the features characterising a TSP instance and the information obtained through them. In particular, computing the features that cannot be determined directly from the description of the instance itself is expensive, but does help learn better algorithm selection models.

In this section, we propose a new set of features that allows us to investigate the trade-off of cost of feature computation vs. information in a much more fine-grained and principled manner. We harness one of the solvers from our portfolio and analyse its progress when run for a small amount of time. We can control the amount of time directly – the longer the solver is run, the more information we get, but the more expensive the feature computation becomes. This information is then used to derive novel features.

Our single best solver, EAX+restart, provides the user with a trace of its execution as it progresses through the different generations. For each generation, the evolutionary algorithm outputs the best and average tour length found over the individuals of the current population. This gives an indication of how the solver progresses. By comparing the tour lengths of successive generations to the initial one, we get information on how quickly the solver is able to improve on initial solutions.

We consider the information obtained during the first n generations. Each best and average tour length is normalised by the best and average tour lengths of the initial population to obtain the improvement over these. We compute the minimum, maximum, mean, and median of both best and average improvements over the n generations. As the solving trajectory varies between different executions, we compute the median values of these numbers over m runs of EAX+restart with different random seeds.

This feature computation can be seen as a pre-solving step, during which we are running the actual algorithm used to find a solution. If the solver finds the solution during the first n generations, no further work needs to be done. Presolving is an effective means of quickly solving easy instances without incurring the overhead of feature computation costs. It is used with great success in the SATzilla system [20] for example.

5.1   Determining the Number of Generations and Probing Runs

We first investigated the impact of the parameters n and m on selector performance. The results of these preliminary experiments were somewhat inconclusive, but led us to choose $n = 10$ generations and $m = 1$ algorithm run for computing our probing features. This keeps the cost of feature computation low, while still providing us with valuable information that can be used effectively to decide which solver to use.

The results vary not only with n and m, but also between different probing runs. As our probing algorithm is stochastic, we obtain different feature values for different random seeds. The resulting performance differences can be quite high, especially for easy instances that are solved almost instantaneously if the solver starts its search process with a good set of initial tours. This means that not only the computed feature values, but also the cost of feature computation is different for different runs. This introduces additional stochasticity and noise into our evaluation.

We therefore average feature costs and values over 10 independent algorithm runs with different random seeds, and the results reported below are averages over those runs. In each of these runs, we extract the features as described above and build and evaluate the models. Averaging the results in this manner makes our conclusions statistically more robust.

The mean cost of computing this set of features (mean over all instances that are not solved during feature computation, and median over 10 independent runs per instance) is 2.81 s, and the mean number of presolved instances over all independent runs is 26.60, all from TSPLIB.

5.2   Results

In the subsequent evaluation, we focus on the approaches that we have identified as the most promising in the previous experiments, namely random forest and MARS models for regression and regression on pairs of algorithms. Table 3 shows the results we were able to achieve with selectors using only our new features. The overall best model is random forest regression and achieves better performance than the single best solver on average.

We note that the performance of the virtual best solver is very slightly worse than that observed in our experiments from Sect. 4, although the difference is less than the two significant digits we round to. This is because for the instances that are solved during feature computation, we take the runtime of the solver used to compute those features, even though a different solver may be faster.

The selector performance obtained using our new models is worse than the single best solver when taking into account the full cost of feature computation; however, because of the nature of our new probing features, this is not necessary: If the solver used for the feature computation is chosen as the solver to be run on the given TSP instance, the features are obtained at no additional cost, by simply continuing the probing run. The performance results for this 'accelerated' feature computation are shown in the third column of Table 3.

On average over all probing runs with different random seeds, the selectors trained using the new features perform worse than the selectors trained using features from the literature. However, there are clear indications for potential to obtain much better performance. In Table 3, we report, in parentheses, the first quartiles of the distributions of mean PAR10 scores over the 10 independent runs per TSP instance. According to these results, the MARS models for pairs of solvers, using our new probing features, can yield better performance than any of the models we have studied previously, using instance features from the literature.

The performance variation between the 10 independent runs underlying the results in Table 3 is quite high, considering the relatively small difference in performance to the single best algorithm; for our accelerated random forest models and our accelerated MARS models for pairs of solvers, we observe standard deviations of 21.77 and 28.09, respectively. The best performance achieved over the 10 independent runs is up to $\approx$30 % better than that of the single best solver. While these results indicate the potential inherent in our new probing features, statistically robust ways to exploit this potential will be investigated in future work.

Figure 3 illustrates the performance of our new algorithm selectors, based on EAX probing features, in more detail. In contrast to the situation when using features from the literature, illustrated in Fig. 2, we are now able to match the performance of the single best solver for the vast majority of easy instances. This is in part due to the fact that the very easy instances are now solved during feature computation. Furthermore, there are no more cases where our selector chooses a solver that times out while the single best solver does not. On the contrary, there are three instances where the single best solver times out, but our selector chooses a solver that does not. This fact further illustrates the potential of our new probing features, which enable us to make better predictions, especially in extreme cases, where incorrect decisions are particularly detrimental.

When comparing the left- and right-hand plots in Fig. 3, we see the impact of the feature computation costs. There is no difference in the top and right-hand parts of the plots, as the instances in these areas take longer to solve, and the time for feature computation is insignificant. In the centre part, however, a small

Table 3. Summary of algorithm selector results using our new EAX probing features. The numbers represent the mean of the mean PAR10 scores over the entire set of instances (including the ones solved during feature computation) and 10 independent runs per instance, rounded to two digits. The numbers in parentheses represent the first quartiles over ten independent runs. The 'accelerated' column denotes the average PAR10 score where the cost of computing the features was added only if the chosen solver was different from the one used for computing those features. We show the scores for the virtual best and single best solver for comparison. The scores for the models that are better than the single best algorithm are shown in **bold face**.

		Without costs	With costs	Accelerated
Virtual best		18.52		
Single best		104.01		
Regression	RF	**103.42 (95.04)**	106.24 (**97.86**)	**103.83 (95.46)**
	MARS	126.20 (116.93)	129.02 (119.73)	126.53 (117.24)
Regression pairs	RF	128.74 (119.76)	131.56 (122.58)	129.28 (120.27)
	MARS	107.13 (**85.91**)	109.95 (**88.74**)	107.51 (**86.30**)

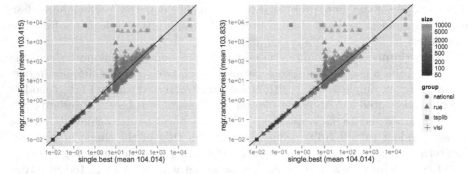

Fig. 3. Algorithm selector performance for the best model trained with the new EAX probing features, random forest regression, without (left) and with (right) feature cost, where in the latter case, the 'accelerated' feature computation method was used. The x- and y- axes show the log PAR10 scores of the single best solver and the selector, respectively. Each point represents one TSP instance. Points on the diagonal correspond to cases where the selector chooses the single best solver, and points below the diagonal to cases where the selector chooses a solver with even better performance for that instance.

shift of points towards the top of the plot can be observed – there is no shift to the right, as the single best solver can be determined statically and does not require features. Easy instances are not affected as much, as the solver used to compute the probing features is also chosen as the solver to continue solving.

Since our use of EAX probing features effectively combines feature computation with presolving, we see considerable benefits over other algorithm selection approaches for relatively easy instances. With further optimised feature compu-

Table 4. Summary of algorithm selector results using the combined set of all features from the literature and our own. The numbers represent the mean of the mean PAR10 scores over the entire set of instances (including the ones solved during feature computation) and all 10 random seeds rounded to two digits. The numbers in parentheses are the first quartiles over 10 independent runs. The 'accelerated' column denotes the average PAR10 score where the full cost of computing the features was added only if the chosen solver was different from the one used for computing those features. If the same solver was chosen, only the cost for the features not derived during the probing run was added. We show the scores for the virtual best and single best solver for comparison. The scores for the models that are better than the single best algorithm are shown in **bold face**.

		Without costs	With costs	Accelerated
Virtual best		18.52		
Single best		104.01		
Regression	RF	**103.89 (95.93)**	163.16 (161.73)	160.76 (159.32)
	MARS	216.89 (190.69)	277.84 (260.15)	275.73 (257.83)
Regression pairs	RF	125.73 (119.59)	180.63 (174.48)	178.29 (172.13)
	MARS	159.13 (145.45)	221.65 (210.95)	219.41 (208.64)

tation and presolving strategies, it should be possible for the selector to focus on improving performance on difficult instances and thus to obtain additional overall performance improvements.

5.3 Combining with Features from the Literature

As we have seen above, our new features have the potential to give rise to better selectors than those obtained by using only features from the literature. For our final set of experiments, we combined the feature sets from the literature with our new EAX probing features to assess whether this could result in even better selectors.

Table 4 shows the performance results for selectors using the combined set of features. Overall, when accounting for the cost of determining the features, performance is worse than for the individual sets in isolation. This is mostly caused by the high cost of feature computation. However, even when ignoring this cost, the selectors do not perform better than before. In particular, while the best selector achieved performance similar to the single best algorithm on average, it appears to be unable to capitalise on the additional information contained in the larger feature set. We believe that the redundant information contained in the set of all features has a detrimental effect on selector performance.

6 Conclusions

The Travelling Salesman Problem is one of the most iconic NP-hard optimisation problems. It has been extensively studied over the years, and many approaches for solving it have been developed. Until recently, a single solver, LKH, has defined the state of the art for inexact TSP solving. With the recent introduction of a new state-of-the-art inexact TSP algorithm, EAX, this picture has changed.

In this work, we have extensively studied the empirical performance of LKH, EAX, and improved variants of these base solvers on a large set of TSP instances ranging from trivial to hard. We have demonstrated the huge potential for algorithm selection in this context. We then successfully applied algorithm selection techniques to improve the state of the art in inexact TSP solving.

On the large set of instances we consider in this paper, we have computed features defined in the literature. We empirically investigated how informative these features are with respect to choosing the best solver for a specific instance. The initial results are very encouraging. Even with features that are relatively cheap to compute, we are able to build algorithm selection models that outperform the current state of the art – the single best solver over the entire set of instances, EAX+restart.

Motivated by this observation, we proposed a new set of features based on information gleaned from the execution trace of one of the solvers in our portfolio. Controlling the trade-off between the amount of information and the cost of computing it, we were able to show that the quality of the selector can improve significantly over selectors that use existing features. Our approach to feature computation combines the extraction of instance characteristics with presolving, which has the additional benefit that trivial instances are solved during this phase and the selector does not have to consider them.

In future work, we will further investigate our new EAX probing features, with the goal of obtaining additional, statistically robust performance improvements. We will also endeavour to add additional features and investigate the impact of cost-sensitive feature selection methods. Our data is available on the Algorithm Selection Benchmark Repository ASlib (http://aslib.net, beta datasets) as scenario TSP-LION2015.

Acknowledgements. We thank Thomas Stützle for letting us use the restart version of LKH 1.3 he implemented in the context of a different project and for helpful comments on earlier versions of this work. Holger Hoos acknowledges support from an NSERC Discovery Grant. Lars Kotthoff is supported by EU FP7 FET project 284715 (ICON) and an IRC "New Foundations" grant. Pascal Kerschke and Heike Trautmann acknowledge support from the European Center of Information Systems (ERCIS).

References

1. Applegate, D.L., Bixby, R.E., Chvatal, V., Cook, W.J.: The Traveling Salesman Problem: A Computational Study. Princeton University Press, Princeton (2007)
2. Bischl, B., Mersmann, O., Trautmann, H., Preuss, M.: Algorithm selection based on exploratory landscape analysis and cost-sensitive learning. In: Proceedings of the 14th Annual Conference on Genetic and Evolutionary Computation, GECCO 2012. ACM, New York (2012)
3. Gomes, C.P., Selman, B.: Algorithm portfolios. Artif. Intell. **126**(1–2), 43–62 (2001)

4. Helsgaun, K.: General k-opt submoves for the LinKernighan TSP heuristic. Math. Program. Comput. 1(2–3), 119–163 (2009)
5. Huberman, B.A., Lukose, R.M., Hogg, T.: An economics approach to hard computational problems. Science 275(5296), 51–54 (1997)
6. Hutter, F., Xu, L., Hoos, H.H., Leyton-Brown, K.: Algorithm runtime prediction: methods and evaluation. Artif. Intell. 206, 79–111 (2014)
7. Kotthoff, L.: LLAMA: leveraging learning to automatically manage algorithms. Technical report, June 2013. arXiv:1306.1031
8. Kotthoff, L.: Algorithm selection for combinatorial search problems: a survey. AI Mag. 35(3), 48–60 (2014)
9. Lacoste, J.D., Hoos, H.H., Stützle, T.: On the empirical time complexity of state-of-the-art inexact tsp solvers. (manuscript in preparation)
10. Malitsky, Y., Sabharwal, A., Samulowitz, H., Sellmann, M.: Algorithm portfolios based on cost-sensitive hierarchical clustering. In: IJCAI, August 2013
11. Mersmann, O., Bischl, B., Trautmann, H., Preuss, M., Weihs, C., Rudolph, G.: Exploratory landscape analysis. In: Proceedings of the 13th Annual Conference on Genetic and Vvolutionary Computation, GECCO 2011, pp. 829–836. ACM, New York (2011). http://doi.acm.org/10.1145/2001576.2001690
12. Mersmann, O., Bischl, B., Trautmann, H., Wagner, M., Bossek, J., Neumann, F.: A novel feature-based approach to characterize algorithm performance for the traveling salesperson problem. Ann. Math. Artif. Intell. 69(2), 151–182 (2013)
13. Nagata, Y., Kobayashi, S.: A powerful genetic algorithm using edge assembly crossover for the traveling salesman problem. INFORMS J. Comput. 25(2), 346–363 (2013)
14. O'Mahony, E., Hebrard, E., Holland, A., Nugent, C., O'Sullivan, B.: Using case-based reasoning in an algorithm portfolio for constraint solving. In: Proceedings of the 19th Irish Conference on Artificial Intelligence and Cognitive Science, January 2008
15. Pihera, J., Musliu, N.: Application of machine learning to algorithm selection for TSP. In: Fogel, D., et al. (eds.) Proceedings of the IEEE 26th International Conference on Tools with Artificial Intelligence (ICTAI). IEEE press (2014)
16. Rice, J.R.: The algorithm selection problem. Adv. Comput. 15, 65–118 (1976)
17. Roussel, O.: Controlling a solver execution with the runsolver tool. JSAT 7(4), 139–144 (2011)
18. Seipp, J., Braun, M., Garimort, J., Helmert, M.: Learning portfolios of automatically tuned planners. In: ICAPS (2012)
19. Smith-Miles, K., van Hemert, J.: Discovering the suitability of optimisation algorithms by learning from evolved instances. Ann. Math. Artif. Intell. 61(2), 87–104 (2011)
20. Xu, L., Hutter, F., Hoos, H.H., Leyton-Brown, K.: SATzilla: portfolio-based algorithm selection for SAT. J. Artif. Intell. Res. (JAIR) 32, 565–606 (2008)
21. Xu, L., Hutter, F., Hoos, H.H., Leyton-Brown, K.: Hydra-MIP: automated algorithm configuration and selection for mixed integer programming. In: RCRA Workshop on Experimental Evaluation of Algorithms for Solving Problems with Combinatorial Explosion at the International Joint Conference on Artificial Intelligence (IJCAI), pp. 16–30 (2011)

A Biased Random-Key Genetic Algorithm for the Multiple Knapsack Assignment Problem

Eduardo Lalla-Ruiz[1]($\boxtimes$) and Stefan Voß[2]

[1] Department of Computer Engineering and Systems, University of La Laguna,
San Cristóbal de La Laguna, Spain
elalla@ull.es
[2] Institute of Information Systems, University of Hamburg, Hamburg, Germany
stefan.voss@uni-hamburg.de

Abstract. The Multiple Knapsack Assignment Problem (MKAP) is an extension of the Multiple Knapsack Problem, a well-known $\mathcal{NP}$-hard combinatorial optimization problem. The MKAP is a hard problem even for small-sized instances. In this paper, we propose an approximate approach for the MKAP based on a biased random key genetic algorithm. Our solution approach exhibits competitive performance when compared to the best approximate approach reported in the literature.

1 Introduction

The Multiple Knapsack Assignment Problem (MKAP) is presented by Kataoka and Yamada [7] as an extension of the well-known Multiple Knapsack Problem (MKP). In the MKAP, we are given a set of items $N = \{1, ..., n\}$, each item having a profit and a weight, and a set of knapsacks $M = \{1, ..., m\}$, where each indivual knapsack has a given capacity. The items may be packed into m knapsacks considering that the items are divided into K mutually disjoint subsets of items N_k ($k = 1, ..., K$), where $N = \bigcup_{k=1}^{K} N_k, n_k := |N_k|$, and $n = \sum_{k=1}^{K} n_k$. The goal of this problem is to determine the assignment of knapsacks to each subset, and fill the knapsacks with the items belonging to those subsets in such a way that the profit of picked items is maximized.

As discussed in [7], the MKAP is $\mathcal{NP}$-hard, since the special case of $K = 1$ is a MKP. Thus, the MKAP may require expensive computational effort when using exact approaches. Therefore, the usage of approximate algorithms, such as Genetic Algorithms (GAs), for finding high-quality solutions is advisable. GAs are known as bio-inspired algorithms (see Holland [6], Goldberg [4]) that use the concepts of biological evolution and survival of the fittest for (hopefully) obtaining optimal or near optimal solutions in optimization problems.

The contributions of this paper are, on the one hand, to propose a Biased Random Key Genetic Algorithm (BRKGA) approach for solving the MKAP. As indicated by Forrest [3], a domain independent problem representation is not always successful on hard optimization problems and a combination with domain-specific knowledge is advised. Thus, we combine the method with a

© Springer International Publishing Switzerland 2015
C. Dhaenens et al. (Eds.): LION 9 2015, LNCS 8994, pp. 218–222, 2015.
DOI: 10.1007/978-3-319-19084-6_19

domain specific solver for the MKP proposed in the related literature by Pisinger [8]. Through the computational experience reported in this work, we have verified that our algorithm reports high quality solutions in terms of objective function value and short computational times. In this regard, on average, it exhibits a better performance than the best approximate algorithm proposed in the literature (with respect to the quality of the solutions).

The remainder of this paper is organized as follows. Section 2 describes the BRKGA proposed to solve the MKAP. Afterwards, Sect. 3 is devoted to analyse the performance of our algorithm over problem instances proposed in the related literature. Lastly, Sect. 4 provides the main conclusions extracted from the work and suggests several directions for further research.

2   Biased Random Key Genetic Algorithm

One of the major drawbacks of GAs is the difficulty to maintain the feasibility of solutions from parents to offspring. Moreover, another drawback is the possible need of specialized representations for each problem variation. To overcome these difficulties, Bean [1] introduced the concept of random keys. A random key is a real-valued number in the interval $[0,1]$ and a solution is a random key vector sampled from the $[0,1]^n$ space, where n depends on the optimization problem considered. The points in this space are mapped and associated to the solution space of the optimization problem via a deterministic procedure called *decoder*. It is a deterministic algorithm that takes as input a random key vector and returns a feasible solution of the optimization problem along with its objective function value.

The idea of the BRKGA [2,5] extends this concept in the way the crossover is performed. That is, given an elite population, for generating the offspring, BRKGA selects one parent from the elite population and the other parent from the rest of the population. Moreover, in the crossover process, for giving more probability to the elite parent genes, a biased coin favoring the elite parent is tossed, so the child would have more probability of inheriting the keys of its elite parent. This strategy implies an implicit learning from the best solutions along the generations.

In the BRKGA, we have a fixed-size population, Pop, consisting of $|Pop|$ individuals, where each individual is a sequence of randomly generated numbers (random keys) in the real interval $[0,1]$. Through a deterministic procedure, termed as decoder, a vector of random keys r is translated to a solution of the optimization problem at hand with a fitness value $f(r)$ for that vector. At each iteration of the algorithm, the population is partitioned into a smaller set Pop_e of elite individuals and a larger set Pop_c with the remaining individuals of Pop. The evolutionary process within BRKGA is as follow. First, all elite individuals are copied, without change, to the population of the next generation. Then, a set Pop_m of mutant individuals, generated in the same way as an individual of the initial population is inserted into the population of the next generation. Finally, the rest of the population is filled with the offspring obtained through a parametrized uniform crossover with a crossover probability ς. This crossover follows an elitist strategy, *i.e.*, one parent from Pop_e and one from Pop_c are selected at random. For a detailed explanation the reader is referred to [5].

Next we specify the BRKGA for the MKAP. That is, we describe the configuration of the initial population, the way in which the solutions are encoded, the decoding process and the stopping criterion of the BRKGA for the MKAP.

Initial Population: The initial population is composed of $|Pop|$ chromosomes, each randomly generated and satisfying the feasibility criteria imposed by the problem.

Coding: The solutions for the MKAP are encoded with an n-dimensional array $R = (r_1, ..., r_n)$, where n is the number of knapsacks. Each component r_i ($i = 1, ..., n$) is a real number in the interval $[0, 1]$.

Decoding: In the decoding process, a random key vector is mapped in the solution space of the MKAP. Each random key, R, is used to determine the assignment of knapsacks to groups. For this purpose, the interval $[0, 1]$ is equally divided into the number of groups, resulting in K subintervals corresponding to each group. If a random key value is found in one of these intervals, then the corresponding knapsack is assigned to that group. Once we assign each knapsack to a group, we solve K MKPs. In doing so, we use a truncated version of the exact algorithm proposed by Pisinger [8]. The reduced version of this recursive branch and bound allows us to provide high-quality solutions in shorter computational times than the complete exact approach. The reduced version differs from the complete version in that it avoids to check the bound again when a feasible solution is obtained, and hence, the recursive part of the algorithm is avoided. Note, that this version does not guarantee the optimality of the solutions provided.

Furthermore, it should be mentioned that the truncated algorithm is used to solve the K reduced problems. In this sense, we may delimit two procedures that are used together but for different decisions, namely one is used for assigning knapsacks to groups and the other to determine how to fill those knapsacks. Therefore, the performance of the complete solution approach proposed in this work should be assessed as a whole.

3   Computational Results

The proposed optimization technique has been implemented in C++ and executed on a computer equipped with an Intel 3.16 GHz and 4 GB of RAM. By preliminary experiments, we identified the following parameters. A population, Pop, consisting of 20 individuals is used, within which an elite population, Pop_e, of 3 and a mutation population, Pop_m, of 3 individuals are considered. The crossover rate, ς, is set to 0.8. Furthermore, regarding the termination criterion, a maximum number of 200 generations or 20 generations without improvement of the best solution found is established.

To assess our algorithm, we use a representative group of the set of instances from the benchmark suite proposed by [7]. In this regard, we selected those

Table 1. Average computational results provided by BRKGA for the instances sets proposed by [7]. Best values are given in bold.

K	m	n	Gurobi [7]			Heuristic [7]			BRKGA		
			Obj.	Time (s.)	n_{opt}	Obj.	Gap (%)	Time (s.)	Obj.	Gap (%)	Time (s.)
2	10	20	**7134.40**	0.03	10	6596.90	7.47	< 1	**7134.40**	0.00	0.05
		40	**15378.80**	371.22	9	15062.80	2.05	< 1	15375.30	0.08	0.02
		60	23163.30	1200.00	0	23018.80	0.62	< 1	**23194.00**	-0.13	0.14
	20	20	**4150.30**	0.00	10	3799.60	7.53	< 1	**4150.30**	0.00	0.06
		40	**14891.90**	0.77	10	14252.20	4.32	< 1	14849.60	0.18	0.29
		60	**23098.00**	943.53	3	22478.10	2.68	< 1	23054.90	0.19	0.29
5	10	20	**6978.50**	0.02	10	5766.40	17.25	< 1	6977.60	0.01	0.14
		40	**15146.60**	8.10	10	14273.60	5.75	< 1	15083.00	0.17	0.42
		60	**23006.10**	1116.37	1	22483.50	2.27	< 1	22949.90	0.24	0.23
	20	20	**4150.30**	0.02	10	3579.80	12.91	< 1	**4150.30**	0.00	0.10
		40	**14781.30**	0.45	10	12677.40	14.19	< 1	14643.10	0.29	0.93
		60	**22932.80**	780.47	4	21369.30	6.82	< 1	22754.20	0.78	0.30
	Avg		14567.69	368.41	7.25	13779.87	6.99	< 1	14526.38	0.15	0.25

sets that are difficult to be solved by their heuristic in terms of solution quality. Each set is composed of 10 instances each. Furthemore, the computational results presented from [7] were conducted on a computer with CPU: Xeon X5482 Quad-Core × 2 3.20 GHz × 2, RAM: 64 GB.

Table 1 shows the average computational results provided by (i) a general purpose solver, Gurobi taken from [7], (ii) the best approximate approach for this problem, Heuristic (see [7]; this method consists of various components including a greedy approach appended by local search with lagrangian relaxation plus solving some MKP), and (iii) our approach, the BRKGA. The first columns correspond to the size of the instance set. Since all the sets are composed of 10 instances each, in the table we report average values. Additionally, we include the number of optimal solutions, n_{opt}, Gurobi reached for each instance set as reported in [7]. Concerning the computational performance of Heuristic, the authors only reported that they are below 1 second.

As can be checked in Table 1, the computational results indicate that BRKGA presents a competitive performance in terms of, both, computational time and solution quality. It requires, on average, about 0.25 seconds to provide a high-quality solution (about 0.15 % gap). Moreover, our proposed approach, in comparison with Gurobi and Heuristic, shows a better performance in terms of computational time. Regarding, the quality of the solutions BRKGA provides better quality solutions than Heuristic in terms of objective function value. On the other hand, when compared to Gurobi, BRKGA provides slightly worse quality solutions. Nevertheless, it should be noted that for the instance set (K=2, m=10 and n=60) BRKGA reports a better performance than Gurobi, this is due to the fact that Gurobi reaches the time limit.

4   Conclusions and Further Research

In this work, a Biased Random Key Genetic Algorithm for solving the Multiple Knapsack Assignment Problem is proposed. From the computational experiments it can be deduced that our algorithm provides high quality solutions in terms of objective function value and short computational times. In this regard, on average, it exhibits a better performance than the best approximate algorithm reported in the literature.

Despite the fact that additional experimentation is still needed, our results point out that the BRKGA is a suitable method for solving the MKAP. In this regard, a more detailed analysis of our algorithm and its capability for learning from the best solutions within its crossover strategy will be a topic for more indepth investigation. We also strive to use a matheuristic approach for solving the MKAP based on ideas from the POPMUSIC approach [10] or the corridor method [9]. For this purpose it might be interesting to check for the effect of rearranging the constraints of a standard mathematical formulation for the MKAP.

Acknowledgements. This work has been partially funded by the European Regional Development Fund, the Spanish Ministry of Economy and Competitiveness (project TIN2012-32608). Eduardo Lalla-Ruiz thanks the Canary Government for the financial support he receives through his doctoral grant.

References

1. Bean, J.C.: Genetic algorithms and random keys for sequencing and optimization. ORSA J. Comput. **6**, 154–160 (1994)
2. Ericsson, M., Resende, M.G.C., Pardalos, P.M.: A genetic algorithm for the weight setting problem in OSPF routing. J. Comb. Optim. **6**, 299–333 (2002)
3. Forrest, S.: Genetic algorithms: principles of natural selection applied to computation. Science **261**(5123), 872–878 (1993)
4. Goldberg, G.E.: Genetic Algorithms in Search Optimization and Machine Learning. Addison Wesley, Reading (1989)
5. Gonçalves, J.F., Resende, M.G.C.: Biased random-key genetic algorithms for combinatorial optimization. J. Heuristics **17**, 487–525 (2011)
6. Holland, J.H.: Adaptation in Natural and Artificial Systems. MIT Press, Boston (1975)
7. Kataoka, S., Yamada, T.: Upper and lower bounding procedures for the multiple knapsack assignment problem. Eur. J. Oper. Res. **237**(2), 440–447 (2014)
8. Pisinger, D.: An exact algorithm for large multiple knapsack problems. Eur. J. Oper. Res. **114**(3), 528–541 (1999)
9. Sniedovich, M., Voß, S.: The corridor method: a dynamic programming inspired metaheuristic. Control Cybern. **35**(3), 551–578 (2006)
10. Taillard, E., Voß, S.: POPMUSIC - Partial optimization metaheuristic under special intensification conditions. In: Ribeiro, C.C., Hansen, P. (eds.) Essays and Surveys in Metaheuristics, pp. 613–629. Kluwer, Boston (2002)

DYNAMOP Applied to the Unit Commitment Problem

Sophie Jacquin[1,2]([✉]), Laetitia Jourdan[1,2], and El-Ghazali Talbi[1,2]

[1] DOLPHIN Project-team, Inria Lille - Nord Europe, 59650 Villeneuve d'ascq, France
sophie.jacquin@inria.fr
[2] LIFL, Université Lille 1, UMR CNRS 8022, 59655 Villeneuve d'ascq Cedex, France
{laetitia.jourdan,el-ghazali.talbi}@lifl.fr

Abstract. In this article, we propose to apply a hybrid method called DYNAMOP (DYNAmic programming using Metaheuristic for Optimization Problems) to solve the Unit Commitment Problem (UCP). DYNAMOP uses a representation based on a path in the graph of states of dynamic programming, which is adapted to the dynamic structure of the problem and facilitates the hybridization between evolutionary algorithms and dynamic programming. Experiments indicate that the proposed approach outperforms the best known approach in literature.

1 Introduction

DYNAMOP is a hybrid method that has been introduced in [1] and applied to a hydro scheduling problem with success. The main idea of DYNAMOP is to use a genetic algorithm (GA) to run though a graph of states defined as in the dynamic programming (DP) method [2]. The GA handles solutions modeled as paths in the graph of states of DP from the initial state to one terminal state. Then a genotype is a valid sequence of states in which a gene is a state and the objective is to find the shortest path. This choice of representation facilitates the proposition of some intelligent evolutionary operators using DP.

In this paper DYNAMOP is applied to solve the Unit Commitment Problem (UCP) [3]. The UCP is a strategic optimization problem in power system operation. The task is to schedule the generating units on-line or off-line over a scheduling horizon. The objective is to minimize the power production cost with the load demand fully met and the operation constraints satisfied. It is a problem that has been extensively studied. Especially with methods based on DP (i.e. [4,5]) and metaheuristics [3,6–9].

Applying DYNAMOP to the UCP has several motivations. Firstly, the representation used in DYNAMOP can be interesting for a problem such as UCP. Indeed, the representation used in the most of the proposed metaheuristics is a binary vector or a binary matrix, giving the on/off scheduling of each unit, so the neighborhood of a solution is the set of solutions whose representation differs only by one bit. However, there is a high dependence between the decision variables: a unit has to stay turned on or off a minimal time and the production cost

© Springer International Publishing Switzerland 2015
C. Dhaenens et al. (Eds.): LION 9 2015, LNCS 8994, pp. 223–228, 2015.
DOI: 10.1007/978-3-319-19084-6_20

depending on how long a unit stays in the same state. With this representation, the neighborhood of a good solution could contain a lot of bad solutions and conversely the neighborhood of a bad solution could be interesting to explore. This is called the *"problem of bad locality property"* as defined in [10]. But the representation proposed in DYNAMOP helps mitigate this problem. Indeed, the neighborhood of a path consists of paths sharing a maximum number of edges with this path, as the fitness function is the sum of the edges values, it implies a positive impact on the locality property. Secondly, the choice of UCP, a well studied problem, will help evaluate the quality of DYNAMOP.

In the following section, the UCP is explained in details. Then, DYNAMOP and its application to the UCP are explained. Finally, our results are compared to the best results found in literature and in conclusion we discuss the potential of this method and future perspectives.

2   Unit Commitment Problem

UCP is to schedule generating units on-line or off-line over a scheduling horizon. UCP is usually modeled as a mixed integer non-linear problem. It consists of binary variables $u_{i,t}$ that take value 1 if a unit i is turned on at time t and 0 otherwise, and continuous variables $p_{i,t}$ that denote their production amounts. The objective is to minimize the cost of production. This production cost is divided into two components, the fuel cost (FC) and the start up cost (CS). For a system of N units and a time horizon of T periods, the objective function can be described as follows:

$$f(u,p) = \sum_{t=1}^{T} \sum_{i=1}^{N} [\mathrm{FC}_i(p_{i,t}) \times u_{i,t} + \mathrm{CS}_i(T_{i,t-1}^{\mathrm{off}}) \times (1 - u_{i,t-1})u_{i,t}],$$

where:

- FC$_i$ is the fuel cost function of the unit i, modeled by a quadratic function.
- CS$_i$ is the start-up cost for unit i, which depends on $T_{i,t-1}^{\mathrm{off}}$, which is the time the unit i has been turned off at time $t-1$:

$$\mathrm{CS}_i(T_{i,t-1}^{\mathrm{off}}) = \begin{cases} \mathrm{CS}_{\mathrm{cold}} & \text{if} \quad T_{i,\mathrm{min}}^{\mathrm{off}} + T_{\mathrm{cs},i} \leq T_{i,t-1}^{\mathrm{off}} \\ \mathrm{CS}_{\mathrm{hot}} & \text{else} \end{cases},$$

where $T_{i,\mathrm{min}}^{\mathrm{off}} + T_{\mathrm{cs},i}$ is the time it takes the unit i to become cold.

The minimization of the objective is subject to the following system and unit constraints:

1. Power Balance Constraints:

$$\sum_{i=1}^{N} p_{i,t} u_{i,t} = D_t \quad \forall t$$

where D_t is a real number giving the load demand at time t.

2. Unit Output Constraints:

$$p_{i,\min} \times u_{i,t} \leq p_{i,t} \leq p_{i,\max} \times u_{i,t} \quad \forall t, i,$$

where $p_{i,\min}$ and $p_{i,\max}$ are the lower and upper bounds on the energy production of unit i respectively.

3. Spinning Reserve Constraints:

$$\sum_{i=1}^{N} p_{i,\max} u_{i,t} \geq D_t + R_t \quad \forall t$$

where R_t is a real number giving the minimal reserve at time t.

4. Minimum Uptime Limit:

$$T_{i,t-1}^{\text{on}} \geq T_{i,\min}^{\text{on}} \times (1 - u_{i,t}) u_{i,t-1} \quad \forall t, i,$$

where $T_{i,t-1}^{\text{on}}$ is the time from which the unit i is turned on at time $t-1$ and $T_{i,\min}^{\text{on}}$ is the minimal time during which unit i has to stay turned on.

5. Minimum Downtime Limit:

$$T_{i,t-1}^{\text{off}} \geq T_{i,\min}^{\text{off}} \times (1 - u_{i,t-1}) u_{i,t} \quad \forall t, i,$$

where $T_{i,t-1}^{\text{off}}$ is the time from which the unit i is turned off at time $t-1$ and $T_{i,\min}^{\text{off}}$ is the minimal time during which unit i has to stay turned off.

3 DYNAMOP

In this section we explain how DYNAMOP is applied to the UCP.

Representation: The main idea of DYNAMOP [1] is to use a hybrid GA to find the shortest path in the graph of states of a DP problem. The GA handles solutions modeled as paths in the graph of states of DP from the initial to the terminal state. A genotype is a valid sequence of states, in which a gene is a state. In the case of the UCP, a state is characterized by a time period and for each unit if it is turned on or off, and for how long. In practice, a state is represented by an integer vector $S_t = (S_{t,i})_i$, with $(S_{t,i}) \in [\![-T_{i,\min}^{\text{off}} - T_{\text{cs},i}, T_{i,\min}^{\text{on}}]\!] - \{0\}$. $S_{t,i}$ is negative if the unit i is turned off at time t, positive else. The absolute value of $S_{t,i}$ gives the time for which the unit remains turned off or on. If $S_{t,i} = T_{i,\min}^{\text{on}}/(T_{i,\min}^{\text{off}} + T_{\text{cs},i})$ it is that the unit i is turned on/off for at least $T_{i,\min}^{\text{on}}/(T_{i,\min}^{\text{off}} + T_{\text{cs},i})$ hours. To be valid, a sequence must be such that it is possible to pass from a state to the next one while respecting the time constraints.

Evaluation: The fitness function is the sum of the edges values. The value of an edge $S_{t-1} \rightarrow S_t$ is the fuel cost generated in S_t plus the start up cost generated by moving from S_{t-1} to S_t. To compute this value, it is necessary to solve a dispatching problem. This problem is to determine which quantity is produced by which turned on unit of production, such that the constraints (1,2,3) are met, and the production cost is minimized. It is solved using the λ-iteration method [11]. As the fitness function is separable with to the edges, a Δ-evaluation is applied. After applying the evolutionary operators, only the values of the modified edges are computed.

Mutation: The mutation proposed in DYNAMOP randomly chooses a state and replaces it by another state. This new state is linked to the current path by modifying the minimum number of its edges. More precisely in the case of the UCP if $(S_t)_t$ is mute on k into S_k^*, S_k^* is attached to $(S_t)_t$ in considering the unit one by one: For each unit it is to find the maximum $j < k$ such that it is possible to reach $S_{k,i}^*$ from $S_{j,i}$ and the minimum $l > k$ such that it is possible to reach $S_{l,i}$ from $S_{k,i}^*$. The way to go from $S_{j,i}$ and $S_{k,i}^*$ then from $S_{k,i}^*$ to $S_{l,i}$ is uniquely defined.

Crossover: Crossover constructs a new path by introducing a portion of the path of one parent into the path of another parent. After having selected a portion of the path in one parent, the idea is to attach this portion to the path of the other parent by using a minimum number of edges (composing transition paths). The advantage is that the fitness of the obtained offspring is the sum of the values of the portion of paths from both parents plus the values of the transition paths. Then, as the transition paths are constructed to be as small as possible, this leads to good inheritance properties [10] at the phenotypic level. The crossover process is illustrated in Fig. 1. The two crossover points are the states with thick borders.

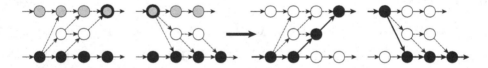

Fig. 1. Illustration of crossover: recombine the two parents using two transition paths.

Intelligent Mutation: The idea of this mutation is to construct a corridor around the considered path and to replace the path by the best path found (by DP) in this corridor. This is exactly a step of Discrete Differential Dynamic Programming methodology [12] which is a hybridization between DP and local search, which has been generalized in [13] to a hybridization between local search and any method. Here a corridor is constructed by fixing the scheduling for some units.

Boosting Crossover: $P^1 = (S_t^1)_t$ and $P^2 = (S_t^2)_t$ are crossed as follows: Let $(t_i)_i$ the increasing serie of the locus of the common states of P^1 and P^2. The off-spring is composed of the best sub-paths between $(S_t^1)_{t_i \leq t < t_{i+1}}$ and $(S_t^2)_{t_i \leq t < t_{i+1}}$.

Table 1. Results and comparison with the best known method in literature

N	DYNAMOP		BGA		Best known (BK)		GAP	
	Best	t (s)	Best	t (s)	Best	t (s)	$\frac{BK-DYNAMOP}{BK}$	$\frac{BK-BGA}{BK}$
10	**563 938**	11	**563 938**	32	**563 938**	19	0.000 %	0.00 %
20	1 123 297	37	1 129 460	97	**1 123 003**	19	−0.026 %	−0.57 %
40	2 242 596	45	2 287 463	358	**2 242 167**	42	−0.019 %	−2.02 %
60	**3 360 320**	88	3 430 710	677	3 361 980	328	0.050 %	−2.04 %
80	**4 480 630**	133	4 891 180	795	4 481 860	113	0.027 %	−9.13 %
100	**5 599 150**	166	5 848 670	1514	5 602 039	162	0.050 %	−4.40 %

4 Results and Conclusion

First, a sensitivity analysis has been carried out in order to determine the effect of algorithm parameters. This analysis is performed using Irace [14]. Irace is a package for R (a statistical software). Its main purpose is to automatically configure optimization algorithms by finding the most appropriate settings for an optimization problem using statistical comparisons. The parameters to fix are the evolutionary operators rates, the population size, and the number of units that are randomly selected to keep their scheduling fixed in the intelligent mutation. Once the parameters had been fixed, the performance of the proposed DYNAMOP method was tested on UCP instances with the number of units from 10 to 100 taken from [6].

Table 1 summarizes the study results. Column "DYNAMOP" shows the results obtained with DYNAMOP. Column "BGA" shows the results obtained with a basic genetic algorithm. This algorithm is implemented using ParadisEO software [15] and parameterized with Irace [14]. Column "Best known" gives the best result found in literature for each instance. These results come from methods: [3, 7–9]. In each case, the stopping criteria are 100 iterations without improvement. 20 independent trials are performed for each test case. Columns "Best" give for each instance the best result obtained over the 20 runs. Columns "t" give the mean time of convergence. The results given in column "Best known" are directly taken from the articles, so in this case, the time is for informational purpose only.

For each case of study, the statistical Friedman test was performed to compare DYNAMOP and BGA and it leads to the conclusion that DYNAMOP is statistically better with a significance level of 1 %. This can be observed in the result table. On top of that it can be observed that the convergence of DYNAMOP is much more faster. It should also be noted that on instances with many units, DYNAMOP outperformed the results found in literature. In the case were the best result of literature is better than the one obtained by DYNAMOP, DYNAMOP is very near to this best. Therefore we can say that the effectiveness and validity of DYNAMOP have been demonstrated through this application. On top that DYNAMOP has the advantage to be adaptable to any problem which holds Bellman's property and the proposed idea of representation can be reused to construct hybridization with a different metaheuristic.

To conclude, we believe that the extension DYNAMOP to stochastic and multi-objective problems could be an interesting line of research.

References

1. Jacquin, S., Jourdan, L., Talbi, E.: Dynamic programming based metaheuristic for energy planning problems. In: EvoStar (2014)
2. Bellman, R.E.: Dynamic Programming. Princeton University Press, Princeton, NJ (1957)
3. Jeong, Y.W., Park, J.B., Shin, J.R., Lee, K.Y.: A thermal unit commitment approach using an improved quantum evolutionary algorithm. Electr. Power Compon. Syst. **37**, 770–786 (2009)
4. Pang, C., Chen, H.: Optimal short-term thermal unit commitment. IEEE Trans. Power Apparatus Syst. **95**, 1336–1346 (1976)
5. Su, C.C., Hsu, Y.Y.: Fuzzy dynamic programming: an application to unit commitment. IEEE Trans. Power Syst. **6**, 1231–1237 (1991)
6. Kazarlis, S.A., Bakirtzis, A., Petridis, V.: A genetic algorithm solution to the unit commitment problem. IEEE Trans. Power Syst. **11**, 83–92 (1996)
7. Lau, T., Chung, C., Wong, K., Chung, T., Ho, S.: Quantum-inspired evolutionary algorithm approach for unit commitment. IEEE Trans. Power Syst. **24**, 1503–1512 (2009)
8. Jeong, Y.W., Park, J.B., Jang, S.H., Lee, K.Y.: A new quantum-inspired binary pso: application to unit commitment problems for power systems. IEEE Trans. Power Syst. **25**, 1486–1495 (2010)
9. Chen, P.H.: Two-level hierarchical approach to unit commitment using expert system and elite pso. IEEE Trans. Power Syst. **27**, 780–789 (2012)
10. Talbi, E.G.: Metaheuristics: from design to implementation, vol. 74. John Wiley & Sons, Hoboken, NJ (2009)
11. Saramourtsis, A., Damousis, J., Bakirtzis, A., Dokopoulos, P.: Genetic algorithm solution to the economic dispatch problemapplication to the electrical power grid of crete island. In: Proceedings of the Workshop Machine Learning Applications to Power Systems (ACAI), pp. 308–317 (2001)
12. Heidari, M., Chow, V.T., Kokotović, P.V., Meredith, D.D.: Discrete differential dynamic programing approach to water resources systems optimization. Water Resour. Res. **7**, 273–282 (1971)
13. Sniedovich, M., Viß, S.: The corridor method: a dynamic programming inspired metaheuristic. Control Cybern. **35**, 551–578 (2006)
14. López-Ibáñez, M., Dubois-Lacoste, J., Stützle, T., Birattari, M.: The irace package, iterated race for automatic algorithm configuration. IRIDIA, Université Libre de Bruxelles, Belgium, Technical report TR/IRIDIA/2011-004 (2011)
15. Humeau, J., Liefooghe, A., Talbi, E.G., Verel, S.: Paradiseo-mo: From fitness landscape analysis to efficient local search algorithms. J. Heuristics **19**, 881–915 (2013)

Scalarized Lower Upper Confidence Bound Algorithm

Mădălina M. Drugan[✉]

Artificial Intelligence Lab, Vrije Universiteit Brussels,
Pleinlaan 2, 1050-B Brussels, Belgium
Madalina.Drugan@vub.ac.be

Abstract. Multi-objective evolutionary optimisation algorithms and stochastic multi-armed bandits techniques are combined in designing stochastic multi-objective multi-armed bandits ($MOMAB$) with an efficient exploration and exploitation trade-off. Lower upper confidence bound ($LUCB$) focuses on sampling the arms that are most probable to be misclassified (i.e., optimal or suboptimal arms) in order to identify the set of best arms aka the Pareto front. Our scalarized multi-objective LUCB (sMO-$LUCB$) is an adaptation of LUCB to reward vectors. Preliminary empirical results show good performance of the proposed algorithm on a bi-objective environment.

1 Introduction

Multi-armed bandits [Auer et. al., 2002] (MAB) is a machine learning paradigm used to study and analyse resource allocation in stochastic and noisy environments. The *multi-objective multi-armed bandits* (MOMAB) [Drugan and Nowe, 2013] algorithms are MABs with reward vectors that import techniques from multi-objective optimisation for an efficient exploration / exploitation trade-off. There are important differences between the MOMAB and the standard MAB algorithms that arise mainly because: (1) there are sets of arms that can be considered to be the best, i.e. the Pareto front, and (2) the number of arms in the Pareto front is unknown.

MOMABs are optimisation algorithms that use a dominance relation to order estimated reward vectors. *The goal of multi-objective multi-armed bandits algorithms is to identify the Pareto front.* A main dominance relation that is imported in MOMABs [Drugan and Nowe, 2013] is the *scalarization functions* that transform the reward vectors into scalar rewards using weight vectors.

Section 2 studies the scalarized MOMAB problem with a new formulation of the scalarized dominance relations for uncertain environments (SDU). SDU is currently used by stochastic MAB to show that the best arm will be selected with a certain accuracy. The linear scalarization function weights each objective from the reward vector and the result is the sum of these weighted values. Due to its simplicity, linear scalarization is the most popular scalarization function in designing both multi-objective optimisation and reinforcement learning algorithms [Roijers et al., 2013].

© Springer International Publishing Switzerland 2015
C. Dhaenens et al. (Eds.): LION 9 2015, LNCS 8994, pp. 229–235, 2015.
DOI: 10.1007/978-3-319-19084-6_21

Section 3 introduces the *scalarized multi-objective lower upper confidence bound* algorithm (*sMO-LUCB*). This is a translation of the adaptive sampling algorithm *lower upper confidence bound* [Kalyanakrishnan et al., 2012] to reward vectors and scalarization functions. For each scalarization, each iteration, the arms that are most probable to be misclassified (optimal or suboptimal arms) are selected. The algorithm stops when the confidence in arm classification (i.e. suboptimal and Pareto optimal arms) is high.

In Sect. 4, we test the proposed scalarization based MOMAB algorithm on a stochastic environment generated with a bi-objective normal distribution. Section 5 concludes the paper.

2   Scalarized Dominance for Uncertain Environments

Let's consider, like in the standard definition of MAB, a fixed set of arms $\mathcal{I}$ with cardinality K, where $K \geq 2$. The vector reward space is defined as the D-dimensional hypercube $[0, 1]^D$, where D is the number of objectives. When an arm i is played, a random vector of rewards is received, one component per objective. The random vectors have a stationary distribution with support in the D-dimensional hypercube $[0, 1]^D$ but the vector of true expected rewards $\boldsymbol{\mu}_i = (\mu_i^1, \ldots, \mu_i^D)$ is unknown. At time steps $t_1, t_2, \ldots$, the corresponding reward vectors $\mathbf{X}_i^{t_1}$, $\mathbf{X}_i^{t_2}$, $\ldots$ are independently and identically distributed according to an unknown law with unknown expectation vector $\boldsymbol{\mu}_i = (\mu_i^1, \ldots, \mu_i^D)$. Reward values obtained from different arms are also assumed to be independent. To bound the performance of MABs, i.e. upper and lower regret bounds, we assume that the rewards are almost surely bounded random vectors with support belonging to the D-dimensional hypercube $[0, 1]^D$ so that we can apply the Hoeffding inequality. A *policy* π is an algorithm that selects the next arm to play based on the list of past plays and obtained reward vectors.

A linear scalarized reward for an arm i is $f_{\boldsymbol{\omega}}(\boldsymbol{\mu}_i) = \boldsymbol{\omega} \cdot \boldsymbol{\mu}_i = \sum_{d=1}^D \omega^d \cdot \mu_i^d$, where $\boldsymbol{\omega} = (\omega^1, \ldots, \omega^D)$ is weight vector selected from a set of predefined weight vectors and $\sum_{d=1}^D \omega^d = 1$.

We say that $\boldsymbol{\mu}_i$ is *better* than $\boldsymbol{\mu}_j$ given a scalarization function $f_{\boldsymbol{\omega}}$, iff $f_{\boldsymbol{\omega}}(\boldsymbol{\mu}_j + \varepsilon_j) < f_{\boldsymbol{\omega}}(\boldsymbol{\mu}_i - \varepsilon_i)$. Note that the scalarization of the uncertainty vector $\varepsilon_i = (\varepsilon_i, \ldots, \varepsilon_i)$ for an arm i is equal to the uncertainty in a single objective, $\varepsilon_i \leftarrow f_{\boldsymbol{\omega}}(\varepsilon_i)$, for any weight vector $\boldsymbol{\omega}$ with the sum of weights equal to 1. The linear scalarization function is also additive, thus $f_{\boldsymbol{\omega}}(\boldsymbol{\mu}_j + \varepsilon_j) = f_{\boldsymbol{\omega}}(\boldsymbol{\mu}_j) + f_{\boldsymbol{\omega}}(\varepsilon_j) = f_{\boldsymbol{\omega}}(\boldsymbol{\mu}_j) + \varepsilon_j$.

If $\varepsilon_i = 0$, then we obtain the definition of the standard scalarization function, and, like all single objective functions, usually, there is a single optimal arm for a scalarization function and that arm belongs to the Pareto front. If $\varepsilon_i > 0$, then there could be multiple optimal arms for $f_{\boldsymbol{\omega}}$ and the number of such arms is not known beforehand.

Let $\boldsymbol{\omega}$ be a scalarization vector with positive weight values in all objectives $\omega^d > 0$. We define an indexing of arms according to their order wrt the scalarization function $f_{\boldsymbol{\omega}}$ such that

$$f_{\boldsymbol{\omega}}(\widehat{\boldsymbol{\mu}}_{i(1)}) \geq f_{\boldsymbol{\omega}}(\widehat{\boldsymbol{\mu}}_{i(2)}) \geq \ldots \geq f_{\boldsymbol{\omega}}(\widehat{\boldsymbol{\mu}}_{i(K)})$$

By definition, an arm with the j-th index is an ε-optimal arm iff

$$f_{\boldsymbol{\omega}}(\widehat{\boldsymbol{\mu}}_{i(1)}) - \varepsilon_{i(1)} < f_{\boldsymbol{\omega}}(\widehat{\boldsymbol{\mu}}_{i(j)}) + \varepsilon_{i(j)}$$

Let's consider that are m such ε-optimal arms, $\{1, \ldots, m\}$. The rest of the arms, with indexes $\{m+1, \ldots, K\}$, are suboptimal. Thus, $\forall j$, $m+1 \leq j \leq K$, we have $f_{\boldsymbol{\omega}}(\widehat{\boldsymbol{\mu}}_{i(j)}) + \varepsilon_{i(j)} < f_{\boldsymbol{\omega}}(\widehat{\boldsymbol{\mu}}_{i(1)}) - \varepsilon_{i(1)}$. The learner is not aware of the indexing $\boldsymbol{\omega}(\cdot)$ and its goal is to identify all the ε-optimal arms for a given $\boldsymbol{\omega}$.

Table 1. Twenty bi-dimensional reward vectors labelled from 1 to 20. The first ten ones are labelled from $\boldsymbol{\mu}_1^*$ till $\boldsymbol{\mu}_{10}^*$ and are Pareto optimal arms, while the last ten ones are labelled from $\boldsymbol{\mu}_{11}$ till $\boldsymbol{\mu}_{20}$ and they are suboptimal.

$\boldsymbol{\mu}_1^* = (0.562, 0.493)$	$\boldsymbol{\mu}_2^* = (0.552, 0.515)$	$\boldsymbol{\mu}_3^* = (0.543, 0.527)$	$\boldsymbol{\mu}_4^* = (0.535, 0.535)$
$\boldsymbol{\mu}_5^* = (0.525, 0.555)$	$\boldsymbol{\mu}_6^* = (0.523, 0.557)$	$\boldsymbol{\mu}_7^* = (0.515, 0.563)$	$\boldsymbol{\mu}_8^* = (0.506, 0.568)$
	$\boldsymbol{\mu}_9^* = (0.503, 0.571)$	$\boldsymbol{\mu}_{10}^* = (0.497, 0.573)$	
$\boldsymbol{\mu}_{11} = (0.498, 0.567)$	$\boldsymbol{\mu}_{12} = (0.502, 0.563)$	$\boldsymbol{\mu}_{13} = (0.505, 0.495)$	$\boldsymbol{\mu}_{14} = (0.508, 0.555)$
$\boldsymbol{\mu}_{15} = (0.512, 0.533)$	$\boldsymbol{\mu}_{16} = (0.514, 0.525)$	$\boldsymbol{\mu}_{17} = (0.522, 0.554)$	$\boldsymbol{\mu}_{18} = (0.531, 0.531)$
	$\boldsymbol{\mu}_{19} = (0.542, 0.523)$	$\boldsymbol{\mu}_{20} = (0.547, 0.513)$	

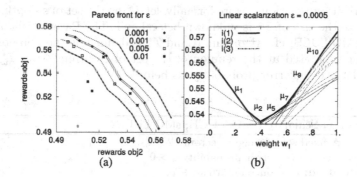

Fig. 1. (a) A bi-objective example with ten Pareto optimal arms and ten suboptimal arms. (b) Scalarization dominance for uncertain environments on a bi-objective example with the uncertainty value $\varepsilon = 5 \cdot 10^{-4}$ and 11 scalarization values, where $\omega^1 = \{0.0, 0.1, 0.2, \ldots, 0.9, 1.0\}$.

We identify multiple Pareto optimal arms using multiple scalarization functions $\boldsymbol{\omega}$.

Example 1. Consider the example from Table 1. Figure 1(a) shows that, when the standard Pareto dominance relation is considered, there are ten Pareto optimal reward vectors in the Pareto front and ten suboptimal arms. Figure 1(b) shows the ε - optimal arms for ten scalarization functions, where $\varepsilon = 5 \cdot 10^{-4}$.

There are only 6 arms identified as the best arms, i.e. with the index $i(1)$, and there is one extra arm identified as second best μ_6^*, with the index $i(2)$, and two extra arms identified as the third best μ_3^* and μ_8^*. Note that the number of ε optimal arms increases for weights vectors with small differences between weight values meaning that the difference in fitness values is also smaller. In our example, for $\omega^1 = 1.0$, there is only one ε-optimal arm, whereas, for $\omega^1 = 0.5$, there are 5 ε-optimal arms.

3   Scalarized Multi-Objective LUCB

Our scalarized MOMAB algorithm's practical value comes from re-usage of arm pulls of different scalarization functions. A common approach in multi-objective optimisation is to consider a fixed set of scalarization functions $W \leftarrow \{\omega_1, \ldots, \omega_{|W|}\}$ that are uniform randomly spread in the weight space in order to identify multiple Pareto optimal arms. The *scalarized MO LUCB* algorithm, sMO-LUCB cf Algorithm 1, is a collection of LUCBs [Kaufmann and Kalyanakrishnan, 2013] with different weight vectors $\omega \in W$.

We assume that for each scalarization function $\omega \in W$, a single objective LUCB identifies the best arm within the confidence interval $[-\varepsilon, \varepsilon]$, where the confidence value ε decreases with the number of arm pulls. We are interested in the Pareto front resulting from the reunion of all ε-optimal arms identified by these scalarization functions. More formally, let $\mathcal{I}_\omega^*$ be the set of ε - optimal arms for f_ω, where the size of $\mathcal{I}_\omega^*$ is unknown beforehand. The Pareto front $\mathcal{I}^*$ is the output of sMO-LUCB, cf. Algorithm 1, and $\mathcal{I}^*$ is defined as the non-dominated Pareto front generated at the reunion of the set of ε-optimal arms $\mathcal{I}_\omega^*$, for all $\omega \in W$ and $\delta > 0$ the error probability, as before.

Algorithm 1. Scalarized MO LUCB, sMO-LUCB

Require: A fixed set of weight vectors W
Require: Accuracy ϵ, error probability $\delta > 0$
  Initialisation: Pull once all arms $i \in \mathcal{I}$;
  Compute the critical arms $u_\omega(1)$ and $\ell_\omega(1)$, $\forall \omega \in W$;
  $\mathcal{I}^*_\omega(1) \leftarrow \emptyset$ and compute the difference $B_\omega(1)$
  while $\max_{\omega \in W} B_\omega(1) > \epsilon$ **do**
    Set $v(n) \leftarrow \arg \max_{\omega \in W} B_\omega(n)$;
    Pull the critical arms $u_v(n)$ and $\ell_v(n)$;
    for all $\omega \in W$ **do**
      Update the set of ε-optimal arms $\mathcal{I}_\omega^*(n)$;
      Update $u_\omega(n)$, $\ell_\omega(n)$, and $B_\omega(n)$
    end for
    $n \leftarrow n + 1$;
  end while
  Compute the Pareto front $\mathcal{I}^* \leftarrow \cup_{\omega \in W} \mathcal{I}_\omega^*$
  return $\mathcal{I}^*$

Let's denote the confidence value of arm i for the scalarization function ω after n arm pulls with $\varepsilon_i(n) \leftarrow \sqrt{\frac{\beta(n,\delta)}{2n_i}}$, and $\beta(n,\delta) \leftarrow \log\left(\frac{k_1 K|\mathcal{W}|n^\alpha}{\delta}\right) + \log\log\left(\frac{k_1 K|\mathcal{W}|n^\alpha}{\delta}\right)$ an exploration parameter and α a positive scalar value. To adapt the original exploration parameter of $LUCB$, we have considered that Hoeffding inequality and union bound sum up over the number of scalarization functions.

For each ω, we consider two arms $\ell_\omega(n)$ and $u_\omega(n)$ to be critical after n arm pulls as the arms that are the most likely to be misclassified. Thus,

$$u_\omega(n) \leftarrow \arg\min_{i \in \mathcal{I}_\omega^*} f_\omega(\widehat{\boldsymbol{\mu}}_i) - \varepsilon_i(n)$$

$$\ell_\omega(n) \leftarrow \arg\max_{j \notin \mathcal{I}_\omega^*} f_\omega(\widehat{\boldsymbol{\mu}}_j) + \varepsilon_j(n)$$

sMO-LUCB pulls each round the two critical arms $u(n)$ and $\ell(n)$ and it stops when

$$\max_{\omega \in \mathcal{W}} B_\omega(n) < \delta$$

where $B_\omega(n) \leftarrow (f_\omega(\widehat{\boldsymbol{\mu}}_\ell) + \varepsilon_\ell(n)) - (f_\omega(\widehat{\boldsymbol{\mu}}_u) - \varepsilon_u(n))$. When the critical arms of the scalarization function ω are pulled, the corresponding difference $B_\omega(n)$ decreases with the decrease in confidence values. The next round, another scalarization function with a larger difference between the upper and lower bounds will be pulled. The algorithm stops when all the scalarization functions have identified with high confidence the corresponding sets of ε-optimal arms.

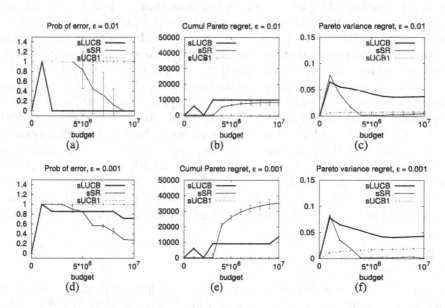

Fig. 2. The numerical simulations for: scalarized multi-objective lower upper confidence bound (sMO-$LUCB$), scalarized successive rejects (sSR), and scalarized upper confidence bound ($sUCB1$). The scalarized dominance relation has (top) $\varepsilon = 0.01$ and (bottom) $\varepsilon = 0.001$.

Consider a set of scalarization functions with the same optimal arm. The arms pulls' for one scalarization function will influence, decrease, the confidence interval of the other scalarization functions with the same optimal value. In the ideal case, when there is a single scalarization function per Pareto optimum arm, sMO-LUCB is fair in pulling arms from the Pareto front. Thus, the algorithm is computational efficient by reusing arm pulls for different scalarization functions.

Unlike for the Pareto version of the MO-LUCB algorithm, the behaviour of scalarized MO-LUCB cannot be straightforwardly bounded using state of the art theorems.

The advantage of scalarized LUCB is that the number of scalarization functions $|\mathcal{W}|$ can be chosen by the user and thus, the $sMO\text{-}LUCB$ algorithm does not depend on the size of the Pareto front. In addition, sMO-LUCB does not need to assume that the optimum value of these scalarization functions are evenly spread over the Pareto front.

4   Preliminary Numerical Simulations

We experimentally compare three performance indexes of three scalarization MOMAB algorithms: (1) $sMO\text{-}LUCB$, cf. Algorithm 1, and (2) scalarized successive rejects sSR from [Drugan and Nowe, 2014], and (3) scalarized UCB1 from [Drugan and Nowe, 2013]. The bi-objective problem from Example 1 is the test problem, where the mean vector rewards are generated using a bi-objective normal distribution with variance 0.1. For all algorithms we take 11 scalarization functions and $N = 10^7$ arm pulls for each of the 100 independent runs of the algorithms, the error probability $\delta = 0.01$ and $\alpha = 2$. In Fig. 2, we compare the performance of the three algorithms in terms of: (1) empirical probability of error in identifying the Pareto front, (2) cumulative Pareto regret [Drugan and Nowe, 2013], and (3) cumulative Pareto variance regret. In the scalarized dominance relation, we consider $\varepsilon \in \{0.01, 0.001\}$. Figure 2(a) shows that $sMO\text{-}LUCB$ has a lower probability of error than sSR whereas in Fig. 2(d) sSR outperforms $sMO\text{-}LUCB$. $sUCB1$ has the largest probability of error. Note the relation between the performance of the algorithms and their goal. $sMO\text{-}LUCB$ and sSR have the goal of identifying the Pareto front whereas $sUCB1$ minimizes the Pareto regret. Therefore, the cumulative Pareto regret, cf Fig. 2(b) and (e) and the Pareto variance regret Fig. 2(c) and (f), is the smallest for $sUCB1$. We conclude that in these preliminary simulations, $sMO\text{-}LUCB$ has a low probability of error.

5   Conclusions

This paper translates a successful multi-armed bandits algorithm, i.e. adaptive sampling, to reward vectors using scalarization functions. $sMO\text{-}LUCB$'s goal is to identify the Pareto front with high accuracy. Our preliminary experiments show good performance for the scalarized MO-LUCB algorithm.

Acknowledgements. Madalina M. Drugan was supported by the IWT-SBO project PERPETUAL (gr. nr. 110041) and FWO project "Multi-criteria RL" (gr. nr. G. 087814N).

References

Auer, P., Cesa-Bianchi, N., Fischer, P.: Finite time analysis of the multiarmed bandit problem. J. Mach. Learn. **47**(2/3), 235–256 (2002)

Drugan, M., Nowe, A.: Designing multi-objective multi-armed bandits: a study. In: Proceedings of International Joint Conference of Neural Networks (IJCNN) (2013)

Drugan, M., Nowe, A.: Scalarization based pareto optimal set of arms identification algorithms. In: Proceedings of International Joint Conference of Neural Networks (IJCNN) (2014)

Kalyanakrishnan, S., Tewari, A., Auer, P., Stone, P.: PAC subset selection in stochastic multi-armed bandits. In: Proceedings of International Conference on Machine Learning (ICML) (2012)

Kaufmann, E., Kalyanakrishnan, S.: Information complexity in bandit subset selection. In: Proceedings of COLT, pp. 228–251 (2013)

Roijers, D.M., Vamplew, P., Whiteson, S., Dazeley, R.: A survey of multi-objective sequential decision-making. J. Artif. Intell. Res. (JAIR) **48**, 67–113 (2013)

Generating Training Data for Learning Linear Composite Dispatching Rules for Scheduling

Helga Ingimundardóttir[(✉)] and Thomas Philip Rúnarsson

School of Engineering and Natural Sciences,
University of Iceland, Reykjavik, Iceland
{hei2,tpr}@hi.is

Abstract. A supervised learning approach to generating composite linear priority dispatching rules for scheduling is studied. In particular we investigate a number of strategies for how to generate training data for learning a linear dispatching rule using preference learning. The results show, that when generating a training data set from only optimal solutions, it is not as effective as when suboptimal solutions are added to the set. Furthermore, different strategies for creating preference pairs is investigated as well as suboptimal solution trajectories. The different strategies are investigated on 2000 randomly generated problem instances using two different problem generator settings.

When applying learning algorithms, the training set is of paramount importance. A training set should have sufficient knowledge of the problem at hand. This is done by the use of features which are supposed to capture the essential measures of a problem's state. For this purpose, the job-shop scheduling problem (JSP) is used as a case study to illustrate a methodology for generating meaningful training data which can be successfully learned.

JSP deals with the allocation of tasks of competing resources where the goal is to minimise a schedule's maximum completion time, i.e., the makespan denoted $C_{\max}$. In order to find good solutions, heuristics are commonly applied in research, such as the simple priority based dispatching rules (SDR) from [11]. Composites of such simple rules can perform significantly better [6]. As a consequence, a linear composite of dispatching rules (LCDR) was presented in [3]. The goal there was to learn a set of weights, $\mathbf{w}$, via logistic regression such that

$$h(\mathbf{x}_j) = \langle \mathbf{w} \cdot \phi(\mathbf{x}_j) \rangle, \tag{1}$$

yields the preference estimate for dispatching job J_j that corresponds to post-decision state $\mathbf{x}_j$, where $\phi(\mathbf{x}_j)$ denotes its feature mapping. The job dispatched is the following,

$$j^* = \arg\max_j \{h(\mathbf{x}_j)\}. \tag{2}$$

The approach was to use supervised learning to determine which feature states are preferable to others. The training data was created from optimal solutions of randomly generated problem instances.

© Springer International Publishing Switzerland 2015
C. Dhaenens et al. (Eds.): LION 9 2015, LNCS 8994, pp. 236–248, 2015.
DOI: 10.1007/978-3-319-19084-6_22

An alternative would be minimising the expected $C_{\max}$ by directly using a brute force search such as CMA-ES [2]. Preliminary experiments were conducted in [5], which showed that optimising the weights in Eq. (1) via evolutionary search actually resulted in a better LCDR than the previous approach. The nature of the CMA-ES is to explore suboptimal routes until it converges to an optimal route. This implies that the previous approach, of restricting the training data only to *one* optimal route, may not produce a sufficiently rich training set. That is, the training set should incorporate a more complete knowledge of *all* possible preferences, i.e., it should make the distinction between suboptimal and sub-suboptimal features, etc. This approach would require a Pareto ranking of preferences which can be used to make the distinction of which feature sets are equivalent, better or worse – and to what degree, e.g. by giving a weight to the preference. This would result in a very large training set, which of course could be re-sampled in order to make it computationally feasible to learn. In this study we will investigate a number of different ranking strategies for creating preference pairs.

Alternatively, training data could be generated using suboptimal solution trajectories. For instance [7] used decision trees to 'rediscover' largest processing time (LPT, a single priority based dispatching rule) by using LPT to create its training data. The limitations of using heuristics to label the training data is that the learning algorithm will mimic the original heuristic (both when it works poorly and well on the problem instances) and does not consider the real optimum. In order to learn heuristics that can outperform existing heuristics, then the training data needs to be correctly labelled. This drawback is confronted in [8,10,15] by using an optimal scheduler, computed off-line. In this study, we will both follow optimal and suboptimal solution trajectories, but for each partial solution the preference pair will be labelled correctly by solving the partial solution to optimality using a commercial software package [1]. For this study most work remaining (MWR), a promising SDR for the given data distributions [4], and the CMA-ES optimised LCDRs from [5] will be deemed worthwhile for generating suboptimal trajectories.

To summarise, the study considers two main aspects of the generation of training data: (a) how preference pairs are added at each decision stage, and (b) which solution trajectorie(s) should be sampled. That is, optimal, random, or suboptimal trajectories, based on a good heuristic, etc.

The outline of the paper is as follows, first we illustrate how JSP can be seen as a decision tree where the depth of the tree corresponds to the total number of job-dispatches needed to form a complete schedule. The feature space is also introduced and how optimal dispatches and suboptimal dispatches are labelled at each node in the tree. This is followed by detailing the strategies investigated in this study by selecting preference pairs ranking and sampling solution trajectories. The authors then perform an extensive study comparing these strategies. Finally, this paper concludes with discussions and a summary of main results.

Table 1. Problem space distributions, $\mathcal{P}$.

Name	Size $(n \times m)$	N_{train}	N_{test}	Note
$\mathcal{P}_{j.rnd}$	6×5	500	500	Random
$\mathcal{P}_{j.rndn}$	6×5	500	500	Random-narrow

Table 2. Feature space, $\mathcal{F}$.

ϕ	Feature description
ϕ_1	Job processing time
ϕ_2	Job start-time
ϕ_3	Job end-time
ϕ_4	When machine is next free
ϕ_5	Current makespan
ϕ_6	Total work remaining for job
ϕ_7	Most work remaining for all jobs
ϕ_8	Total idle time for machine
ϕ_9	Total idle time for all machines
ϕ_{10}	ϕ_9 weighted w.r.t. number of assigned tasks
ϕ_{11}	Time job had to wait
ϕ_{12}	Idle time created
ϕ_{13}	Total processing time for job

1   Problem Space

In this study synthetic JSP data instances are considered with the problem size $n \times m$, where n and m denotes number of jobs and machines, respectively. Problem instances are generated stochastically. By fixing the number of jobs and machines while processing time are i.i.d. samples from a discrete uniform distribution from the interval $I = [u_1, u_2]$, i.e., $p \sim \mathcal{U}(u_1, u_2)$. Two different processing time distributions are explored, namely $\mathcal{P}_{j.rnd}$ where $I = [1, 99]$ and $\mathcal{P}_{j.rndn}$ where $I = [45, 55]$ are referred to as random and random-narrow, respectively. The machine order is a random permutation of all of the machines in the job-shop.

For each data distribution N_{train} and N_{test} problem instances were generated for training and testing, respectively. Values for N are given in Table 1. Note, that difficult problem instances are not filtered out beforehand, such as the approach in [16].

2   JSP Tree Representation

When building a complete JSP schedule $\ell = n \cdot m$ dispatches must be made consecutively. A job is placed at the earliest available time slot for its next machine, whilst still fulfilling constraints that each machine can handle, which is at most one job at each time, and jobs need to have finished their previous machines according to its machine order. Unfinished jobs, referred to as the job-list denoted $\mathcal{L}$, are dispatched one at a time according to a heuristic. After each dispatch, the schedule's current features are updated based on its resulting partial schedule. For each possible post-decision state the temporal features, $\mathcal{F}$, applied in this study are given in Table 2. These features are based on SDRs which are widespread in practice. For example if $\mathbf{w}$ is zero, save for $w_6 = 1$, then Eq. (1) gives $h(\mathbf{x}_j) > h(\mathbf{x}_i)$, $\forall i$ which are jobs with less work remaining than job J_j, namely Eq. (2) yields the job with the highest ϕ_6 value, i.e., equivalent to dispatching rule most work remaining (MWR).

Figure 1 illustrates how the first two dispatches could be executed for a 6×5 JSP with the machines $a \in \{M_1, ..., M_5\}$ on the vertical axis and the horizontal axis yields the current makespan, $C_{\max}$. The next possible dispatches are denoted as dashed boxes with the job index j within and its length corresponding to processing time p_{ja}. In the top layer one can see an empty schedule. In the middle layer one of the possible dispatches from the layer above is fixed (depicted solid) and one can see the resulting schedule (i.e., what are the next possible dispatches given this new scenario?). Finally, the bottom layer depicts all outcomes if job J_3 on machine M_3 would be dispatched. This sort of tree representation is similar to *game trees* [9] where the root node denotes the initial (i.e., empty) schedule and the leaf nodes denote the complete schedule. Therefore, the distance k from an internal node to the root yields the number of operations already dispatched. Traversing from root to leaf node, one can obtain a sequence of dispatches that yielded the resulting schedule, i.e., the sequence indicates in which order the tasks should be dispatched for that particular schedule.

However, one can easily see that this sequence of task assignments is by no means unique. Inspecting a partial schedule further along in the dispatching process such as in Fig. 1 (top layer), then let's say J_1 would be dispatched next, and in the next iteration J_2. This sequence would yield the same schedule as if J_2 would have been dispatched first and then J_1 in the next iteration (since these are non-conflicting jobs). This indicates that some of the nodes in the tree can merge despite states of the partial schedules being different in previous layers. In this particular instance one can not infer that choosing J_1 is better and J_2 is worse (or vice versa) since they can both yield the same solution.

Furthermore, in some cases there can be multiple optimal solutions to the same problem instance. Hence not only is the sequence representation 'flawed' in the sense that slight permutations on the sequence are in fact equivalent w.r.t. the end-result, but varying permutations on the dispatching sequence (given the same partial initial sequence) can result in very different complete schedules with the same makespan, and thus same deviation from optimality, ρ defined by Eq. (4), which is the measure under consideration. Care must be taken in this case that neither resulting features are labelled as undesirable or suboptimal. Only the resulting features from a dispatch resulting in a suboptimal solution should be labelled undesirable.

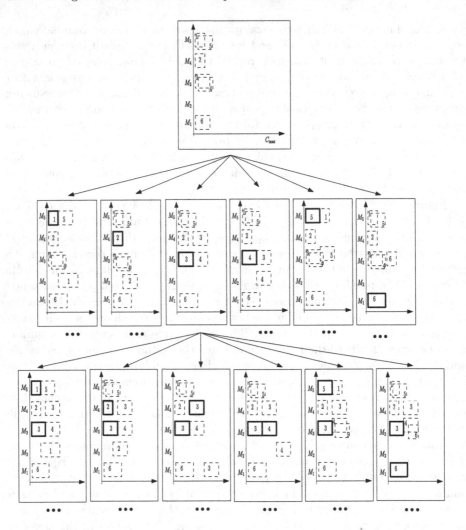

Fig. 1. Partial Tree for JSP for the first two dispatches. Executed dispatches are depicted solid, and all possible dispatches are dashed.

The creation of the tree for job-shop scheduling can be done recursively for all possible permutation of dispatches in the manner described above, resulting in a full n-ary tree of height $\ell = n \cdot m$. Such an exhaustive search would yield at the most n^ℓ leaf nodes (worst case scenario being that no sub-trees merge). Now, since the internal vertices (i.e., partial schedules) are only of interest to learn,[1] the number of those can be at the most $n^{\ell-1}/_{n-1}$ [12]. Even for small dimensions of n and m the number of internal vertices are quite substantial and thus computationally expensive to investigate them all.

---

[1] The root is the empty initial schedule and for the last dispatch there is only one option left to dispatch, so there is no preferred 'choice' to learn.

The optimum makespan is known for each problem instance. At each time step (i.e., layer of the tree) a number of feature pairs are created. The feature pairs consist of the features ϕ_o resulting from optimal dispatches $o \in \mathcal{O}^{(k)}$, versus features ϕ_s resulting from suboptimal dispatches $s \in \mathcal{S}^{(k)}$ at time k. Note, $\mathcal{O}^{(k)} \cup \mathcal{S}^{(k)} = \mathcal{L}^{(k)}$ and $\mathcal{O}^{(k)} \cap \mathcal{S}^{(k)} = \emptyset$. In particular, each job is compared against another job from the job-list, $\mathcal{L}^{(k)}$, and if the makespan differs, i.e., $C_{\max}^{(s)} \gtrless C_{\max}^{(o)}$, an optimal/suboptimal pair is created. However, if the makespan would be unaltered the pair is omitted since they give the same optimal makespan. This way, only features from a dispatch resulting in a suboptimal solution is labelled undesirable.

The approach taken in this study is to verify analytically, at each time step, whether it can indeed *somehow* yield an optimal schedule by manipulating the remainder of the sequence, while maintaining the current temporal schedule fixed as its initial state. This also takes care of the scenario that having dispatched a job resulting in a different temporal makespan would have resulted in the same final makespan even if another optimal dispatching sequence would have been chosen. That is to say the data generation takes into consideration when there are multiple optimal solutions to the same problem instance.

3   Selecting Preference Pairs

At each dispatch iteration k, a number of preference pairs are created, which is then iterated over all N_{train} instances available. A separate data set is deliberately created for each dispatch iteration, as the initial feeling is that DRs used in the beginning of the schedule building process may not necessarily be the same as in the middle or end of the schedule. As a result there are ℓ linear scheduling rules for solving a $n \times m$ job-shop specified by a set of preference pairs for each step,

$$S = \big\{ \{\phi_o - \phi_s, +1\}, \{\phi_s - \phi_o, -1\} \big\} \subset \Phi \times Y \qquad (3)$$

for all $o \in \mathcal{O}^{(k)}, s \in \mathcal{S}^{(k)}, k \in \{1, \ldots, \ell\}$ where $Y = \{-1, 1\}$ denotes, suboptimal or optimal preferences, respectively, and $\phi_o, \phi_s \in \Phi \subset \mathcal{F}$ are features from the collected training set Φ. The reader is referred to [3] for a detailed description of how the linear ordinal regression model is trained on preference set S. Defining the size of the preference set as $l = |S|$, then if l is too large re-sampling may be needed to be done in order for the ordinal regression to be computationally feasible.

3.1   Trajectory Sampling Strategies

The following trajectory sampling strategies were explored for adding features to the training set Φ,

Φ^{opt} at each dispatch some (random) optimal task is dispatched.

Φ^{cma} at each dispatch the task corresponding to highest priority, computed with fixed weights $\mathbf{w}$, which were obtained by directly optimising the mean of the performance measure defined in Eq. (4) with CMA-ES.

Φ^{mwr} at each dispatch the task corresponding to most work remaining is dispatched, i.e., following the simple dispatching rule MWR.

Φ^{rnd} at each dispatch some random task is dispatched.

Φ^{all} all aforementioned trajectories are explored, i.e.,

$$\Phi^{all} = \Phi^{opt} \cup \Phi^{cma} \cup \Phi^{mwr} \cup \Phi^{rnd}.$$

In the case of Φ^{mwr} and Φ^{cma} it is sufficient to explore each trajectory exactly once for each problem instance, since they are static DRs. Whereas, for Φ^{opt} and Φ^{rnd} there can be several trajectories worth exploring. However, only one is chosen at random, this is deemed sufficient as the number of problem instances N_{train} is relatively large.

3.2 Ranking Strategies

The following ranking strategies were implemented for adding preference pairs to S,

S_b all optimum rankings r_1 versus all possible suboptimum rankings r_i, $i \in \{2, \ldots, n'\}$, preference pairs are added, i.e., same basic set-up as in [3].

S_f full subsequent rankings, i.e., all possible combinations of r_i and r_{i+1} for $i \in \{1, \ldots, n'\}$, preference pairs are added.

S_p partial subsequent rankings, i.e., sufficient set of combinations of r_i and r_{i+1} for $i \in \{1, \ldots, n'\}$, are added to the preference set – e.g. in the cases that there are more than one operation with the same ranking, only one of that rank is needed to compared to the subsequent rank. Note that $S_p \subset S_f$.

S_a all rankings, i.e., all possible combinations of r_i and r_j for $i, j \in \{1, \ldots, n'\}$, $i \neq j$, preference pairs are added.

where $r_1 > r_2 > \ldots > r_{n'}$ $(n' \leq n)$ are the rankings of the job-list, $\mathcal{L}^{(k)}$, at time step k.

4 Experimental Study

To test the validity of different rankings and strategies, the problem spaces outlined in Table 1 were used. The optimum makespan is denoted $C_{\max}^{\text{opt}}$, and the makespan obtained from the heuristic model is $C_{\max}^{\text{model}}$. Since the optimal makespan varies between problem instances the performance measure is the following,

$$\rho = \frac{C_{\max}^{\text{model}} - C_{\max}^{\text{opt}}}{C_{\max}^{\text{opt}}} \cdot 100\,\% \tag{4}$$

which indicates the percentage relative deviation from optimality.

The preference set, S, across varying trajectories and ranking strategies is depicted in Fig. 2, where the figure is divided vertically by problem space and horizontally by trajectory scheme.

A linear ordinal regression model (PREF) was created for each preference set, S, for problem spaces $\mathcal{P}_{j.rnd}$ and $\mathcal{P}_{j.rndn}$. A box-plot with the results of percentage relative deviation from optimality, ρ, defined by Eq. (4), is presented in Fig. 3. The box-plots are grouped w.r.t. trajectory strategies and colour-coded w.r.t. ranking schemes. Moreover, the simple priority dispatching rule MWR and the weights obtained by the CMA-ES optimisation used to obtain the training sets Φ^{mwr} and Φ^{cma} respectively are shown in black in the far left of the group comparison. From Fig. 3 it is apparent there can be a performance edge gained by implementing a particular ranking or trajectory strategy. Moreover, the behaviour is analogous across different disciplines. Main statistics are reported in Table 3a and b for $\mathcal{P}_{j.rnd}$ and $\mathcal{P}_{j.rndn}$, respectively. Models are sorted w.r.t. mean relative error.

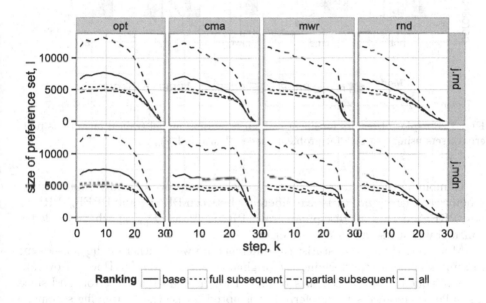

Fig. 2. Size of preference set, $l = |S|$, for different trajectories and ranking strategies obtained from the training set for problem spaces $\mathcal{P}_{j.rnd}$ and $\mathcal{P}_{j.rndn}$.

4.1   Ranking Strategies

There is no statistical difference between PREF_f and PREF_p ranking-models across all trajectory disciplines (cf. Fig. 3), which is expected since S_p is designed to contain the same preference information as S_f. The results hold for both problem spaces.

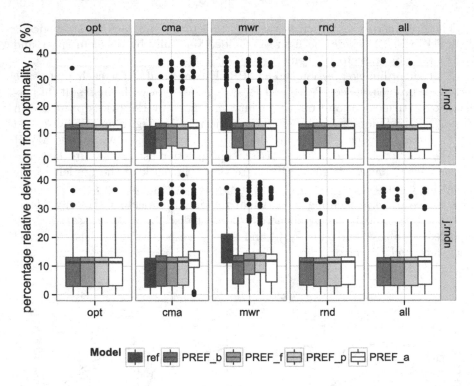

Fig. 3. Box-plot of results for linear ordinal regression model trained on various preference sets using test sets for problem spaces $\mathcal{P}_{j.rnd}$ and $\mathcal{P}_{j.rndn}$.

Combining the ranking schemes, S_a, does not improve the individual ranking-schemes as there is no statistical difference between PREF_a and PREF_b, PREF_f nor PREF_p across all disciplines, save PREF_a^{cma} for $\mathcal{P}_{j.rndn}$ which yielded a considerably worse mean relative error.

Moreover, there is no statistical difference between either of the subsequent ranking-schemes outperforming the original S_b set-up from [3]. However overall, the subsequent ranking schemes results in lower mean relative error, and since a smaller preference set is preferred, it is opted to use the S_p ranking scheme.

Furthermore, it is noted that PREF^{mwr} is able to significantly outperform the original heuristic (MWR) used to create its training data Φ^{mwr}, irrespective of the ranking schemes. Whereas the fixed weights found via CMA-ES outperform the PREF^{cma} models for all ranking schemes. This implies that ranking scheme is relatively irrelevant. The results hold for both problem spaces.

4.2   Trajectory Sampling Strategies

Learning preference pairs from good scheduling policies, as done in PREF^{cma} and PREF^{mwr}, can give favourable results. However, tracking optimal paths yield generally a lower mean relative error.

Table 3. Main statistics of percentage relative deviation from optimality, ρ, defined by Eq. (4) for various models.

(a) $\mathcal{P}_{j.rnd}$ test set							(b) $\mathcal{P}_{j.rndn}$ test set						
model	track	rank	mean	med	sd	max	model	track	rank	mean	med	sd	max
CMA			8.84	10.59	6.14	28.18	CMA			9.13	10.91	6.16	26.23
PREF	all	p	9.63	11.16	6.32	35.97	PREF	rnd	b	9.82	11.36	6.07	33.05
PREF	all	f	9.68	11.11	6.38	35.97	PREF	rnd	f	9.87	11.22	6.57	33.92
PREF	opt	a	9.92	11.22	6.49	27.39	PREF	opt	b	9.94	11.31	6.52	36.32
PREF	all	b	9.98	11.27	6.61	37.36	PREF	opt	f	9.98	11.36	6.58	26.84
PREF	opt	b	10.05	11.45	6.53	34.23	PREF	rnd	p	9.99	11.35	6.42	32.33
PREF	opt	p	10.13	11.33	6.74	27.39	PREF	opt	a	10.01	11.34	6.31	36.60
PREF	all	a	10.15	11.38	6.30	27.57	PREF	all	f	10.05	11.33	6.53	36.60
PREF	opt	f	10.31	11.54	6.87	27.39	PREF	opt	p	10.06	11.42	6.52	26.84
PREF	rnd	b	10.51	11.55	6.86	37.87	PREF	all	p	10.08	11.39	6.49	34.15
PREF	rnd	p	10.75	11.49	6.70	35.60	PREF	all	b	10.12	11.34	6.73	36.60
PREF	cma	p	10.78	11.52	6.89	36.60	PREF	rnd	a	10.14	11.49	6.25	33.05
PREF	rnd	a	10.82	11.59	6.73	28.65	PREF	all	a	10.39	11.45	6.69	36.60
PREF	cma	f	10.90	11.55	6.89	36.60	PREF	cma	f	10.56	11.38	7.28	38.31
PREF	cma	b	10.90	11.55	7.10	36.91	PREF	cma	b	10.73	11.47	7.62	36.60
PREF	mwr	p	10.95	11.46	7.20	37.47	PREF	cma	p	10.74	11.51	7.43	41.60
PREF	mwr	f	11.07	11.48	7.35	37.47	PREF	mwr	b	11.33	11.52	7.72	36.41
PREF	rnd	f	11.09	11.58	6.92	35.60	PREF	mwr	a	11.70	11.82	7.88	37.20
PREF	mwr	a	11.09	11.44	7.21	44.55	PREF	mwr	f	12.07	11.93	8.07	39.17
PREF	mwr	b	11.30	11.54	7.63	36.26	PREF	mwr	p	12.14	11.84	8.32	39.12
PREF	cma	a	11.39	11.74	7.59	38.38	PREF	cma	a	12.59	12.02	7.94	38.27
MWR			13.76	12.72	7.41	38.27	MWR			14.16	12.74	7.59	37.25

It is particularly interesting there is no statistical difference between PREF^{opt} and PREF^{rnd} for both $\mathcal{P}_{j.rnd}$ and $\mathcal{P}_{j.rndn}$ ranking-models. That is to say, tracking optimal dispatches gives the same performance as completely random dispatches. This indicates that exploring only optimal trajectories can result in a training set where the learning algorithm is inept to determine good dispatches in the circumstances when newly encountered features have diverged from the learned feature set labelled to optimum solutions.

Finally, PREF^{all} and PREF^{opt} gave the best combination for $\mathcal{P}_{j.rnd}$ and $\mathcal{P}_{j.rndn}$. However, in the latter case PREF^{rnd} had the best mean relative error although not statistically different from PREF^{all} and PREF^{opt}.

For $\mathcal{P}_{j.rnd}$ the best mean relative error was for PREF^{all}. In that case adding random suboptimal trajectories with the optimal trajectories gave the learning algorithm a greater variety of preference pairs for getting out of local minima. Therefore, a general trajectory scheme would explore both optimal with suboptimal paths.

4.3 Following CMA-ES Guided Trajectory

The rational for using the Φ^{cma} strategy was mostly due to the fact that a linear classifier created the training data (using the weights found via CMA-ES optimisation). Hence the training data created should be linearly separable, which in turn should boost the training accuracy for a linear classification learning model. However, this is not the case since PREF^{cma} does not improve the

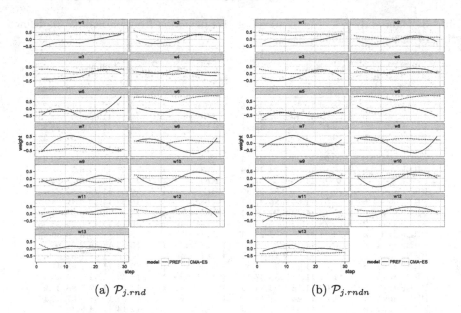

(a) $\mathcal{P}_{j.rnd}$ (b) $\mathcal{P}_{j.rndn}$

Fig. 4. Linear weights (w_1 to w_{13} from left to right, top to bottom) found via CMA-ES optimisation (dashed), and weights found via learning classification PREF_p^{cma} model (solid).

original CMA-ES heuristic which was used to guide its training set Φ^{cma}. However, the PREF^{cma} approach is preferred to that of PREF^{mwr}, so there is some information gained by following the CMA-ES obtained weights instead of simple priority dispatching rules, such as MWR. Inspecting the CMA-ES guided training data more closely, in particular the linear weights for Eq. (1). The weights are depicted in Fig. 4 for problem spaces $\mathcal{P}_{j.rnd}$ (left) and $\mathcal{P}_{j.rndn}$ (right). The original weights found via CMA-ES optimisation that are used to guide the collection of training data are depicted dashed whereas weights obtained by the linear classification PREF_p^{cma} model are depicted solid.

From the CMA-ES experiments it is clear that a lot of weight is applied to decision variable w_6 which corresponds to implementing MWR, yet the existing weights for other features directs the evolutionary search to a "better" training data to learn than the PREF models. Arguably, the training data could be even better, however implementing CMA-ES is rather costly. In [5] the optimisation had not fully converged given its allocated 288 hrs of computation time.

It might also be an artefact because the sampling of the feature space during CMA-ES search is completely different to the data generation described in this study. Hence the different scaling parameters for the features might influence the results. Moreover, the CMA-ES is minimising the makespan directly, whereas the PREF models are learning to discriminate optimal versus suboptimal features sets that are believed to imply a better deviation from optimality later on. However, in that case, the process is very vulnerable when it comes to any divergence from the optimal path. Ideally, it would be best to combine both methodologies: Collect training data from the CMA-ES optimisation which optimises w.r.t. the ultimate performance measure used, and in order to improve upon those weights even further, use a preference based learning approach to deter from any local minima.

5   Summary and Conclusion

The study presents strategies for how to generate training data to be used in supervised learning of linear composite dispatching rules for job-shop scheduling. The experimental results provide evidence of the benefit of adding suboptimal solutions to the training set apart from optimal ones. The subsequent rankings are not of much value, since they are disregarded anyway, but the classification of optimal[2] and suboptimal features are of paramount importance. However, the trajectories to create training instances have to be varied to boost performance. This is due to the fact that sampling only states that correspond to optimal or close-to optimal schedules isn't of much use when the model has diverged too far. Since we are dealing with sequential decision making, all future observations are dependent on previous operations. Therefore, to account for this drawback, an imitation learning approach by [13,14] could fruitful. In that case, we could continue with our $PREF^{opt}$ model and collect a new training set by following the learned policy and use that to create a new model similar to the Φ^{all} scheme. In short, using the model to update itself. This can be done several times until the weights converge. The benefit of this approach is that the states that are likely to occur in practice are investigated and as such used to dissuade the model from making poor choices. Alas, due to the computational cost[3] of collecting the training set Φ, this sort of methodology isn't suitable for high dimensionality of job-shops.

Unlike [8,10,15] learning only optimal training data was not fruitful. However, inspired by the original work of [7], having heuristics guide the generation of training data (while using optimal labelling based on a solver) gave meaningful preference pairs which the learning algorithm could learn. In conclusion, henceforth, the training data will be generated with $PREF_p^{all}$ scheme for the authors' future work. Based on these preliminary experiments, we continue to test on a greater variety of problem data distributions for scheduling, namely job-shop and permutation flow-shop problems. Once training data has been carefully created, global dispatching rules can finally be learned with the hope of implementing them for a greater number of jobs and machines. This is the focus of our current work.

---

[2] Here the tasks labelled 'optimal' do not necessarily yield the optimum makespan (except in the case of following optimal trajectories), instead these are the optimal dispatches for the given partial schedule.

[3] Note, each partial schedule corresponding to a feature in Φ is optimised to obtain its correct labelling.

References

1. Gurobi Optimization Inc: Gurobi optimization (version 5.6.2) [software] (2013). http://www.gurobi.com/
2. Hansen, N., Ostermeier, A.: Completely derandomized self-adaptation in evolution strategies. Evol. Comput. **9**(2), 159–195 (2001)
3. Ingimundardottir, H., Runarsson, T.P.: Supervised learning linear priority dispatch rules for job-shop scheduling. In: Coello, C.A.C. (ed.) LION 2011. LNCS, vol. 6683, pp. 263–277. Springer, Heidelberg (2011)
4. Ingimundardottir, H., Runarsson, T.P.: Determining the characteristic of difficult job shop scheduling instances for a heuristic solution method. In: Hamadi, Y., Schoenauer, M. (eds.) LION 2012. LNCS, vol. 7219, pp. 408–412. Springer, Heidelberg (2012)
5. Ingimundardottir, H., Runarsson, T.P.: Evolutionary learning of weighted linear composite dispatching rules for scheduling. In: International Conference on Evolutionary Computation Theory and Applications (ECTA) (2014)
6. Jayamohan, M., Rajendran, C.: Development and analysis of cost-based dispatching rules for job shop scheduling. Eur. J. Oper. Res. **157**(2), 307–321 (2004)
7. Li, X., Olafsson, S.: Discovering dispatching rules using data mining. J. Sched. **8**, 515–527 (2005)
8. Malik, A.M., Russell, T., Chase, M., Beek, P.: Learning heuristics for basic block instruction scheduling. J. Heuristics **14**(6), 549–569 (2008)
9. von Neumann, J., Morgenstern, O.: Theory of Games and Economic Behavior (Commemorative Edition). Princeton University Press, Princeton Classic Editions, Princeton (2007)
10. Olafsson, S., Li, X.: Learning effective new single machine dispatching rules from optimal scheduling data. Int. J. Prod. Econ. **128**(1), 118–126 (2010)
11. Panwalkar, S.S., Iskander, W.: A survey of scheduling rules. Oper. Res. **25**(1), 45–61 (1977)
12. Rosen, K.H.: Discrete Mathematics and its Applications, Chap. 9, 5th edn. McGraw-Hill Inc, New York (2003)
13. Ross, S., Bagnell, D.: Efficient reductions for imitation learning. In: Teh, Y.W., Titterington, D.M. (eds.) Proceedings of the Thirteenth International Conference on Artificial Intelligence and Statistics (AISTATS 2010), vol. 9, pp. 661–668 (2010). http://www.jmlr.org/proceedings/papers/v9/ross10a/ross10a.pdf
14. Ross, S., Gordon, G.J., Bagnell, D.: A reduction of imitation learning and structured prediction to no-regret online learning. In: Gordon, G.J., Dunson, D.B. (eds.) Proceedings of the Fourteenth International Conference on Artificial Intelligence and Statistics (AISTAT 2011), vol. 15, pp. 627–635, Journal of Machine Learning Research - Workshop and Conference Proceedings (2011). http://www.jmlr.org/proceedings/papers/v15/ross11a/ross11a.pdf
15. Russell, T., Malik, A.M., Chase, M., van Beek, P.: Learning heuristics for the superblock instruction scheduling problem. IEEE Trans. Knowl. Data Eng. **21**(10), 1489–1502 (2009)
16. Watson, J.P., Barbulescu, L., Whitley, L.D., Howe, A.E.: Contrasting structured and random permutation flow-shop scheduling problems: search-space topology and algorithm performance. INFORMS J. Comput. **14**, 98–123 (2002)

A Practical Case of the Multiobjective Knapsack Problem: Design, Modelling, Tests and Analysis

Brahim Chabane[1,2(✉)], Matthieu Basseur[1], and Jin-Kao Hao[1]

[1] LERIA, Université d'Angers, 2 Bd Lavoisier, 49045 Angers Cedex 01, France
{chabane,basseur,hao}@info.univ-angers.fr
[2] GePI Conseil, 14 Place de la Dauversire, 49000 Angers, France

Abstract. In this paper, we present a practical case of the multiobjective knapsack problem which concerns the elaboration of the optimal action plan in the social and medico-social sector. We provide a description and a formal model of the problem as well as some preliminary computational results. We perform an empirical analysis of the behavior of three metaheuristic approaches: a fast and elitist multiobjective genetic algorithm (NSGA-II), a Pareto Local Search (PLS) algorithm and an Indicator-Based Multi-Objective Local Search (IBMOLS).

1 Introduction

During the last decades, combinatorial optimization has received great interest and takes an important and even strategic place in industrial settings. Multiobjective metaheuristics have proven their efficiency for solving many practical problems, which usually consist in handling simultaneously several conflicting objectives [2].

The aim of this paper is to present a practical problem, proposed by the company "GePI" which works in the social and medico-social domain. This study is unique in the sector because even if this sector is increasingly computerized these last years, it remains among the sectors where optimization is not yet used as a tool for decision support.

The problem considered in this paper consists in elaborating action plans in order to improve the overall management of the considered structure. The aim is to choose a subset of actions among many possible actions while optimizing several objectives. Each action has a realization cost and can influence other objectives (positively or negatively). The global cost of the solution should not exceed a predefined budget. Our problem is a multiobjective knapsack problem [5,8], which is well known in the literature. The action plan represents the knapsack and the selected actions represent the items to put in the knapsack respecting the budget constraint.

The considered problem can include more than one thousand possible actions and involve up to eight objectives. Here, we are interested in providing efficient techniques in terms of solutions quality and response time.

© Springer International Publishing Switzerland 2015
C. Dhaenens et al. (Eds.): LION 9 2015, LNCS 8994, pp. 249–255, 2015.
DOI: 10.1007/978-3-319-19084-6_23

In the following, a description and a formal model of the problem are first introduced. Then, the ways of generating problem instances is provided. Next, we present the first results using three metaheuristic algorithms: PLS (Pareto Local Search) [7,9], IBMOLS (Indicator-Based Multi-Objective Local Search) [1] and NSGA-II [4]. Finally, we end with a conclusion and the future work.

2   Problem Modeling and Description

This project is a part of "*MSQualité*" software developed by the company GePI which is dedicated specifically to the social and medico-social sector that includes 34000 different structures (rest houses, accommodation and rehabilitation centers, work-based support centers, etc.) [10]. Even if the use of computer resources has made considerable progress in recent years in this sector, they are basically employed for the daily management of the structures. In particular, optimization tools are completely absent. In this context of lack of advanced models and tools, GePI has decided to set up this project to develop a multiobjective decision support system to assist managers in their action plan elaboration.

We can define an action plan p as a subset of actions selected among a set of feasible actions A, in order to maximize or minimize a set F of conflicting objectives. p can be represented by a vector $p = (a_1, a_2, ..., a_n)$ with n equal to the size of A. $a_i = 1$ if the action a_i is selected and $a_i = 0$ otherwise. The set of the possible action plans (solutions) is denoted by P. The origins of the actions are either issued from action plans already made in the structure itself or other similar structures, or are decided by the managers for continuous improvements.

The objectives can be of varied nature, namely qualitative (such as "improve resident's quality of life") or quantitative (such as "increase the resident's autonomy"). In both cases, each objective is represented by an objective function f_j which associates to every action $a_i \in A$ its impact on the objective j.

An action $a_i \in A$ can have a positive or a negative impact on an objective $f_j \in F$. This impact is evaluated by the function $f_j(a_i) = v_{ij}$ which assigns to any action a_i an integer value $v_{ij} \in [-100, +100]$ that represents the contribution of the action a_i to the achievement of the objective j ($v_{ij} > 0$) or the degradation of the action a_i for the objective j ($v_{ij} < 0$). $v_{ij}=0$ when the action a_i has no effect on the objective j. Thus, we can associate to each action a_i an objective vector $v=(f_1(a_i), f_2(a_i), ..., f_m(a_i))$ with m equal to the number of objectives to optimize. We define m in the interval [2,8] because in practice, the projects can have up to eight objectives (otherwise the project management and evaluation[1] will be difficult).

Considering an action plan $p = (a_1, a_2, ..., a_n) \in \{0, 1\}^n$, the impact that p^* has on an objective j is obtained by:

---

[1] In social and medico-social structures, a project is defined for a period of five years. At the sixth year, the evaluation of the project is carried out and the attainment of each objective is measured. Therefore, the more there are objectives, the more the evaluation is difficult.

$$f_j(p) = \sum_{i=1}^{n} a_i f_j(a_i) \tag{1}$$

Thus, an objective vector $z = (f_1(p), f_2(p), ..., f_m(p))$ is associated to each solution $p \in P$. A constraint c_j is added for every objective j determining the minimal threshold accepted for f_j. In the following, we consider that all the objectives must be improved:

$$f_j(p) \geq c_j \geq 0 \tag{2}$$

An additional constraint concerns the realization cost of the solution which should not exceed some budget β fixed by the decision maker. Indeed, each action a_i has a realization cost ω_i which can take negative values since there may be actions with negative cost when it is about selling of objects or services. Actions with no cost are also to be taken into account. The global cost of a solution p corresponds to the following cost sum of the actions of p:

$$\begin{cases} W(p) = \sum_{i=1}^{n} a_i \omega_i \\ W(p) \leq \beta \end{cases} \tag{3}$$

So, the optimization goal aims to find $p^* \in \arg\max_{p \in P} F(p)$ verifying:

$$\begin{cases} p^* \in \{0, 1\}^n \\ \forall_j \in \{1, m\}, f_j(p^*) \geq c_j \\ \sum_{i=1}^{n} a_i \omega_i \leq \beta \end{cases} \tag{4}$$

Since we deals with a multiobjective case, p^* is not unique. Instead, we obtain a set of non-dominated solutions (in Pareto optimality sens). The aim is to approximate the Pareto front effectively.

3 Instance Generation

Based on the above model, we have randomly generated a number of instances with different sizes (actions) {50,100,250,500,750,1000} and different number of objectives $m \in \{2, ..., 8\}$. We have also generated several partially structured instances which are more representative of real cases. To be as close as possible to the real problem, for each objective function, an action has a chance of 50 % to be neutral, 40 % to have a positive impact and 10 % to have a negative impact. Moreover, the cost of 40 % of the actions is set to 0. The non-null action values are uniformly taken from the interval [0,100] (positively or negatively). The non-null action costs are uniformly taken in the interval [-10000,10000].

4   Preliminary Results

We have tested, on random instances, three metaheuristic algorithms: NSGA-II, IBMOLS and PLS. For NSGA-II, we have used a population of size 100, a mutation probability of $1/n$ (where n is the number of the actions). The initial population is generated randomly while verifying that the cost of the individuals do not exceed the budget β. For IBMOLS, we have used the iterative version with a population of size 10 (the initial population is generated in the same way as NSGA-II) and the epsilon indicator as realized in [1] and in [12]. The fitness of each individual in the population is evaluated, with respect to the rest of the population, using the following formula:

$$I_\epsilon(P\backslash\{x\}, x) = \sum_{z \in P\backslash\{x\}} - \exp^{-I_\epsilon(z,x)/k} \tag{5}$$

where $k > 0$ represents the scaling factor [1] (k is set to 0.01 in our experiments).

NSGA-II is compared with IBMOLS and shows to be inferior to IBMOLS on the large size problems. Indeed, both algorithms use a bounded population and the same selection strategy: one random neighbor of each individual of the current population is selected to be a member of the child population in NSGA-II or to integrate the current population in IBMOLS.

PLS [7] is used with an archive of unbounded size and an initial population of one individual. The neighborhood generation is the same as for PLS and IBMOLS. The ith neighbor of the solution $p = (a_1, a_2, ..., a_n)$ is obtained by flipping the value of a_i and only the neighbors verifying the constraint β are accepted. The budget constraint β is fixed to one million e for the three methods.

For the quality assessment, we have performed 30 runs of each method to solve each instance. For IBMOLS and NSGA-II, a run time of $n^2 * m$ milliseconds is used for each run (where n is the number of actions and m is the number of objectives). But for PLS, the run time is limited to one hour because the size of the archive and the response time increase exponentially with the instance size, making PLS inefficient for large size problems. Our experiments are realized on an Intel core i5-2400 CPU machine with 2 x 3.10 Ghz and 16 Gb of RAM. Then, we have evaluated our outputs using the R and ϵ indicators and computed their average values over the 30 runs for each algorithm and each tested instance. For the statistical analysis, we have used the Mann-Whitney test. In our experiments, we say that algorithm A outperforms algorithm B if the Mann-Whitney test provides a confidence level greater than 95 %. To calculate the indicator values and the Mann-Whitney test, we have used the performance assessment package (PISA) [6] which is available at: http://www.tik.ee.ethz.ch/sop/pisa/?page=assessment.php.

Table 1 shows a comparison of NSGA-II and IBMOLS in terms of the mean values obtained for R and ϵ indicators over 30 runs, using 30 instances with different sizes. The first column presents the instance size "m_n" where m and n are the number of objectives and actions respectively. The values in bold mean that the corresponding algorithm is at least 95 % statistically better than the other one for the considered instance and indicator.

Table 1. Comparison of mean values of I_ϵ and I_R of IBMOLS and NSGA-II

Instance	I_ϵ		I_R	
	NSGA-II	IBMOLS	NSGA-II	IBMOLS
2_50	**0.520**	0.135	**0.160**	0.030
2_100	**0.520**	0.135	**0.160**	0.030
2_150	**0.491**	0.200	**0.159**	0.055
2_250	**0.521**	0.283	**0.174**	0.074
2_500	**0.558**	0.358	**0.191**	0.097
2_1000	**0.567**	0.306	**0.191**	0.072
3_50	**0.415**	0.229	**0.112**	0.044
3_100	**0.412**	0.368	**0.119**	0.096
3_150	**0.461**	0.386	0.122	0.108
3_250	0.442	0.411	0.108	0.109
3_500	0.451	**0.482**	0.121	**0.143**
3_1000	0.480	**0.654**	0.119	**0.086**
4_50	0.401	0.398	0.094	0.086
4_100	0.351	**0.438**	0.083	**0.108**
4_150	0.451	0.450	0.099	**0.133**
4_250	0.405	**0.537**	0.100	**0.189**
2_500	0.347	**0.594**	0.089	**0.230**
4_1000	0.396	**0.709**	0.085	**0.278**
5_50	**0.364**	0.292	**0.073**	0.047
5_100	0.375	**0.459**	0.081	**0.111**
5_150	0.396	**0.556**	0.092	**0.165**
5_250	0.316	**0.650**	0.077	**0.246**
5_500	0.374	**0.700**	0.086	**0.285**
5_1000	0.307	**0.788**	0.084	**0.353**
6_50	**0.381**	0.350	0.091	0.085
6_100	0.267	**0.385**	0.046	**0.084**
5_150	0.291	**0.564**	0.066	**0.204**
6_250	0.219	**0.664**	0.042	**0.278**
6_500	0.336	**0.748**	0.104	**0.368**
6_1000	0.199	**0.846**	0.058	**0.459**

From Table 1 we can conclude that on the whole NSGA-II is more efficient on the small instances (instances with 50 actions or no more than 3 objectives) but IBMOLS performs better than NSGA-II as soon as we exceed 4 objectives. It still remains that the diversity of the compromise solutions is reduced with IBMOLS and should be improved.

5   Conclusion

In this paper, we presented an application of the multiobjective knapsack problem encountered in the structures of the social and medico-social sector. A formal model of the problem has been provided. The efficiency of IBMOLS and its superiority to NSGA-II on a large size problems has been shown. However, the epsilon indicator of IBMOLS does not always maintain naturally the diversity of the population in the objective space. It should be interesting to consider a modified version of IBMOLS or to evaluate the effectiveness of other quality indicators. In [3], the hypervolume contribution indicator has shown a high performance level and outperforms the I_ϵ indicator on different multiobjective combinatorial problems. However, it cannot be applied to the present problem since when the number of objective function is greater than three, the high computational cost of the hypervolume contribution calculation tends to drastically reduce the convergence speed of the algorithm. An interesting idea should be to consider the R2 indicator [11], which can be a good trade-off between a reduced computation cost and an efficient indicator.

Acknowledgments. We are grateful to the reviewers for their useful comments. This work was partially supported by the French Ministry for Research and Education through a CIFRE grant (number 0450/2013). We thank M. Rachid Naitali, the Director of GePI Conseil, for his support.

References

1. Basseur, M., Liefooghe, A., Khoi, L., Burke, E.K.: The efficiency of indicator-based local search for multi-objective combinatorial optimisation problems. J. Heuristics **18**, 263–296 (2012)
2. Basseur, M., Talbi, E., Nebro, A., Alba, E.: Metaheuristics for multiobjective combinatorial optimization problems: review and recent issues. INRIA Research Report 5978 (2006)
3. Basseur, M., Zeng, R.-Q., Hao, J.-K.: Hypervolume-based Multi-objective Local Search. Neural Computing and Applications, pp. 1917–1929. Springer, London (2012)
4. Deb, K., Pratap, A., Agarwal, S., Meyarivan, T.: A fast and elitist multiobjective genetic algorithm: NSGA-II. IEEE Trans. Evol. Comput. **6**, 182–197 (2002)
5. Kellerer, H., Pferschy, U., Pisinger, D.: Knapsack Problems. Springer Science & Business Media, Heidelberg (2004)
6. Knowles, J.D., Thiele, L., Zitzler, E.: A Tutorial on the performance assessment of stochastic multiobjective optimizers. Technical report, Computer Engineering and Networks Laboratory (TIK) 214, ETH Zurich (2006)
7. Liefooghe, A., Humeau, J., Mesmoudi, S., Jourdan, L., Talbi, E.-G.: On dominance-based multiobjective local search: design, implementation and experimental analysis on scheduling and traveling salesman problems. J. Heuristics **18**, 317–352 (2012)
8. Lust, T., Teghem, J.: The multiobjective multidimensional knapsack problem: a survey and a new approach. Int. Trans. Oper. Res. **19**, 495–520 (2012)

9. Paquete, L., Chiarandini, M., Sttzle, T.: Pareto local optimum sets in the biob-jective traveling salesman problem: an experimental study. In: Gandibleux, X., Sevaux, M., Sörensen, K., T'kindt, V. (eds.) Metaheuristics for Multiobjective Optimisation, pp. 177–199. Springer, Heidelberg (2004)
10. The ministry of social affairs and health: Le champ social et médico-social: une activité en forte croissance, des métiers qui se développent et se diversifient. Repères Anal. 44 (2012)
11. Trautmann, H., Wagner, T., Brockhoff, D.: R2-EMOA: focused multiobjective search using R2-indicator-based selection. In: Nicosia, G., Pardalos, P. (eds.) LION 7. LNCS, vol. 7997, pp. 70–74. Springer, Heidelberg (2013)
12. Zitzler, E., Künzli, S.: Indicator-based selection in multiobjective search. In: Yao, X., Burke, E.K., Lozano, J.A., Smith, J., Merelo-Guervós, J.J., Bullinaria, J.A., Rowe, J.E., Tiño, P., Kabán, A., Schwefel, H.-P. (eds.) PPSN 2004. LNCS, vol. 3242, pp. 832–842. Springer, Heidelberg (2004)

A Bayesian Approach to Constrained Multi-objective Optimization

Paul Feliot[1,2]($\boxtimes$), Julien Bect[1,2], and Emmanuel Vazquez[1,2]

[1] IRT SystemX, Palaiseau, France
{paul.feliot,julien.bect,emmanuel.vazquez}@irt-systemx.fr
[2] SUPELEC, Gif-sur-yvette, France
{paul.feliot,julien.bect,emmanuel.vazquez}@supelec.fr

Abstract. This paper addresses the problem of derivative-free multi-objective optimization of real-valued functions under multiple inequality constraints. Both the objective and constraint functions are assumed to be smooth, nonlinear, expensive-to-evaluate functions. As a consequence, the number of evaluations that can be used to carry out the optimization is very limited. The method we propose to overcome this difficulty has its roots in the Bayesian and multi-objective optimization literatures. More specifically, we make use of an extended domination rule taking both constraints and objectives into account under a unified multi-objective framework and propose a generalization of the expected improvement sampling criterion adapted to the problem. A proof of concept on a constrained multi-objective optimization test problem is given as an illustration of the effectiveness of the method.

1 Introduction

This paper addresses the problem of derivative-free multi-objective optimization of real-valued functions under multiple inequality constraints:

$$\begin{cases} \text{Minimize} & f(x) \\ \text{Subject to } x \in \mathbb{X} \text{ and } c(x) \leq 0 \end{cases}$$

where $f = (f_j)_{1 \leq j \leq p}$ is a vector of objective functions to be minimized, $\mathbb{X} \subset \mathbb{R}^d$ is the search domain and $c = (c_i)_{1 \leq i \leq q}$ is a vector of constraint functions. Both the objective functions f_j and the constraint functions c_i are assumed to be smooth, nonlinear functions that are expensive to evaluate. As a consequence, the number of evaluations that can be used to carry out the optimization is very limited. This setup typically arises when the values $f(x)$ and $c(x)$ for a given $x \in \mathbb{X}$ correspond to the outputs of a computationally expensive computer program.

In this work, we consider a Bayesian approach to this optimization problem. The objective and constraint functions are modelled using a vector-valued Gaussian process and $\mathbb{X}$ is explored using a sequential Bayesian design of experiments approach. More specifically, we focus on the Expected Improvement (EI)

C. Dhaenens et al. (Eds.): LION 9 2015, LNCS 8994, pp. 256–261, 2015.
DOI: 10.1007/978-3-319-19084-6_24

sampling criterion. This criterion was originally introduced in the context of single-objective, unconstrained optimization [10,13]. It was later extended to handle constraints [7,16,18,20,21] and to address unconstrained multi-objective problems [4,9,17,23]. However, to the best of our knowledge, the general case of a constrained multi-objective problem has only been addressed very recently by [22]. In their paper, Shimoyama et al. consider three different Bayesian criteria for unconstrained multi-objective optimization and study the effect of multiplying the criteria by a probability of feasibility in order to handle the constraints.

The approach we propose to handle the constraints is based on an extended domination rule, in the spirit of [6,15,19], which takes both objectives and constraints into account under a unified framework. The extended domination rule makes it possible to derive a new expected improvement criterion to deal with constrained multi-objective optimization problems. Section 2 introduces the proposed method, while Sect. 3 presents a proof of concept on a classical test case from the literature. Results and future works are briefly discussed at the end of Sect. 3.

2 An Expected Improvement Criterion for Constrained Multi-objective Optimization

In this section, we present our extended domination rule and introduce a new expected improvement criterion suitable for constrained and unconstrained multi-objective problems. The new criterion is equivalent to the original EI on unconstrained single-objective problems and to Schonlau's extension to the constrained case [21] once a feasible point has been found. It is also similar to the formulation of [23] for unconstrained multi-objective problems and to that of [22] in the constrained case once a feasible point has been found. As such, it can be seen as a generalization of the above-mentioned criteria.

Denote by $\mathbb{F} \subset \mathbb{R}^p$ and $\mathbb{C} \subset \mathbb{R}^q$ the objective and constraint spaces respectively, and let $\mathbb{Y} = \mathbb{F} \times \mathbb{C}$. We shall say that $y_1 \in \mathbb{Y}$ dominates $y_2 \in \mathbb{Y}$, which will be denoted by $y_1 \lhd y_2$, if $\psi(y_1)$ dominates $\psi(y_2)$ in the usual Pareto sense, where

$$\psi : \mathbb{F} \times \mathbb{C} \rightarrow \overline{\mathbb{R}}^p \times \mathbb{R}^q$$
$$(y_f, y_c) \mapsto \begin{cases} (y_f, 0) & \text{if } y_c \leq 0, \\ (+\infty, \max(y_c, 0)) & \text{otherwise,} \end{cases}$$

In the above system of equations, $\overline{\mathbb{R}}$ denotes the extended real line. For unconstrained problems, we simply take the usual domination rule on $\mathbb{F}$. Figure 1 illustrates this extended domination rule in different cases.

Assume now that $\mathbb{Y}$ is bounded. Much like [4,17,23], we define the improvement yielded by a new observation as the increase of the dominated hypervolume:

$$I_N(x_{N+1}) = |H_{N+1}| - |H_N|,$$

where H_N is the subset of $\mathbb{Y}$ dominated by the solutions observed so far $(f(x_1), c(x_1)), \ldots, (f(x_N), c(x_N))$ and $|\cdot|$ denotes the usual (Lebesgue) volume

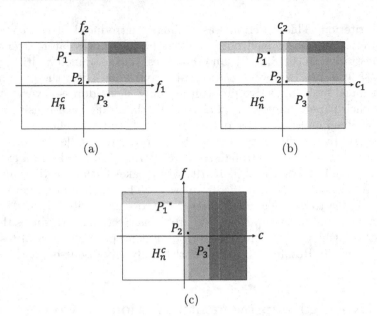

Fig. 1. Illustration of the extended domination rule in different situations. The region dominated by each point is represented by a shaded area. Darker shades of gray indicate overlapping regions. (a) Feasible solutions are compared with respect to their objective values using the usual domination rule in the objective space. (b) Non-feasible solutions are compared component-wise with respect to their constraint violations using the usual domination rule applied in the constraint space. (c) Feasible solutions always dominate non-feasible solutions; other cases are handled as in the first two figures.

measure in $\mathbb{R}^{p+q}$. The corresponding expected improvement criterion can be written as

$$\begin{aligned}
\mathrm{EI}_N(x_{N+1}) &= \mathbb{E}_N\left((I_N(x_{N+1}))\right) \\
&= \mathbb{E}_N\left(\int_{\mathbb{Y}\setminus H_N} \mathbb{1}_{\xi(x_{N+1})\vartriangleleft y}\,\mathrm{d}y\right) \\
&= \int_{\mathbb{Y}\setminus H_N} \mathbb{P}_N(\xi(x_{N+1})\vartriangleleft y)\,\mathrm{d}y
\end{aligned}$$

where $\mathbb{P}_N$ denotes the probability conditional to the observations and ξ is a vector-valued Gaussian model for (f,c).

Even though the integrand of the EI formula can be readily computed analytically, its integration is not trivial due to the combinatorial nature of the problem [2,5,8]. To overcome this difficulty, we propose to use a Sequential Monte Carlo (SMC) approximation [1,3,11,12]:

$$\mathrm{EI}_N(x_{N+1}) \approx \sum_{i=1}^{n} w_i\,\mathbb{P}_N(\xi(x_{N+1})\vartriangleleft y_i),$$

where $\mathcal{Y}_N = (w_i, y_i)_{1\le i\le n}$ is a weighted sample that targets the uniform density on $\mathbb{Y}\setminus H_N$.

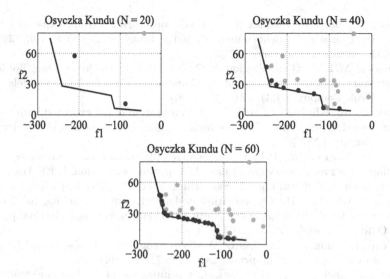

Fig. 2. Test results on Osyczka and Kundu test problem with, from left to right, $N = 20$, 40 and 60 evaluations. Only feasible points are shown on the figures. The dark dots represent non-dominated observations while the light gray dots represent dominated ones. The dark curve represents the target Pareto front.

3   Proof of Concept

In this paper, we illustrate the behavior of our new optimization strategy using the Osyczka and Kundu test problem [14] for constrained multi-objective optimization ($d = 6$, $p = 2$, $q = 6$). The algorithm is initialized using a Latin Hypercube sample of 18 samples and proceeds using the above mentionned criterion. Figure 2 shows the convergence of the algorithm at different steps of the optimization.

We are also able to report good results on other challenging test cases from the literature and future communications will include a comparison of our method to reference optimization methods. More details about the SMC procedure will also be proposed.

Acknowledgements. This research work has been carried out in the frame of the Technological Research Institute SystemX, and therefore granted with public funds within the scope of the French Program *Investissements d'Avenir*.

References

1. Au, S.K., Beck, J.L.: Estimation of small failure probabilities in high dimensions by subset simulation. Probab. Eng. Mech. **16**(4), 263–277 (2001)
2. Bader, J., Zitzler, E.: Hype: An algorithm for fast hypervolume-based many-objective optimization. Evol. Comput. **19**(1), 45–76 (2011)

3. Benassi, R., Bect, J., Vazquez, E.: Bayesian optimization using sequential Monte Carlo. In: Hamadi, Y., Schoenauer, M. (eds.) LION 2012. LNCS, vol. 7219, pp. 339–342. Springer, Heidelberg (2012)
4. Emmerich, M.T.M., Giannakoglou, K.C., Naujoks, B.: Single- and multi-objective evolutionary optimization assisted by Gaussian random field metamodels. IEEE Trans. Evol. Comput. **10**(4), 421–439 (2006)
5. Emmerich, M., Klinkenberg, J.W.: The computation of the expected improvement in dominated hypervolume of Pareto front approximations. Leiden University, Rapport Technique (2008)
6. Fonseca, C.M., Fleming, P.J.: Multiobjective optimization and multiple constraint handling with evolutionary algorithms. I. A unified formulation. IEEE Trans. Syst. Man Cybern. B Cybern. Part A: Syst. Hum. **28**(1), 26–37 (1998)
7. Gramacy, R.L., Lee, H.: Optimization under unknown constraints. In: Bayesian Statistics 9. In: Proceedings of the Ninth Valencia International Meeting, pp. 229–256. Oxford University Press (2011)
8. Hupkens, I., Emmerich, M., Deutz, A.: Faster computation of expected hypervolume improvement. arXiv preprint arXiv:1408.7114 (2014)
9. Jeong, S., Minemura, Y., Obayashi, S.: Optimization of combustion chamber for diesel engine using kriging model. J. Fluid Sci. Technol. **1**(2), 138–146 (2006)
10. Jones, D.R., Schonlau, M., Welch, W.J.: Efficient global optimization of expensive black-box functions. J. Global Optim. **13**(4), 455–492 (1998)
11. Li, L., Bect, J., Vazquez, E.: Bayesian Subset Simulation: a kriging-based subset simulation algorithm for the estimation of small probabilities of failure. In: Proceedings of PSAM 2011 & ESREL 2012, 25–29 June 2012, Helsinki, Finland. IAPSAM (2012)
12. Liu, J.S.: Monte Carlo strategies in scientific computing. Springer, Heidelberg (2008)
13. Mockus, J.: Application of bayesian approach to numerical methods of global and stochastic optimization. J. Global Optim. **4**(4), 347–365 (1994)
14. Osyczka, A., Kundu, S.: A new method to solve generalized multicriteria optimization problems using the simple genetic algorithm. Struct. Optim. **10**(2), 94–99 (1995)
15. Oyama, A., Shimoyama, K., Fujii, K.: New constraint-handling method for multi-objective and multi-constraint evolutionary optimization. Trans. Jpn. Soc. Aeronaut. Space Sci. **50**(167), 56–62 (2007)
16. Parr, J.M., Keane, A.J., Forrester, A.I.J., Holden, C.M.E.: Infill sampling criteria for surrogate-based optimization with constraint handling. Eng. Optim. **44**(10), 1147–1166 (2012)
17. Picheny, V.: Multiobjective optimization using Gaussian process emulators via stepwise uncertainty reduction. Stat. Comput. 16 p. (2014). doi:10.1007/s11222-014-9477-x
18. Picheny, V.: A stepwise uncertainty reduction approach to constrained global optimization. In: Proceedings of the 17th International Conference on Artificial Intelligence and Statistics (AISTATS), Reykjavik, Iceland. vol. 33, pp. 787–795. JMLR: W&CP (2014)
19. Ray, T., Tai, K., Seow, K.C.: Multiobjective design optimization by an evolutionary algorithm. Eng. Optim. **33**(4), 399–424 (2001)
20. Sasena, M.J., Papalambros, P., Goovaerts, P.: Exploration of metamodeling sampling criteria for constrained global optimization. Eng. Optim. **34**(3), 263–278 (2002)

21. Schonlau, M., Welch, W.J., Jones, D.R.: Global versus local search in constrained optimization of computer models. In: New Developments and Applications in Experimental Design: Selected Proceedings of a 1997 Joint AMS-IMS-SIAM Summer Conference. IMS Lecture Notes-Monographs Series, vol. 34, pp. 11–25. Institute of Mathematical Statistics (1998)
22. Shimoyama, K., Sato, K., Jeong, S., Obayashi, S.: Updating kriging surrogate models based on the hypervolume indicator in multi-objective optimization. J. Mech. Des. **135**(9), 094503 (2013)
23. Wagner, T., Emmerich, M., Deutz, A., Ponweiser, W.: On expected-improvement criteria for model-based multi-objective optimization. In: Schaefer, R., Cotta, C., Kołodziej, J., Rudolph, G. (eds.) PPSN XI. LNCS, vol. 6238, pp. 718–727. Springer, Heidelberg (2010)

Solving Large MultiZenoTravel Benchmarks with Divide-and-Evolve

Alexandre Quemy[1]([✉]), Marc Schoenauer[1], Vincent Vidal[2], Johann Dréo[3], and Pierre Savéant[3]

[1] TAO Project, INRIA Saclay and LRI Paris-Sud
University and CNRS, Orsay, France
alexandre.quemy@gmail.com, marc.schoenauer@inria.fr
[2] ONERA-DCSD, Toulouse, France
Vincent.Vidal@onera.fr
[3] THALES Research and Technology, Palaiseau, France
{pierre.saveant,johann.dreo}@thalesgroup.com

Abstract. A method to generate various size tunable benchmarks for multi-objective AI planning with a known Pareto Front has been recently proposed in order to provide a wide range of Pareto Front shapes and different magnitudes of difficulty. The performance of the Pareto-based multi-objective evolutionary planner DAE$_{\text{YAHSP}}$ are evaluated on some large instances with singular Pareto Front shapes, and compared to those of the single-objective aggregation-based approach.

1 Introduction

Multi-Objectives Problems (MOP) involves several contradictory criteria to be optimized. The Pareto Set of a MOP is the set of the best trade-offs between these objectives, i.e., solutions that cannot be improved w.r.t. one objective without deteriorating at least another one. The projection of the Pareto Set on the objective space is called the Pareto Front.

Many benchmark suites exist for continuous multi-objective optimization, for which the exact Pareto Front can be analytically computed, and with known difficulties (e.g. dimensionality, shape of the Pareto Fronts, existence of local Pareto-optima, ...). For combinatorial optimization, the situation is not yet so clear, and whereas there exist famous benchmark problems of all sizes, their Pareto Fronts are generally not exactly known except the simplest ones (see e.g., MOCOLIB at http://www.mcdmsociety.org/MCDMlib.html).

The context of the present work is that of *AI planning*: a planning domain D is defined by (i) a set of predicates, that define the state of the system when instantiated, and (ii) a set of possible actions that can be triggered in states where their pre-conditions are satisfied, resulting in a new state. A planning problem instance $\mathcal{P}_D(I, G)$ is defined on a given planning domain D by a list of objects, used to instantiate the predicates to define the states, an initial state I and a goal state G. The aim is to come up with a *feasible plan*, i.e., a set of actions that, when applied in turn to the initial state, lead the system to

© Springer International Publishing Switzerland 2015
C. Dhaenens et al. (Eds.): LION 9 2015, LNCS 8994, pp. 262–267, 2015.
DOI: 10.1007/978-3-319-19084-6_25

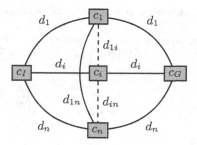

Fig. 1. A schematic view of a general MultiZenoTravel problem.

the goal state, that is optimal w.r.t. a given measure: the number of actions, or the total cost of the plan when actions have non-uniform costs, or the total *makespan* (total duration of the plan) when actions have durations, and can be run in parallel, as in the present work.

The present work presents the first results of DAE$_{YAHSP}$, an Evolutionary Pareto-based multi-objective planner [5] on large instances of MultiZeno-Travel domain with known Pareto Fronts, as proposed in [6]. The paper is organized as follows: Sect. 2 introduces the MultiZenoTravel benchmark suite, and the ZenoSolver algorithm that can derive the true Pareto Front for these instances. Sample very diverse experimental Pareto Fronts illustrate its versatility. Experimental results on some of the large MultiZenoTravel instances obtained by Divide-and-Evolve, the only Pareto-based evolutionary AI planner to-date [5], are compared with those of its single-objective version using the weighted sum aggregation on problems with non-convex fronts in Sect. 3.

2 MultiZenoTravel Benchmarks and ZenoSolver

MiniZenoTravel is a simple temporal planning domain related to logistics, inspired by the well-known ZenoTravel problem of IPC series[1]. It involves cities, passengers, and planes (see e.g., Fig. 1); Planes can fly from one city to another when a link exists; Planes fly either empty, or carrying a unique passenger – and these are the only possible actions. In a MiniZenoTravel instance (Fig. 1), there are n central cities C_i, linked as a clique, and all are linked to the initial city C_I and the goal city C_G; the flight durations are d_i from city C_i to city C_I or C_G, and d_{ij} between cities C_i and C_j. There are t passengers and p planes, and all are in C_I in the initial state I, and all passengers must be in city C_G in the goal state G. The single objective version aims at minimizing the total makespan. Previous work [5,7] proposed a multi-objective version of these benchmarks called MultiZenoTravel, by adding a *cost* c_i for landing in city C_i: the second objective is to minimize the *total cost* of the plan. More recent work [6] extended these benchmarks to problems of various complexity, and proved that such problems could provide Pareto Fronts of various shapes and difficulties, thanks to ZenoSolver, an exact Pareto solver.

---

[1] http://ipc.icaps-conference.org/.

Table 1. Large instances: parameters and generation statistics.

Inst.	n	t	p	Generating functions		Pareto#	h	PPP(k)	Iter.(k)	Time
1	20	6	2	$\frac{5}{2}i + \frac{(i \bmod 2)}{10}$	$\frac{5}{2}i + \frac{(i \bmod 2)}{10}$	409	4015	1568220	3317140	16 h 46
2	3	21	2			61	861	53	233	2006 s
3	10	10	3			383	7205	1056804	7918940	51 h
4	8	26	25	$\sqrt{i}$	i	15	190	34176	60457	4240 s

ZENOSOLVERis a C++11 software dedicated to generate and exactly solve MULTIZENOTRAVEL instances in cases where $d_i + d_j < d_{ij}$ for all (i, j). Firstly, it allows to tune the problem parameters in order to adjust the difficulty or to obtain different shapes of Pareto Fronts. In particular, vectors c and d are generated using two user-defined functions, f and g, such that $c_i = x_c f(i) + y_c$ and $d_i = x_d g(n - i) + y_d$, ensuring that both objectives are conflicting. ZENOSOLVER outputs the corresponding PDDL file (Planning Domain Definition Language, universally used to describe planning problems), that can be directly used by any standard AI planner, and computes the true Pareto Front (see all details in [6]).

We identified some large instances with very diverse front shapes and complexities, that could become a basic set of representative instances for MULTIZENOTRAVEL, allowing fair comparisons between various solvers and approaches. Table 1 gives the parameters used by ZENOSOLVER to build some of them, as well as some statistics about their complexity: The generating time strongly varies, from some minutes for Instance 1 up to 51h for Instance 3. The choice of the generating functions was purely empirical, guided by the fact that we wanted to obtain mainly piecewise concave fronts with uneven point distributions. This is why none of these fronts is linear, and most contain concave parts, i.e., parts where all points are above the segment made of the two extreme points. Unfortunately, this is not obvious on Fig. 2, due to the large scale used here (but see [6] for some zooms). Note the small number of Pareto points of Instance 4, in spite of the complexity of this instance (26 persons), due to the small ratio $\frac{p}{t}$.

3   Multi-objective Experiments

3.1   Divide-and-Evolve

Based on the Divide-and-Conquer paradigm, this generic hybrid evolutionary approach has been originally introduced in [7]. The main idea to solve a planning problem $\mathcal{P}_D(I, G)$ is to find a sequence of states $S_1, \ldots, S_n$, and to use some embedded planner to solve in turn the series of planning problems $\mathcal{P}_D(S_k, S_{k+1})$, for $k \in [0, n]$ (with the convention that $S_0 = I$ and $S_{n+1} = G$). The generation and optimization of the sequence of states $(S_i)_{i \in [1,n]}$ is driven by an evolutionary algorithm, and each subproblems $\mathcal{P}_D(S_k, S_{k+1})$ is handled to an external 'embedded' planner. The concatenation of the corresponding plans (possibly with some compression step) is a solution of the initial problem. A more detailed presentation is given in [1].

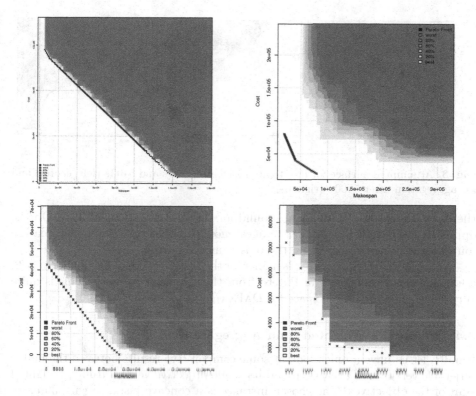

Fig. 2. Attainment surfaces for the Instances 1, 2, 3 and 4.

3.2 Experimental Conditions

The MOEA used here is IBEA_{H^-} [10], the Indicator Base Evolutionary Algorithm [10] using the Hypervolume Difference Indicator, that was demonstrated the best choice in previous work [5]. $\text{DAE}_{\text{YAHSP}}$ internal parameters have been tuned with PARAMILS [3], also using H^-. For each instance, 20 independent runs limited to 5400 s (1800 for instance 4) have been performed. This limit is arbitrary but early experiments on small MULTIZENOTRAVEL instances not shown here have empirically demonstrated (stagnation of the hypervolume for all runs) that indeed the algorithm had reached a stationary state within this limit. All performance assessments and comparisons have been done using the PISA platform [2].

3.3 DAE_YAHSP on Large Instances

Attainment surfaces are displayed on Fig. 2: the darker the region in objective space, the higher the probability to reach it (full white meaning that none of the 20 runs ever reached it). The attainment surfaces for the Instance 1 are uniformly distributed close to the true Pareto Front, even though very few Pareto optima were actually reached. The surfaces for Instances 2 and 3 are much further from

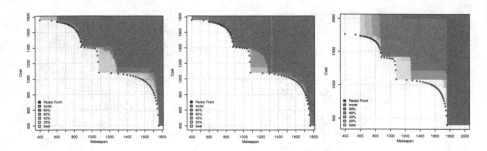

Fig. 3. Attainment surface for Pareto approach after 900 s and 5400 s (left, center) and for aggregation after 5400 s (right).

the exact front (only 2 points are found for the Instance 3 out of 383). On the opposite, even if with a smaller budget, most of the actual Pareto optima are found for Instance 4, except on the most concave part.

We can notice that, even if n is higher for the Instance 4 than for the Instance 2, adding planes results here in Pareto front that is easier to reach. This is quite surprising since the search space for $\text{DAE}_{\text{YAHSP}}$ is increasing with p.

3.4 Pareto Vs Weighted Sum Aggregation

Finally, let us have a quick look at some comparative results between the multi-objective version of $\text{DAE}_{\text{YAHSP}}$ and its single-objective version using a weighted sum of the objectives. The chosen instance is a concave instance with 30 cities (resulting in a Pareto Front made of 66 points) not displayed here. All experimental conditions are the same than in [4]. One aggregated run amounts to 11 independent runs, the weight α taking values from 0 to 1 by step of 0.1.

The attainment surfaces (Fig. 3) show that in the case of Pareto approach, the exact Pareto Front is already delineated after 900 s, even considering only the worst run. On the opposite, even the best of the 20 runs is still far from the Pareto Front apart from a few points that lie in the convex parts. This trend, though preliminary here, nevertheless confirms the well-known fact that weight sum aggregation has difficulties to reach the concave parts of Pareto fronts. However, using an archive shows that several non-dominated plans where found all over the Pareto front, strongly reducing the impact of the weight parameter. We hypothesize that this is due to the highly stochastic nature of YAHSP, that seems to be able to reach good results without the help of the genetic algorithm: A single individual can lead to several different objective vectors depending on YAHSP strategy and random choices. The causality between the good structure of an individual and its fitness is thus very weak. Further work will study more deeply this hypothesis, and try to learn the relation between the individual structure and its ability to provide good plans.

4 Conclusion and Perspectives

This paper has proposed some first experiments with the recently proposed MULTIZENOTRAVEL test suite for multi-objective AI planning [6], where the instance generator comes with ZENOSOLVER, an exact solver that is able to identify the true Pareto front for even very large instances. The complete code is publicly available at https://descarwin.lri.fr, making it easy for everyone to generate his/her own benchmark instances. However, we hope that the few typical instances that have been provided here, and that exhibit very different shapes of Pareto Fronts for very different levels of complexity, could be the starting point for a general benchmark for AI planning.

The results of DAE$_{YAHSP}$ on some of these instances show the need for further improvement of the multi-objective search efficiency of MO-DAE. The results on the aggregation approach raise interesting issues regarding the respective usefulness of the evolutionary (DAE$_X$) and the AI-planning (YAHSP) parts of DAE$_{YAHSP}$. Further experiments are also needed, in which DAE$_X$ approach is used within other state-of-the-art decomposition algorithms (e.g., from the MOEA/D family [9], or using Tchebychev decomposition), and compared in detail to other non-Pareto multi-objective planners [8].

References

1. Bibaï, J., Savéant, P., Schoenauer, M., Vidal, V.: An evolutionary metaheuristic based on state decomposition for domain-independent satisficing planning. In: Brafman, R., et al. (eds.) 20th ICAPS, pp. 18–25. AAAI Press (2010)
2. Bleuler, S., Laumanns, M., Thiele, L., Zitzler, E.: PISA – a platform and programming language independent interface for search algorithms. In: Fonseca, C.M., Fleming, P.J., Zitzler, E., Deb, K., Thiele, L. (eds.) EMO 2003. LNCS, vol. 2632, pp. 494–508. Springer, Heidelberg (2003)
3. Hutter, F., Hoos, H.H., Leyton-Brown, K., Stützle, T.: ParamILS: an automatic algorithm configuration framework. JAIR **36**, 267–306 (2009)
4. Khouadjia, M.R., Schoenauer, M., Vidal, V., Dréo, J., Savéant, P.: Multi-objective AI planning: comparing aggregation and pareto approaches. In: Middendorf, M., Blum, C. (eds.) EvoCOP 2013. LNCS, vol. 7832, pp. 202–213. Springer, Heidelberg (2013)
5. Khouadjia, M.R., Schoenauer, M., Vidal, V., Dréo, J., Savéant, P.: Pareto-based multiobjective AI planning. In: Rossi, F. (eds.) Proceedings of the IJCAI. AAAI Press (2013)
6. Quemy, A., Schoenauer, M.: True Pareto Fronts for Multi-Objective AI Planning Instances (2015, submitted)
7. Schoenauer, M., Savéant, P., Vidal, V.: Divide-and-Evolve: a new memetic scheme for domain-independent temporal planning. In: Gottlieb, J., Raidl, G.R. (eds.) EvoCOP 2006. LNCS, vol. 3906, pp. 247–260. Springer, Heidelberg (2006)
8. Sroka, M., Long, D.: Exploring metric sensitivity of planners for generation of pareto frontiers. In: Kersting, K., Toussaint, M. (eds.) 6th STAIRS, pp. 306–317. IOS Press (2012)
9. Zhang, Q., Hui, L.: A Multi-objective evolutionary algorithm based on decomposition. IEEE Trans. Evol. Comput. **11**(6), 712–731 (2007)
10. Zitzler, E., Künzli, S.: Indicator-based selection in multiobjective search. In: Yao, X., Burke, E.K., Lozano, J.A., Smith, J., Merelo-Guervós, J.J., Bullinaria, J.A., Rowe, J.E., Tiňo, P., Kabán, A., Schwefel, H.-P. (eds.) PPSN 2004. LNCS, vol. 3242, pp. 832–842. Springer, Heidelberg (2004)

Incremental MaxSAT Reasoning to Reduce Branches in a Branch-and-Bound Algorithm for MaxClique

Chu-Min Li[1,2]($\boxtimes$), Hua Jiang[1], and Ru-Chu Xu[1]

[1] Huazhong University of Sciences and Technology (HUST), Wuhan, China
chu-min.li@u-picardie.fr,jh_hgt@163.com
[2] MIS, Université de Picardie Jules Verne, Amiens, France

Abstract. When searching for a maximum clique of a graph using a branch-and-bound algorithm, it is usually believed that one should minimize the set of branching vertices from which search is necessary. It this paper, we propose an approach called incremental MaxSAT reasoning to reduce the set of branching vertices in three ways, developing three algorithms called DoMC (short for Dynamic ordering MaxClique solver), SoMC and SoMC- (short for Static ordering MaxClique solver), respectively. The three algorithms differ only in the way to reduce the set of branching vertices. To our surprise, although DoMC achieves the smallest set of branching vertices, it is significantly worse than SoMC and SoMC-, because it has to change the vertex ordering for branching when reducing the set of branching vertices. SoMC is the best, because it preserves the static vertex ordering for branching and reduces the set of branching vertices more than SoMC-.

1 Introduction

A clique in an undirected graph $G = (V, E)$, where V is a set of n vertices $\{v_1, v_2, ..., v_n\}$ and E is a set of m edges, is a subset C of V in which every two vertices are adjacent. The maximum clique problem (MaxClique for short) consists in finding a clique of G of the largest size. The size of a maximum clique of G is usually denoted by $\omega(G)$. MaxClique is a very important NP-hard problem, because it is useful in many real-world applications such as bioinformatics and fault diagnosis. A huge amount of effort has been devoted to solve it. In this paper, we focus on exact algorithms for MaxClique based on the Branch-and-Bound (BnB) scheme.

In order to search for a maximum clique in G, a BnB algorithm typically uses a heuristic to order vertices of G to obtain an ordering such as $v_1 < v_2 < v_3 < ... < v_n$, and branches on every vertex v_i for $i = 1, 2, ..., n$ to recursively search for a maximum clique containing v_i in the subgraph G_i induced by $\{v_i, v_{i+1}, ..., v_n\}$. To be efficient, the algorithm maintains a global variable C_{max} to denote the largest clique found so far in G and prunes useless branches in which a clique larger than C_{max} cannot be found. Recent BnB algorithms for MaxClique such

C. Dhaenens et al. (Eds.): LION 9 2015, LNCS 8994, pp. 268–274, 2015.
DOI: 10.1007/978-3-319-19084-6_26

as MCS [5], MaxCliqueDyn [1], and MaxCLQ [3,4] prune useless branches as follows. They first partition the vertices in G into independent sets $D_1, D_2, ..., D_r$ (an independent set is a subset of V in which no two vertices are adjacent). Then if $r > |C_{max}|$, they order the vertices according to their independent set: $v_i < v_j$ if $v_i \in D_p$ and $v_j \in D_q$ and $q < p$. In the vertex ordering $v_1 < v_2 < ... < v_n$ obtained in this way, the vertices in the subset $D_r \cup D_{r-1} \cup ... \cup D_{|C_{max}|+1}$ are the smallest. The algorithms only need to branch on vertices in this subset, since vertices in $D_1, D_2, ...,$ and $D_{|C_{max}|}$ cannot form alone a clique larger than C_{max}.

We call *branching vertices* the vertices that a BnB algorithm needs to branch on. It is a common practice for a state-of-the-art BnB algorithm to reduce as much as possible the number of branching vertices by cleverly ordering vertices. For example, the *Re-NUMBER* procedure in MCS aims at reducing the number of branching vertices by re-organizing the independent sets $D_1, D_2, ...,$ and $D_{|C_{max}|}$ to make them accept more vertices. Consequently, the vertex ordering for branching in the algorithm is dynamic and is different at different search tree nodes.

An exception is the algorithm IncMaxCLQ [2] which uses a static vertex ordering for branching and needs to branch on all vertices of G. Let v_i and v_j be two vertices in G and $v_i < v_j$, the static vertex ordering implies $v_i < v_j$ in any subgraph of G containing v_i and v_j and at every search tree node. The static vertex ordering allows IncMaxCLQ to use an efficient incremental upper bound.

In this paper, we show that deriving the smallest possible set of branching vertices is not necessarily beneficial, that keeping a static vertex ordering prob ably is more important, and that reducing the number of branching vertices by keeping a static vertex ordering is really beneficial. Concretely, we propose an approach called *incremental MaxSAT reasoning* to reduce the number of branching vertices in three ways, developing three algorithms called DoMC (short for Dynamic ordering MaxClique solver), SoMC and SoMC- (short for Static order ing MaxClique solver), respectively. DoMC uses incremental MaxSAT reasoning to reinforce the Re-NUMBER procedure of MCS, reducing the number of branching vertices more than MCS. Nevertheless, this reduction prohibits any static vertex ordering for branching in DoMC as in MCS. SoMC and SoMC- reduce the number of branching vertices using incremental MaxSAT reasoning by preserving a static vertex ordering. Experimental results show that SoMC, SoMC-, and even IncMaxCLQ that preserves a static vertex ordering but does not reduce the number of branching vertices at all, are significantly better than DoMC, in terms of both search tree size and runtime, although the set of branching vertices in DoMC is smaller. SoMC and SoMC- are also faster than the stat-of-the-art algorithms such as MCS, MaxCliqueDyn, MaxCLQ and IncMaxCLQ. SoMC derives smaller sets of branching vertices than SoMC-, and is better than SoMC-.

2 Incremental MaxSAT Reasoning

Let V' be a subset of V, the subgraph of G induced by V' is defined as $G(V') = (V', E')$, where $E' = \{(v_i, v_j) \in E \mid v_i, v_j \in V'\}$. The set of adjacent vertices of a vertex v in G is denoted by $\Gamma(v) = \{v' \mid (v, v') \in E\}$. The cardinality $|\Gamma(v)|$ of

$\Gamma(v)$ is called the degree of v. The density of a graph of n vertices and m edges is $2m/(n(n-1))$.

Recent BnB algorithms such as MCS, MaxCliqueDyn, MaxCLQ and IncMax-CLQ partition G into independent sets by sequentially inserting vertices of G into independent sets. Unfortunately, the upper bound given by the independent set partition, called UB_{IndSet} in this paper, may not be tight, because a set of r independent sets may not form a clique of size r. In this case, these independent sets are said *conflicting*. A recent approach proposed in [3,4] uses MaxSAT reasoning to improve UB_{IndSet} by detecting conflicting independent sets, after (implicitly) encoding a MaxClique problem into a partial MaxSAT problem. MaxSAT reasoning as described in [3,4] is not incremental because it is always done from scratch. In this paper, we propose *incremental MaxSAT reasoning* which, given an induced subgraph G' of G with a known upper bound of $\omega(G')$, successively adds vertices of G into G' and detects a conflict in G' after inserting each vertex. The purpose of incremental MaxSAT reasoning is to show that the upper bound of $\omega(G')$ is not increased after inserting these vertices into G'.

Example 1. Consider the graph in Fig. 1 and its subgraph G' induced by $\{v_1, v_2, v_3, v_4\}$. G' is partitioned into 2 independent sets: $\{v_1, v_4\}$, $\{v_2, v_3\}$, so $\text{UB}_{IndSet}=2$ for G'. When inserting v_5 into G', we have a new independent set $\{v_5\}$. Incremental MaxSAT reasoning detects a conflict as follows: assume that each of the three independent sets contributes a vertex to the maximum clique under construction, then v_5 is in the clique, excluding v_1 and v_2 from the clique because they are not adjacent to v_5, so the only remaining v_4 in the first set and the only remaining v_3 in the second set should be in the clique. However, this is not possible, because v_3 and v_4 are not adjacent. So the three independent sets $\{v_1, v_4\}$, $\{v_2, v_3\}$ and $\{v_5\}$ are conflicting.

We add a new vertex $z_1(z_2, z_3)$ into the first (second, third) independent set. Each z_i is unconnected to z_j (for any $j \neq i$) and the vertices in the same independent set, but is adjacent to all other vertices in G. If the conflicting independent sets can form a clique of size p without the new vertices, they can form a clique of size $p+1$ with the new vertices. So, the new vertices cover exactly one conflict in these independent sets.

Then v_6 is inserted into G'. We have now 4 independent sets: $\{v_1, v_4, z_1\}$, $\{v_2, v_3, z_2\}$, $\{v_5, z_3\}$, and $\{v_6\}$. Incremental MaxSAT reasoning detects a new conflict as follows: the adding of v_6 in the maximum clique under construction excludes v_1 and v_4 from the first set, and v_5 from the third set. However, z_1 and z_3 are not adjacent and cannot both belong to a clique. So, $\{v_1, v_4, z_1\}$, $\{v_5, z_3\}$, and $\{v_6\}$ are conflicting.

The two conflicts detected above for v_5 and v_6 are clearly disjoint because of the adding of z_1, z_2 and z_3, showing that the upper bound of $\omega(G')$ is always 2 after G' includes v_5 and v_6. The second conflict can also be covered by adding a new vertex into each independent set involved in the conflict.

Formally, we define a function $\text{IncMaxSAT}(G, S, B)$, where $G = (V, E)$ is a graph with a vertex ordering, S is a subset of vertices that is partitioned into r

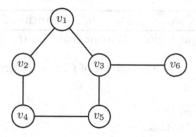

Fig. 1. A simple graph ($\omega(G)$=2) from [3]

Algorithm 1. GetBranches(G, r)

Input: $G=(V, E)$, and r: an imposed lower bound for the size of a MaxClique of G

Output: a set of branching vertices

1 **begin**
2      $G' \leftarrow G$; $P \leftarrow \emptyset$; $S \leftarrow \emptyset$; $B \leftarrow \emptyset$;
3   **while** G' *is not empty* **do**
4          $v \leftarrow$ the biggest vertex of G';
5   remove v from G';
6   **if** P *contains an independent set D in which v is not adjacent to any vertex* **then**
7   insert v into D; insert v into S;
8   **else**
9   **if** $|P|<r$ **then**
10   **create** a new independent set $D = \{v\}$; $P \leftarrow P \cup \{D\}$; **insert** v into S;
11   **else**
12   **if** P *contains an independent set D in which v has only one adjacent vertex u and u can be inserted into another independent set D'* **then**
13   **move** u from D to D'; **insert** v into D; **insert** v into S;
14   **else** $B \leftarrow \{v\} \cup B$;
15      $B \leftarrow$ IncMaxSAT(G, S, B);
16   **return** the set of all vertices of G smaller than or equal to the biggest vertex in B.

---

independent sets, and $B = V \setminus S$ is a set of branching vertices to be reduced. The function successively inserts vertices of B (from the biggest vertex to the smallest one in the predefined vertex ordering) into S, and detects a disjoint conflict in S for each inserted vertex. The detected conflicts show that S with the inserted vertices cannot form a clique of size larger than r. The function stops as soon as it fails to detect a conflict when inserting a vertex v into S, and returns the set of remaining vertices in B (including v).

Algorithm 2. SoMC(G, C, C_{max}), a BnB algorithm for MaxClique

---

Input: $G=(V, E)$, clique C under construction, and the largest clique C_{max} found so far

Output: $C \cup C'$, where C' is a maxclique of G, if $|C \cup C'|>|C_{max}|$; C_{max} otherwise

1 **begin**
2   **if** $|V|=0$ **then** return C;
3    $B \leftarrow$ GetBranches(G, $|C_{max}|-|C|$);
4   **if** $B=\emptyset$ **then** return C_{max};
5    $S \leftarrow V \backslash B$;
6   **for** $i:=|B|$ downto 1 **do**
7      $C_1 \leftarrow$ SoMC($G(\Gamma(b_i) \cap S)$, $C \cup \{b_i\}$, C_{max});
8      $S \leftarrow \{b_i\} \cup S$;
9   **if** $|C_1|>|C_{max}|$ **then** $C_{max} \leftarrow C_1$;
10   **return** C_{max};

Table 1. Median runtimes in seconds and tree sizes in thousands for random graphs (computed by solving 51 graphs at each point). The points where fewer than 26 graphs are solved within 5000 s are marked by "-". "Dyn" stands for MaxCliqueDyn.

N	D	Dyn	MCS	MaxCLQ	IncMaxCLQ		SoMC		DoMC		SoMC-	
		Time	Time	Time	Time	Tree size	Time	Tree size	Time	Tree size	Time	Tree size
200	0.80	4.56	2.26	1.63	1.27	111	**0.92**	**97.1**	1.33	113	1.19	121
200	0.90	61.87	34.18	9.18	6.22	305	**4.64**	**263**	6.86	341	5.98	299
200	0.95	28.45	13.41	1.59	0.74	22.2	**0.54**	**20.7**	0.67	22.4	0.71	22.1
300	0.70	7.91	6.38	6.12	5.62	642	**3.68**	**503**	5.62	567	4.77	646
300	0.80	269.5	203.1	117.3	105.7	8166	**74.12**	**6611**	143.5	10326	99.18	8372
300	0.90	-	-	-	-	-	4169	182516	-	-	-	-
400	0.60	4.70	4.19	5.94	4.99	685	**3.53**	**521**	4.84	556	4.58	697
400	0.70	99.76	96.79	89.96	86.43	8590	**58.15**	**6687**	90.56	8416	75.24	8488
400	0.80	-	-	4877	4986	475219	**2834**	**269528**	-	-	3868	345920
500	0.50	1.85	**1.62**	2.96	2.42	519	1.66	410	1.94	**294**	1.93	470
500	0.60	26.15	23.27	29.93	26.97	3822	**17.31**	**2688**	25.66	2875	21.08	3478
500	0.70	915.8	916.1	766.2	883.8	81697	**646.2**	**61687**	905.5	78280	709.4	79602
1000	0.30	0.75	**0.70**	2.51	1.27	374	1.02	217	1.04	**163**	1.23	354
1000	0.40	8.47	7.57	19.15	8.90	2279	**6.59**	1934	8.68	**1556**	7.20	2083
1000	0.50	176.9	167.2	303.2	214.3	40154	**139.9**	27445	188.8	**25256**	164.6	34645

3   Applying Incremental MaxSAT Reasoning to Reduce the Number of Branching Vertices

A BnB algorithm always searches for a maximum clique of size larger than a given lower bound r in $G = (V, E)$. Assuming V is totally ordered, we define the function GetBranches(G, r) in Algorithm 1 that returns a set of branching vertices B by showing vertices in $V \setminus B$ cannot form a clique of size larger than r. The function works in two phases: in the first phase, r independent sets are

Table 2. Runtimes in seconds and tree sizes in thousands for DIMACS instances that are solved by at least one solver in 10^5 s, excluding the instances solved by all solvers in 10 s. "-" stands for instances that cannot be solved in 10^5 s. "Dyn" stands for MaxCliqueDyn.

Instance	N	D	Dyn Time	MCS Time	MaxCLQ Time	IncMaxCLQ Time	Tree size	SoMC Time	Tree size	DoMC Time	Tree size	SoMC- Time	Tree size
brock400_1	400	0.74	466.0	379.7	339.2	222.3	18906	**147.8**	**14541**	345.2	52167	189.5	18826
brock400_2	400	0.74	192.1	166.2	**105.9**	170.2	14474	109.9	**11160**	303.7	22489	146.7	14367
brock400_3	400	0.74	371.6	256.2	102.4	204.6	17735	**80.17**	**8022**	267.7	8275	108.5	10272
brock400_4	400	0.74	185.7	138.2	125.3	159.2	13319	**105.7**	**10307**	196.5	10746	140.3	13337
brock800_1	800	0.65	5988	5209	4889	8830	890495	**2142**	**233383**	9063	810594	2838	298022
brock800_2	800	0.65	5349	4686	4857	11210	1125211	**2139**	**221625**	8315	538086	2872	287018
brock800_3	800	0.65	3455	3208	3452	4221	398861	**895.8**	**105807**	7628	392722	1161	131127
brock800_4	800	0.65	2691	2259	3441	5832	547564	**1483**	**173233**	4589	251810	1947	217924
C2000.5	2000	0.50	-	-	-	61009	9901896	41222	6606311	46381	**5704024**	48332	8604571
C250.9	250	0.89	2376	2074	298.2	278.9	12066	**202.1**	**10055**	372.6	16361	264.5	11707
DSJC1000.5	1000	0.50	185.2	169.6	295.8	226.3	38431	**134.9**	26826	196.2	**26604**	154.9	33278
gen400_p0.9_55	400	0.90	-	37220	-	1.23	**4.01**	**1.17**	4.03	1.56	5.13	**1.17**	4.03
gen400_p0.9_65	400	0.90	-	96567	26134	0.34	3.09	**0.29**	**3.09**	0.37	3.24	0.30	**3.09**
gen400_p0.9_75	400	0.90	-	-	1372	0.27	6.64	**0.17**	**3.34**	0.18	3.52	**0.17**	**3.34**
hamming10-2	1024	0.99	49.73	0.19	**0.06**	32.65	131	34.49	131	34.59	131	34.43	131
keller5	776	0.75	-	-	5376	**141.6**	**2092**	192.4	7818	344.3	10837	199.3	8106
MANN_a45	1035	0.99	1712	63.09	20.04	115.8	218	15.40	85.4	**13.23**	**75.7**	15.48	86.2
p_hat1000-2	1000	0.49	276.6	131.6	219.9	48.75	1855	**33.87**	**1391**	53.38	1778	44.69	1612
p_hat1000-3	1000	0.75	-	-	-	42244	1028854	**27727**	**618456**	-	-	36521	804780
p_hat1500-2	1500	0.51	-	10448	15138	2165	45143	**1322**	**26354**	4023	77650	1829	35299
p_hat500-3	500	0.75	235.1	79.23	81.04	22.02	941	**15.59**	**704**	23.59	823	19.20	792
p_hat700-3	700	0.75	3946	1586	1009	269.5	7544	**178.9**	**4954**	434.8	10891	233.3	6121
sanr200_0.9	200	0.90	32.56	19.68	6.08	2.71	128	**1.97**	**107**	3.63	170	2.39	119
sanr400_0.7	400	0.70	110.0	99.69	98.56	104.7	10607	**70.06**	**8390**	100.1	15146	90.26	10658

formed using the coloring process of MCS with the Re-NUMBER procedure, and an initial set B of branching vertices is obtained; in the second phase, incremental MaxSAT reasoning is applied to eliminate the biggest vertices of B from which any clique of size larger than r cannot be found.

The BnB algorithm SoMC depicted in Algorithm 2 calls the GetBranches function to obtain a reduced set B of branching vertices $\{b_1, b_2, ..., b_{|B|}\}$ and successively branches on vertex b_i (for $i = |B|, |B|-1, ..., 1$) to search for a maximum clique containing b_i in the subgraph of G induced by $\{b_i, b_{i+1}, ..., b_{|B|}\} \cup S$, where $S = V \setminus B$, and B is ordered as V. Note that for any $b \in B$ and any $v \in S$, we have $b < v$, meaning that the set $\{b_i, b_{i+1}, ..., b_{|B|}\} \cup S$ can never contain a vertex smaller than b_i. This fact is exploited in the implementation of SoMC to speed up search using an incremental upper bound as in IncMaxCLQ. See [2] for details.

The GetBranches function returns the set of all vertices of G smaller than or equal to the biggest vertex in B, which is larger than the set given by IncMaxSAT(G, S, B) in line 15, because some vertices smaller than the biggest vertex in B could be inserted into S by the independent set partition, but are included in the set returned by the GetBranches function in line 16. We can modify GetBranches to make it simply return IncMaxSAT(G, S, B) in line 15, giving another BnB algorithm called DoMC, in which the set of branching vertices is smaller, but the static vertex ordering for branching is not preserved any more, because the set $\{b_i, b_{i+1}, ..., b_{|B|}\} \cup S$ can now contain some vertices smaller than b_i when DoMC branches on b_i.

Another possibility to make S not contain vertices smaller than any vertex in B is to stop the independent set partition in the GetBranches function as soon as a vertex cannot be inserted into the r independent sets (i.e. line 14 is replaced by "**else** *break*", then the function returns IncMaxSAT($G, S, V \setminus S$) in line 15). This modification gives the BnB algorithm SoMC- which also exploits the incremental upper bound as SoMC and IncMaxCLQ, thanks to the preserved static vertex ordering for branching.

Note that SoMC, SoMC- and DoMC are the same except the modifications described above. They share the same implementation and use the same vertex ordering as in IncMaxCLQ to partition G into independent sets in the GetBranches function. We now compare them, as well as MaxCliqueDyn, MCS, MaxCLQ, and IncMaxCLQ on standard MaxClique benchmarks on an Intel Xeon CPU X5460@3.16 GHz under Linux with 16 GB of memory. All solvers were compiled using gcc/g++ -O3.

Tables 1 and 2 show the runtimes (in seconds) of all algorithms and search tree sizes (in thousands) of IncMaxCLQ, SoMC, SoMC- and DoMC. Except few graphs, the search trees of DoMC are larger than IncMaxCLQ, SoMC and SoMC- that keep static vertex ordering for branching, although DoMC derives the smallest set of branching vertices. SoMC- is better than IncMaxCLQ because IncMaxCLQ does not reduce the number of branching vertices at all, while SoMC- does. SoMC is better than SoMC- because SoMC derives smaller sets of branching vertices than SoMC-.

SoMC and SoMC- are faster than MaxCLQ, MaxCliqueDyn, and MCS.

References

1. Konc, J., Janezic, D.: An improved branch and bound algorithm for the maximum clique problem. Commun. Math. Comput. Chem. **58**, 569–590 (2007)
2. Li, C.M., Fang, Z.W., Xu, K.: Combining MaxSAT reasoning and incremental upper bound for the maximum clique problem. In: Proceedings of the 2013 IEEE 25th International Conference on Tools with Artificial Intelligence (ICTAI2013), pp. 939–946 (2013)
3. Li, C.M., Quan, Z.: An efficient branch-and-bound algorithm based on maxsat for the maximum clique problem. In: Proceedings of the 24th AAAI, pp. 128–133 (2010)
4. Li, C.M., Quan, Z.: Combining graph structure exploitation and propositional reasoning for the maximum clique problem. In: Proceedings of the 22th ICTAI, pp. 344–351 (2010)
5. Tomita, E., Sutani, Y., Higashi, T., Takahashi, S., Wakatsuki, M.: A Simple and Faster Branch-and-Bound Algorithm for Finding a Maximum Clique. In: Rahman, M.S., Fujita, S. (eds.) WALCOM 2010. LNCS, vol. 5942, pp. 191–203. Springer, Heidelberg (2010)

Reusing the Same Coloring in the Child Nodes of the Search Tree for the Maximum Clique Problem

Alexey Nikolaev[1]([✉]), Mikhail Batsyn[1], and Pablo San Segundo[2]

[1] Laboratory of Algorithms and Technologies for Network Analysis,
National Research University Higher School of Economics, 136 Rodionova Street,
Nizhny Novgorod 603093, Russia
{ainikolaev,mbatsyn}@hse.ru
[2] Centro de Automática y Robótica (CAR), UPM-CSIC,
C/ Jose Gutiérrez Abascal, 2, 28006 Madrid, Spain
pablo.sansegundo@upm.es

Abstract. In this paper we present a new approach to reduce the computational time spent on coloring in one of the recent branch-and-bound algorithms for the maximum clique problem. In this algorithm candidates to the maximum clique are colored in every search tree node. We suggest that the coloring computed in the parent node is reused for the child nodes when it does not lead to many new branches. So we reuse the same coloring only in the nodes for which the upper bound is greater than the current best solution only by a small value δ. The obtained increase in performance reaches 70 % on benchmark instances.

Keywords: Maximum clique problem · Branch-and-bound algorithm · Reusing coloring

1 Introduction

A complete graph, or a clique, is a graph which vertices are all pairwise adjacent. The *maximum clique problem* (MCP) consists in finding a clique (a complete subgraph) of a given graph with the largest number of vertices. MCP is an important and deeply studied NP-hard problem, because it has many applications in a wide range of fields (e.g. [1,3]). The detailed analysis of heuristics and exact algorithms for MCP can be found in [5]. The comparison survey reports that MCS [4] and MaxCLQ [2] have the best performance among the existing exact algorithms.

In this paper, we show that for MCP solvers that use coloring as the upper bound it is not necessary to color every subproblem. It is efficient to reuse the same coloring in the child nodes. In Sect. 2 we present some preliminaries and give the formulation of the MCP. In Sect. 3 we describe the branch-and-bound algorithm. A description of a new approach and some examples are presented in Sect. 4. In Sect. 5 computational results and comparison with MaxCLQ are provided.

© Springer International Publishing Switzerland 2015
C. Dhaenens et al. (Eds.): LION 9 2015, LNCS 8994, pp. 275–280, 2015.
DOI: 10.1007/978-3-319-19084-6_27

2   Preliminaries

Consider a simple undirected graph $G = (V, E)$ which consists of a finite set of vertices $V = \{v_1, v_2, \ldots, v_n\}$ and edges $E \subseteq V \times V$ that pair distinct vertices. A *clique* Q is a subset of V where all vertices are pairwise adjacent. A *maximal clique* is a clique that cannot be enlarged by adding any other vertex to it. A clique which has the maximum size (number of vertices) in a graph is called *maximum clique*. The number of vertices in a maximum clique of a graph G is the *clique number* $\omega(G)$ of the graph G. The maximum clique problem (MCP) is the problem of finding the maximum clique in a given graph.

Two vertices are said to be *neighbors* (or *adjacent vertices*) if they are connected by an edge. The neighbor set of any vertex $v \in V$ in $G = (V, E)$ is denoted by $N(v)$, i.e. $N(v) = \{w \in V | (v, w) \in E\}$. A *coloring* (*vertex coloring*) of a graph G is an assignment of colors $c(v) : V \to \mathbb{N}$ to every vertex of the graph G so that any two adjacent vertices have different colors. Color set notation $C(G) = \{C_1, C_2, \ldots, C_l\}$ denotes a vertex coloring that employs l different color numbers. $C(G)$ partitions the vertex set into l disjoint color sets C_i, where C_i contains all vertices with color number i. The following proposition is important for MCP solvers because it gives an upper bound for the clique number. The proposition can be easily proved by contradiction.

Proposition 1. *If G can be colored into l colors, then $\omega(G) \leq l$.*

3   The Exact Algorithm for the MCP

The majority of branch-and-bound algorithms for the MCP begin with the initializing of two global variables Q and Q_{max} to $\varnothing$, where Q is the current clique and Q_{max} is the largest clique found so far. Then we use greedy coloring heuristic (GCH function) for coloring. Greedy coloring heuristic is an approximate coloring algorithm that iteratively assigns the smallest positive integer (color number) to vertices, so that any two adjacent vertices have different positive integers. There is an extension of greedy coloring heuristic which is called greedy coloring heuristic with recoloring (RECOLORING function). The details and implementation of the greedy coloring heuristic with recoloring (termed *Re-Number-Sort*) can be found in [4]. The result of GCH and RECOLORING is a vertex coloring C ($C = \{C_1, C_2, \ldots, C_l\}$).

FINDING_OF_MAXIMUM_CLIQUE is an implementation of a branch-and-bound algorithm (Listing 1). A list of candidate vertices sorted according to the initial ordering is denoted by U. At the beginning U is equal to V. We choose vertex v with maximum color (step 2) and if pruning condition (step 3) is ruled out then we add the vertex v to the current clique Q and compute the new set of candidate vertices $newU$ (step 4). We replace Q_{max} with Q (step 10) only if Q is maximal ($Q_{max} = \varnothing$) and $|Q| > |Q_{max}|$. Step 11 is performed to remove the already considered vertex v.

The current branch-and-bound algorithm and MCS [4] are similar but there is a difference in FINDING_OF_MAXIMUM_CLIQUE. The difference is in the condition $|Q| - 1 + c(v) \leq |Q_{max}| + \delta$ (step 6) which is added for integrating of a new approach (REUSING_OF_COLORING procedure). Here δ is a non-negative integer. If the condition holds then we use our approach for computing the upper bound. Otherwise, greedy coloring with recoloring is used.

Listing 1. Branch-and-bound algorithm

```
procedure FINDING_OF_MAXIMUM_CLIQUE (U, C)
initial step: Q:=∅, Qmax:=∅, U:=V, C:=GCH(V)
begin
1. while U≠∅
2.      v:= a vertex with the maximum color number in C;
3.      if (|Q|+c(v)≤|Qmax|) then return
4.      Q:=Q∪{v}, newU:=U∩N(v);
5.      if newU≠∅ then
6.            if (|Q|-1+c(v)≤|Qmax|+δ) then
7.                REUSING_OF_COLORING (newU, C, newC);
8.            else newC:=RECOLORING (newU);
9.            FINDING_OF_MAXIMUM_CLIQUE (newU, newC);
10.     else if (|Q|>|Qmax|) then Qmax:=Q;
11.     Q:=Q\{v}, C:=C\{v}, U:=U\{v};
end
```

Listing 2. Reusing of parent coloring

```
procedure REUSING_OF_COLORING (newU, C, newC)
begin
1. for i:=1 to |C|-1
2.     newCi=Ci∩newU;
3. remove empty colors from newC;
4. for i:=1 to |newC|
5.     if (|newCi|=1) then
6.          w:= the vertex in newCi;
7.          for j:=1 to |newC|
8.              if (newCi∩N(w)=∅ and i≠j) then
9.                  recolor w to color j and remove newCi;
10.                 j:=|newC|+1;
11. sort newC by size of color sets in descending order;
end
```

4 Reusing of Parent Coloring

Reusing of parent coloring is a new approach for branch-and-bound algorithms for the MCP. It is based on the idea that for some node of the search tree we can use the parent coloring instead of computing greedy coloring or greedy coloring with recoloring. So it is not necessary to color candidate vertices every time. The implementation of this idea is REUSING_OF_COLORING procedure (Listing 2). The procedure can be divided into 3 steps: 1. Reusing of parent coloring; 2. Improving the quality of coloring; 3. Sorting colors by size. The following example shows how it works:

Consider a node of the search tree for which $\delta = 2, |Q| = 1$ (the current clique has 1 vertex), $|Q_{max}| = 4$ (the largest clique found so far has 4 vertices), and the subgraph of candidates (neighbors of the vertex in the current clique Q) is shown in Fig. 1. The colors of vertices are shown in brackets in this figure.

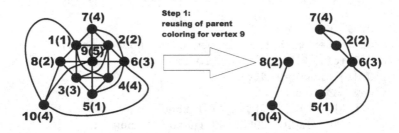

Fig. 1. Reusing of parent coloring

According to the algorithm we consider candidates starting from the vertex 9 because it is the candidate with the largest color. We add this vertex to the current clique, so $|Q| = 2$. The condition $|Q| - 1 + c(9) \leq |Q_{max}| + \delta$ (Listing 2, step 6) holds that is why for the vertex 9 we use our approach. For vertex 9 the new subgraph of candidates is shown in Fig. 1 (right). We do not color this subgraph and reuse the colors from the coloring in the parent node (Listing 2, steps 1–3). It is worth noting that the quality of the obtained coloring is not so high because the greedy coloring heuristic can color this subgraph in three colors.

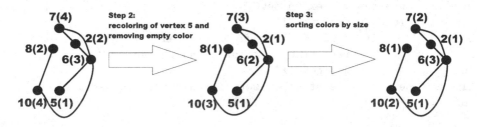

Fig. 2. Improving the quality of coloring and sorting colors by size

Steps 4–10 improve the quality of coloring. At the beginning we find a color which contains only one vertex w (step 6). Then we try to recolor vertex w, i.e. we attempt to add vertex w to another color set (steps 7–10). In our example color 1 contains only one vertex 5. We recolor vertex 5 to color 2 and remove empty color 1 (see Fig. 2).

Table 1. Computational time (in seconds)

Instances	n	p	ω	$\delta = 0$	$\delta = 1$	$\delta = 2$	$\delta = 3$	$\delta = 4$	$MaxCLQ$
C250.9	250	0.909	44	1494.3	1230.1	1125.1	1124.9	1116.9	**344.5**
MANN_a45	1035	0.996	345	72.59	66.92	77.40	103.3	150.7	**34.15**
brock400_1	400	0.748	27	316.6	278.1	255.8	**252.7**	262.1	259.7
brock400_2	400	0.749	29	135.8	117.3	**110.5**	111.0	111.0	118.9
brock400_3	400	0.748	31	214.0	187.6	**172.8**	174.4	178.1	204.2
brock400_4	400	0.749	33	107.9	92.86	88.07	**86.97**	89.70	130.7
brock800_1	800	0.649	23	4993.3	4257.0	3922.6	**3831.3**	3870.4	5606.6
brock800_2	800	0.651	24	4661.6	3871.7	3594.6	3539.9	**3470.8**	4889.0
brock800_3	800	0.649	25	3030.4	2550.2	2342.8	2271.2	**2252.6**	3222.6
brock800_4	800	0.65	26	2218.0	1861.2	1700.3	1653.9	**1645.1**	2438.4
dsjc1000.5	1000	0.5	15	171.3	151.5	**144.7**	144.9	148.8	317.9
frb30-15-1	450	0.824	30	992.6	609.8	463.3	399.4	**365.8**	655.2
frb30-15-2	450	0.823	30	827.2	485.3	343.4	283.0	**239.9**	951.7
frb30-15-3	450	0.824	30	585.1	329.0	226.2	178.6	**152.9**	581.0
frb30-15-4	450	0.823	30	1543.8	902.8	653.4	535.1	**480.1**	1155.6
frb30-15-5	450	0.824	30	893.1	503.7	364.5	296.1	**263.6**	873.7
p_hat500-3	500	0.752	50	70.73	59.02	55.44	56.08	60.08	**49.83**
p_hat700-3	700	0.748	62	1289.5	1055.6	964.1	**949.1**	1019.2	1082.2
p_hat1000-2	1000	0.49	46	141.5	115.5	105.2	**104.0**	109.3	117.8
sanr200_0.9	200	0.898	42	13.79	11.40	10.41	10.19	10.26	**5.604**
sanr400_0.7	400	0.7	21	88.85	78.09	**75.49**	75.91	77.94	97.66
			Total	23862	18814	16796	16182	**16075**	23137

Step 11 changes the order of colors, so that color sets are sorted by the size in descending order. In our example after sorting by size colors become: $C_1 = \{2, 5, 8\}, C_2 = \{7, 10\}, C_3 = \{6\}$. The result of sorting is shown in Fig. 2 (right).

5 Computational Results

We have tested our algorithm with different values of δ and compared it with MaxCLQ [2]. The implementation of MaxCLQ was kindly provided by Li and Quan. Table 1 presents the considered DIMACS instances and the performance of the compared algorithms. Column header n, p and ω stands for the number of vertices, the density and the clique number respectively. It takes more computational time (the total time over all instances is 16917 s) if $\delta = 5$, that is why it is not included in Table 1.

It is interesting that only for four instances MaxCLQ has the best time. The total computational time over all instances for the algorithm with $\delta = 4$ is 33 % and 31 % less than for the original algorithm ($\delta = 0$) and MaxCLQ respectively.

For frb30-15-1, frb30-15-3, frb30-15-5 the new algorithm with $\delta = 4$ gives 63 %, 74 % and 70 % reduction in computational time. The new approach does not look so powerful for MANN_a45. The reason for this fact may be that the density of the instance is high and $\omega \cong 0.3n$.

6   Conclusion

We have proposed a new approach based on reusing of parent coloring. The computational results show that this technique is very efficient. It reduces the computational time by 60–70 % for some instances and by 30 % on average.

The proposed approach is quite flexible and in the future it can be applied to some other state-of-the-art algorithms.

Acknowledgments. The authors would like to thank Chu-Min Li and Zhe Quan for the source code of their MaxCLQ algorithm. We gratefully acknowledge their kindness. The work was conducted at National Research University Higher School of Economics and supported by RSF grant 14-41-00039.

References

1. Butenko, S., Wilhelm, W.E.: Clique-detection models in computational biochemistry and genomics. Eur. J. Oper. Res. **173**, 1–17 (2006)
2. Li, C.M., Quan, Z.: Combining graph structure exploitation and propositional reasoning for the maximum clique problem. In: 2010 22nd IEEE International Conference on Tools with Artificial Intelligence, pp. 344–351. IEEE (2010)
3. San Segundo, P., Rodriguez-Losada, D., Matia, F., Galan, R.: Fast exact feature based data correspondence search with an efficient bit-parallel MCP solver. Appl. Intell. **32**, 311–329 (2010)
4. Tomita, E., Sutani, Y., Higashi, T., Takahashi, S., Wakatsuki, M.: A simple and faster branch-and-bound algorithm for finding a maximum clique. In: Rahman, M.S., Fujita, S. (eds.) WALCOM 2010. LNCS, vol. 5942, pp. 191–203. Springer, Heidelberg (2010)
5. Wu, Q., Hao, J.K.: A review on algorithms for maximum clique problems. Eur. J. Oper. Res. **242**, 693–709 (2015)

A Warped Kernel Improving Robustness in Bayesian Optimization Via Random Embeddings

Mickaël Binois[1,2]($\boxtimes$), David Ginsbourger[3], and Olivier Roustant[1]

[1] Mines Saint-Étienne, UMR CNRS 6158, LIMOS, 42023 Saint-Étienne, France
{mickael.binois,olivier.roustant}@mines-stetienne.fr
[2] Renault S.A.S., 78084 Guyancourt, France
[3] Department of Mathematics and Statistics, University of Bern,
Alpeneggstrasse 22, 3012 Bern, Switzerland
david.ginsbourger@stat.unibe.ch

Abstract. This works extends the Random Embedding Bayesian Optimization approach by integrating a warping of the high dimensional subspace within the covariance kernel. The proposed warping, that relies on elementary geometric considerations, allows mitigating the drawbacks of the high extrinsic dimensionality while avoiding the algorithm to evaluate points giving redundant information. It also alleviates constraints on bound selection for the embedded domain, thus improving the robustness, as illustrated with a test case with 25 variables and intrinsic dimension 6.

Keywords: Black-box optimization · Expected Improvement · Low-intrinsic dimensionality · Gaussian processes · REMBO

1 Introduction

The scope of Bayesian Optimization methods is usually limited to moderate-dimensional problems [2]. To overcome this restriction, [9] recently proposed to extend the applicability of these methods to up to billions of variables, when only few of them are actually influential, through the so-called Random EMbedding Bayesian Optimization (REMBO) approach. In REMBO, optimization is conducted in a low-dimensional domain $\mathcal{Y}$, randomly embedded in the high-dimensional source space $\mathcal{X}$. New points are chosen by maximizing the Expected Improvement (EI) criterion [4] with Gaussian process (GP) models incorporating the considered embeddings via two kinds of covariance kernels proposed in [9]. A first one, $k_{\mathcal{X}}$, relies on Euclidean distances in $\mathcal{X}$. It delivers good performance in moderate dimension, albeit its main drawback is to remain high-dimensional so that the benefits of the method are limited. A second one, $k_{\mathcal{Y}}$, is defined directly over $\mathcal{Y}$ and is therefore independent from the dimension of $\mathcal{X}$. However, it has been shown [9] to possess artifacts that may lead EI algorithms to spend many iterations exploring equivalent points.

C. Dhaenens et al. (Eds.): LION 9 2015, LNCS 8994, pp. 281–286, 2015.
DOI: 10.1007/978-3-319-19084-6_28

Here we propose a new kernel with a warping (see e.g. [7]) inspired by simple geometrical ideas, that retains key advantages of $k_\mathcal{X}$ while remaining of low dimension like $k_\mathcal{Y}$. Its effectiveness is illustrated on a 25-dimensional test problem with 6 effective variables.

2   Background on the REMBO Method and Related Issues

The considered minimization problem is to find $\mathbf{x}^* \in \mathrm{argmin}_{\mathbf{x} \in \mathcal{X}} f(\mathbf{x})$, with $f : \mathcal{X} \subseteq \mathbb{R}^D \to \mathbb{R}$, where $\mathcal{X}$ is a compact subset of $\mathbb{R}^D$, assumed here to be $[-1, 1]^D$ for simplicity. From [9], one main hypothesis about f is that its effective dimensionality is $d_e < D$: there exists a linear subspace $\mathcal{T} \subset \mathbb{R}^D$ of dimension d_e such that $f(\mathbf{x}) = f(\mathbf{x}_\top + \mathbf{x}_\perp) = f(\mathbf{x}_\top)$, $\mathbf{x}_\top \in \mathcal{T}$ and $\mathbf{x}_\perp \in \mathcal{T}^\perp \subset \mathbb{R}^D$ ([9], Definition 1). Given a random matrix $\mathbf{A} \in \mathbb{R}^{D \times d}$ ($d \geq d_e$) with components sampled independently from $\mathcal{N}(0, 1)$, for any optimizer $\mathbf{x}^* \in \mathbb{R}^D$, there exists at least a point $\mathbf{y}^* \in \mathbb{R}^d$ such that $f(\mathbf{x}^*) = f(\mathbf{A}\mathbf{y}^*)$ with probability 1 ([9], Theorem 2). To respect box constraints, f is evaluated at $p_\mathcal{X}(\mathbf{A}\mathbf{y})$, the convex projection of $\mathbf{A}\mathbf{y}$ onto $\mathcal{X}$. The low dimensional function to optimize is then $g : \mathbb{R}^d \to \mathbb{R}$, $g(\mathbf{y}) = f\left(p_\mathcal{X}(\mathbf{A}\mathbf{y})\right)$.

Optimizing g is carried out using Bayesian Optimization, e.g., with the EGO algorithm [1]. It bases on Gaussian Process Regression [5], also known as Kriging [3], to create a surrogate of g. Supposing that g is a sample from a GP with known mean (zero here to simplify notations) and covariance kernel $k(.,.)$, conditioning it on n observations $\mathbf{Z} = f(\mathbf{x}_{1:n}) = g(\mathbf{y}_{1:n})$, provides a GP $Z(.)$ with mean $m(\mathbf{x}) = \mathbf{k}(\mathbf{x})^T K^{-1} \mathbf{Z}$ and kernel $c(\mathbf{x}, \mathbf{x}') = k(\mathbf{x}, \mathbf{x}') - \mathbf{k}(\mathbf{x})^T K^{-1} \mathbf{k}(\mathbf{x}')$, where $\mathbf{k}(\mathbf{x}) = (k(\mathbf{x}, \mathbf{x}_i))_{1 \leq i \leq n}$ and $K = (k(\mathbf{x}_i, \mathbf{x}_j))_{1 \leq i, j \leq n}$. The choice of k is preponderant, since it reflects a number of beliefs about the function at hand. Among the most commonly used are the "squared exponential" (SE) and "Matérn" stationary kernels, with hyperparameters such as length scales or degree of smoothness [6,8]. For REMBO, [9] proposed two versions of the SE kernel with length scales l, namely the low-dimensional $k_\mathcal{Y}(\mathbf{y}, \mathbf{y}') = \exp\left(-\|\mathbf{y} - \mathbf{y}'\|_d^2 / 2l_\mathcal{Y}^2\right)$ and the high-dimensional $k_\mathcal{X}(\mathbf{y}, \mathbf{y}') = \exp\left(-\left\|p_\mathcal{X}(\mathbf{A}\mathbf{y}) - p_\mathcal{X}(\mathbf{A}\mathbf{y}')\right\|_D^2 / 2l_\mathcal{X}^2\right)$ $(\mathbf{y}, \mathbf{y}' \in \mathcal{Y})$.

Selecting the domain $\mathcal{Y} \subset \mathbb{R}^d$ is a major difficulty of the method: if too small, the optimum may not be reachable while a too large domain renders optimizing harder, in particular since $p_\mathcal{X}$ is far from being injective. Distant points in $\mathcal{Y}$ may coincide in $\mathcal{X}$, especially far from the center, so that using $k_\mathcal{Y}$ leads to sample useless new points in $\mathcal{Y}$ corresponding to the same location in $\mathcal{X}$ after the convex projection. On the other hand, $k_\mathcal{X}$ suffers from the curse of dimensionality when $\mathcal{Y}$ is large enough so that most or all of the points of $\mathcal{X}$ belonging to the convex projection of the subspace spanned by $\mathbf{A}$ onto $\mathcal{X}$ have at least one pre-image in $\mathcal{Y}$. Indeed, whereas embedded points $p_\mathcal{X}(\mathbf{A}\mathbf{y})$ lie in a d dimensional subspace when they are inside of $\mathcal{X}$, they belong to a D-dimensional domain when they are projected onto the faces and edges of $\mathcal{X}$. To alleviate these shortcomings, after showing that with probability $1 - \epsilon$ the optimum is contained in the centered ball of radius d_e/ϵ (Theorem 3), the authors of [9] then suggest to set $\mathcal{Y} = [-\sqrt{d}, \sqrt{d}]^d$. In practice, they split the evaluation budget over several random embeddings or set $d > d_e$ to increase the probability for the optimum to actually be inside $\mathcal{Y}$, slowing down the convergence.

3 Proposed Kernel and Experimental Results

Both $k_{\mathcal{Y}}$ and $k_{\mathcal{X}}$ suffering from limitations, it is desirable to have a kernel that retains as much as possible of the actual high dimensional distances between points while remaining of low dimension. This can be achieved by first projecting points orthogonally on the faces of the hypercube to the subspace spanned by $\mathbf{A}$: $\mathrm{Ran}(\mathbf{A})$, with $p_{\mathbf{A}} : \mathcal{X} \mapsto \mathbb{R}^D$, $p_{\mathbf{A}}(\mathbf{x}) = \mathbf{A}(\mathbf{A}^T\mathbf{A})^{-1}\mathbf{A}^T\mathbf{x}$. Note that these back-projections from the hypercube can be outside of $\mathcal{X}$. The calculation of the projection matrix is done only once, inverting a $d \times d$ matrix. This solves the problem of adding already evaluated points: their back-projections coincide. Nevertheless, distant points on the sides of $\mathcal{X}$ from the convex projection can be back-projected close to each other, which may cause troubles with the stationary kernels classically used.

The next step is to respect as much as possible distances on the border of $\mathcal{X}$, denoted $\partial\mathcal{X}$. Unfolding and parametrizing the manifold corresponding to the convex projection of the embedding of $\mathcal{Y}$ with $\mathbf{A}$ would be best but unfortunately it seems intractable with high D. Indeed, it amounts to finding each intersection of the d-dimensional subspace spanned by $\mathbf{A}$ with the faces of the D-hypercube, before describing the parts resulting from the convex projection. Alternatively, we propose to distort the back-projections which are outside of $\mathcal{X}$, corresponding to those convex-projected parts on the sides of $\partial\mathcal{X}$. In more details, from the back-projection of the initial mapping with $p_{\mathcal{X}}$, a pivot point is selected as the intersection between $\partial\mathcal{X}$ and the line $(O; p_{\mathbf{A}}(p_{\mathcal{X}}(\mathbf{Ay})))$. Then the back-projection is stretched out such that the distance between the pivot point and the initial convex projection are equal. It results in respecting the distance *on the embedding* between the center O and the initial convex projection. The resulting warping, denoted Ψ, is detailed in Algorithm 1 and illustrated in Fig. 1. Based on this, any positive definite kernel k on $\mathcal{Y}$ can be used. For example, the resulting SE kernel is $k_\Psi(\mathbf{y}, \mathbf{y}') = \exp\left(- \|\Psi(\mathbf{y}) - \Psi(\mathbf{y}')\|_D^2 / 2l_\Psi^2\right)$. Note that the function value corresponding to $\Psi(\mathbf{y})$ remains $g(\mathbf{y})$.

Like $k_{\mathcal{X}}$, k_Ψ is not hindered by the non-injectivity brought by the convex projection $p_{\mathcal{X}}$. Furthermore, it can explore sides of the hypercube without spending too much budget since belonging to $\mathrm{Ran}(\mathbf{A})$ (all distances between embedded points after warping are d-dimensional instead of D-dimensional, thus smaller, hence limiting the risk of over-exploring sides of $\mathcal{X}$). It is thus possible to extend the size of $\mathcal{Y}$ to avoid the risk of missing the optimum. For instance, one can check that $\mathcal{Y}$ is larger than $[-\gamma, \gamma]^d$ with γ such that $\gamma^{-1} = \min_{j\in[1,...,D]} \sum_{i=1}^d |A_{j,i}|$, with $A_{j,i}$ the components of $\mathbf{A}$, ensuring to span $[-1, 1]$ for each of the D variables.

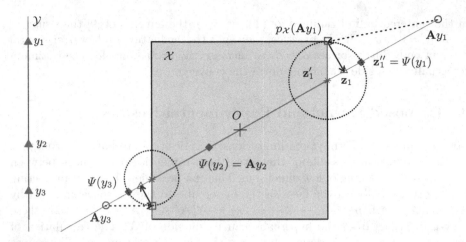

Fig. 1. Illustration of the new warping Ψ, $d = 1$ and $D = 2$, from triangles in $\mathcal{Y}$ to diamonds in $\mathcal{X}$, on three points y_1, y_2, y_3. As for REMBO, the points y_i are first mapped by $\mathbf{A}$ and convexly projected onto $\mathcal{X}$ (if out of $\mathcal{X}$). If the resulting image is strictly contained in $\mathcal{X}$ – as for y_2 – nothing else is done. Otherwise, the new warping is defined in two supplementary steps: back-projection onto Ran($\mathbf{A}$) (giving $\mathbf{z}_i$) and stretching out in the resulting line $[0, \mathbf{z}_i]$ (red solid line) by reporting the distance between the intersection of $[0, \mathbf{z}_i]$ on the frontier of $\mathcal{X}$, $\mathbf{z}_i'$, and the initial convex projection $p_\mathcal{X}(\mathbf{A}y_i)$. The points y_1 and y_3 correspond to cases where such projections are on a corner or a face of $\mathcal{X}$ (Color figure online).

We compare the performances of the usual REMBO method with $k_\mathcal{Y}$, $k_\mathcal{X}$ and the proposed k_Ψ, with a unique embedding. Tests are conducted with the *DiceKriging* and *DiceOptim* packages [6]. We use the isotropic Matérn 5/2 kernel with hyperparameters estimated with Maximum Likelihood and we start optimization with space filling designs of size $10d$. Initial designs are modified such that no points are repeated in $\mathcal{X}$ for $k_\mathcal{Y}$ and $k_\mathcal{X}$. For k_Ψ, we apply Ψ to bigger initial designs before selecting the right number of points, as distant as possible between each other. Experiments are repeated fifty times, taking the same random embeddings for all kernels. To allow a fair comparison, $\mathcal{Y}$ is set to

---

Algorithm 1. Calculation of Ψ.

1: Map $\mathbf{y} \in \mathcal{Y}$ to $\mathbf{A}y$
2: **If** $\mathbf{A}y \in \mathcal{X}$ **Then**
3:    Define $\Psi(\mathbf{y}) = \mathbf{A}y$
4: **Else**
5:    Project onto $\mathcal{X}$ and back-project onto Ran($\mathbf{A}$): $\mathbf{z} = p_\mathbf{A}(p_\mathcal{X}(\mathbf{A}y))$
6:    Compute the intersection of $[O; \mathbf{z}]$ with $\partial\mathcal{X}$: $\mathbf{z}' = (\max_{i=1,\ldots,D} |z_i|)^{-1}\mathbf{z}$
7:    Define $\Psi(\mathbf{y}) = \mathbf{z}' + \|p_\mathcal{X}(\mathbf{A}y) - \mathbf{z}'\|_D \cdot \frac{\mathbf{z}'}{\|\mathbf{z}'\|_D}$
8: **EndIf**

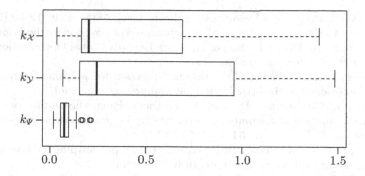

Fig. 2. Boxplot of the optimality gap (best value found minus actual minimum) for kernels $k_{\mathcal{X}}$, $k_{\mathcal{Y}}$ and k_{Ψ} on the Hartmann6 test function (see e.g. [1]) with 250 evaluations, $d = d_e = 6$, $D = 25$.

$[-\sqrt{d}, \sqrt{d}]^d$ for all kernels and the computational efforts on the maximization of the Expected Improvement are the same.

Results in Fig. 2 show that the proposed kernel k_{Ψ} outperforms both $k_{\mathcal{Y}}$ and $k_{\mathcal{X}}$ when $d = 6$. In particular, $k_{\mathcal{Y}}$ loses many evaluations on the sides of $\mathcal{Y}$ for already known points in $\mathcal{X}$ and $k_{\mathcal{X}}$ has a propensity to explore sides of $\mathcal{X}$, while k_{Ψ} avoids both pitfalls.

4   Conclusion and Perspectives

The composition with a warping of the covariance kernel used with REMBO wipes out some of the previous shortcomings. It thus achieved the goal of improving the results with a single embedding, as was shown on the Hartman6 example. Studying the efficiency of splitting the evaluation budget between several random embeddings, compared to relying on a single one along with k_{Ψ}, would be the scope of future research. Of interest is also the study of the embedding itself, such as properties ensuring fast convergence in practice.

Acknowledgments. This work has been conducted within the frame of the ReDice Consortium, gathering industrial (CEA, EDF, IFPEN, IRSN, Renault) and academic (Ecole des Mines de Saint-Etienne, INRIA, and the University of Bern) partners around advanced methods for Computer Experiments.

The authors also thanks the anonymous reviewers as well as Frank Hutter for their helpful suggestions.

References

1. Jones, D., Schonlau, M., Welch, W.: Efficient global optimization of expensive black-box functions. J. Global Optim. **13**(4), 455–492 (1998)
2. Koziel, S., Ciaurri, D.E., Leifsson, L.: Surrogate-based methods. In: Koziel, S., Yang, X.-S. (eds.) Comput. Optimization, Methods and Algorithms. SCI, vol. 356, pp. 33–59. Springer, Heidelberg (2011)

3. Matheron, G.: Principles of geostatistics. Econ. Geol. **58**(8), 1246–1266 (1963)
4. Mockus, J., Tiesis, V., Zilinskas, A.: The application of Bayesian methods for seeking the extremum. In: Dixon, L., Szego, G. (eds.) Towards Global Optimization, vol. 2, pp. 117–129. Elsevier, Amsterdam (1978)
5. Rasmussen, C.E., Williams, C.: Gaussian Processes for Machine Learning. MIT Press, Cambridge (2006). http://www.gaussianprocess.org/gpml/
6. Roustant, O., Ginsbourger, D., Deville, Y.: DiceKriging, DiceOptim: two R packages for the analysis of computer experiments by kriging-based metamodeling and optimization. J. Stat. Softw. **51**(1), 1–55 (2012)
7. Snoek, J., Swersky, K., Zemel, R.S., Adams, R.P.: Input warping for Bayesian optimization of non-stationary functions. In ICML (2014)
8. Stein, M.L.: Interpolation of Spatial Data: Some Theory for Kriging. Springer, New York (1999)
9. Wang, Z., Zoghi, M., Hutter, F., Matheson, D., de Freitas, N.: Bayesian optimization in high dimensions via random embeddings. In: IJCAI (2013)

Making EGO and CMA-ES Complementary for Global Optimization

Hossein Mohammadi[1,2]([✉]), Rodolphe Le Riche[1,2], and Eric Touboul[1,2]

[1] Ecole Nationale Supérieure des Mines de Saint-Etienne, Saint-Étienne, France
[2] CNRS LIMOS, UMR 5168, Saint-Étienne, France
{hossein.mohammadi,leriche,touboul}@emse.fr

Abstract. The global optimization of expensive-to-calculate continuous functions is of great practical importance in engineering. Among the proposed algorithms for solving such problems, *Efficient Global Optimization (EGO)* and *Covariance Matrix Adaptation Evolution Strategy (CMA-ES)* are regarded as two state-of-the-art unconstrained continuous optimization algorithms. Their underlying principles and performances are different, yet complementary: EGO fills the design space in an order controlled by a Gaussian process (GP) conditioned by the objective function while CMA-ES learns and samples multi-normal laws in the space of design variables. This paper proposes a new algorithm, called EGO-CMA, which combines EGO and CMA-ES. In EGO-CMA, the EGO search is interrupted early and followed by a CMA-ES search whose starting point, initial step size and covariance matrix are calculated from the already sampled points and the associated conditional GP. EGO-CMA improves the performance of both EGO and CMA-ES in our 2 to 10 dimensional experiments.

Keywords: Continuous global optimization · CMA-ES · EGO

1 Introduction and Basic Concepts

Continuous numerical optimization problems are at the core of many applications in science and engineering. They are formalized as

$$min_{x \in \mathcal{S} \subset \mathbb{R}^d} f(x).$$

It often happens that the underlying function, f, is not only expensive to evaluate but also mathematically multimodal.

EGO Algorithm. One approach to deal with expensive and multimodal optimization problems is to use GP as (meta)models for the objective function. The deterministic Efficient Global Optimization (EGO) algorithm [7] instanciates this idea and has become a standard for continuous global optimization in less than twenty dimensions when the number of function evaluations is inferior to 1000. The principle of model-based optimizers such as EGO and SMAC [4] is

© Springer International Publishing Switzerland 2015
C. Dhaenens et al. (Eds.): LION 9 2015, LNCS 8994, pp. 287–292, 2015.
DOI: 10.1007/978-3-319-19084-6_29

to build a prediction for $f(x)$ called $m(x)$ and the associated prediction uncertainty $v(x)$. A new point $x' \in S$ is selected which strikes a compromise between the best known regions (low $m(x)$) and the least known regions (large $v(x)$). In EGO, $m(x)$ and $v(x)$ are GP's mean and variance and the compromise is the maximization of the expected improvement below the best observed f value. Once x' has been found, $f(x')$ is calculated, the model ($m()$ and $v()$) is updated and the process is iterated.

CMA-ES Algorithm. Another popular algorithm in continuous global optimization is the Covariance Matrix Adaptation Evolution Strategy (CMA-ES) [3]. CMA-ES relies on the iterative sampling and updating of a multi-normal density

$$\mathbf{x} \sim \mathbf{m}^{(g)} + \sigma^{(g)} \mathcal{N}\left(\mathbf{0}, \mathbf{C}^{(g)}\right), \ i = 1, ..., \lambda, \tag{1}$$

where g is the iteration counter, $\sigma^{(g)} \in \mathbb{R}^+$ is called mutation step size and $\mathbf{C}^{(g)} \in \mathbb{R}^{d \times d}$ is a covariance matrix. $\sigma^{(g)}$ controls the step length and $\mathbf{C}^{(g)}$ governs the ellipsoidal shape of the density function. The effective covariance matrix $\sigma^2 \times \mathbf{C}$ of CMA-ES describes good steps in S. CMA-ES is sometimes interpreted as a robust local search method [3]. Its robustness is related to invariance properties with respect to objective function scaling and coordinate system rotations. This algorithm was consistently found to be highly performing in the Black-Box Optimization Benchmarking (BBOB) workshops for low, moderate and highly multimodal functions for problems dimensions between 5 and 40 [2] if it is coupled with a restart mechanism.

CMA-ES and Models. Past works on global optimization of costly functions have already involved augmenting Evolution Strategies (ESs) with metamodels [6,8,10]. The general idea is to replace some evaluations of the true objective function with metamodel estimates and trigger true evaluations through an error rate measure. Kriging has sometimes been the metamodel added to the ESs. The motivation for using kriging is the availability of a prediction uncertainty. In [12], a pre-selection of the most promising points is done based on a kriging model, which enables sampling more solutions and makes the search more efficient. Two criteria are investigated as performance measures, the (mean) objective function prediction and the probability of improvement over the best observed point. In [1], kriging serves as a local metamodel and various performances are measured by different compromises between search intensification around the current best solution and exploration. In [9], a local kriging enables dealing with noisy objective functions by easing the estimation of the objective function expectation.

The optimization algorithm introduced in this paper differs from previous contributions in the fact that the EGO and CMA-ES search principles are invoked one after each other without iterations. The motivation is that EGO is efficient in the early design of experiments (DoE) stage of the optimization (volume search), while CMA-ES is a converging search process that efficiently transitions from volume to local search.

2   The EGO-CMA Algorithm

2.1   Experimental Setup and Initial Observations

The optimization algorithms compared in the paper (EGO, CMA-ES and later EGO-CMA) are tested with four well-known functions called Sphere, Ackley, Rastrigin, and Michalewicz (cf. [5]). The search spaces of the functions have been rescaled to $[-5, 5]^d$, $d = 2, 5, 10$. The total number of calls to the objective function or *budget* is $70 \times d$. The initial design points of EGO are obtained by Latin Hypercube Samples (LHS) of size $3 \times d$. We repeat EGO three times on each function. CMA-ES being a stochastic optimizer, it arguably exhibits larger performance variation so its runs are repeated ten times from three different starting points. Figure 1 illustrates one typical run of EGO and CMA-ES on the Sphere function in dimension 5. The solid line represents each function value obtained by the optimization algorithm and the dashed-dotted line shows the best observed function value so far. In a characteristic manner, EGO makes early progress and then loses efficiency (left) while CMA-ES steadily converges to the minimum as the number of calls to the objective function increases (right). Such an observation was confirmed on the other test functions and started the idea of combining EGO and CMA-ES.

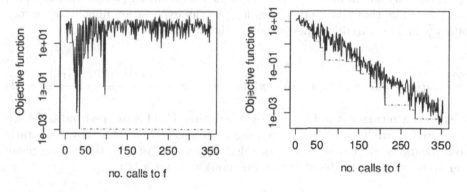

Fig. 1. One typical run of EGO (left) and CMA-ES (right) on the Sphere, $d = 5$. Solid line: f history during optimization. Dash-dotted line: best f.

2.2   Combining EGO and CMA-ES

We now introduce the EGO-CMA algorithm, which first explores the search space with EGO and then switches to CMA-ES in order to converge to the optimum.

The switch occurs after the best observed f has not improved for at least $0.1 \times budget$ analyses and if one of the following conditions is met: (*i*) 50 percent of the *budget* is exhausted or (*ii*) $\overline{EI} < 0.01 \times \left(f_{DoE}^{best} - f^{best} \right)$. $\overline{EI}$ is the average of the maximum expected improvement over the 5 last iterations. f_{DoE}^{best} and f^{best}

are the best f values in the initial design of experiments and the current best point, respectively. When the switch takes place, the best point obtained by EGO, $\mathbf{x}^{best}$, becomes CMA-ES's starting point. Furthermore, EGO-CMA uses of the fitted kriging mean as an approximation to the true function to warm start CMA-ES.

Let us provide some background on CMA-ES initialization. Consider first the optimization of a convex-quadratic function $f_{\mathbf{H}}(\mathbf{x}) = \frac{1}{2}(\mathbf{x} - \mathbf{x}_{\mathbf{H}}^*)^\top \mathbf{H}(\mathbf{x} - \mathbf{x}_{\mathbf{H}}^*)$, where $\mathbf{H}$ is positive definite and $\mathbf{x}_{\mathbf{H}}^*$ is the optimum. $\mathbf{H}$ can be decomposed into $\mathbf{H} = \mathbf{B}\mathbf{D}^2\mathbf{B}^\top$, where $\mathbf{B}$ is made of the eigenvectors of $\mathbf{H}$ as columns ($\mathbf{B}^\top \mathbf{B} = \mathbf{B}\mathbf{B}^\top = \mathbf{I}$) and $\mathbf{D}$ is a diagonal matrix with the square roots of $\mathbf{H}$'s eigenvalues as diagonal elements. The optimal ES covariance matrix has lines of equiprobable mutation aligned with the level sets of the objective function [11]. This happens when the covariance matrix of the search distribution, $\mathbf{C}$ (from (1) without superscript), is proportional to the inverse of $\mathbf{H}$ so we set

$$\mathbf{C} = \mathbf{B}\mathbf{D}^{-2}\mathbf{B}^\top. \tag{2}$$

The step size σ can now be tuned by performing a change of variable to turn to a spherical landscape: define the new variable $\mathbf{t} = \mathbf{D}\mathbf{B}^\top(\mathbf{x} - \mathbf{x}_{\mathbf{H}}^*)$, the objective function becomes $f_{\mathbf{H}}(\mathbf{t}) = \frac{1}{2}\mathbf{t}^\top \mathbf{t}$. In the t-space, the CMA-ES search points distribution (1) becomes $\mathbf{t} \sim \mathbf{D}\mathbf{B}^\top(\mathbf{m} - \mathbf{x}_{\mathbf{H}}^*) + \sigma\mathcal{N}(\mathbf{0}, I)$. In terms of t, one optimizes a spherical function with a spherical distribution, a situation in which one would like that the average step length (the expectation of the square root of a χ_d^2 random variable times σ) equals the distance to the optimum

$$\sigma\sqrt{d - 0.5} = \left\| \mathbf{D}\mathbf{B}^\top(\mathbf{m} - \mathbf{x}_{\mathbf{H}}^*) \right\| \quad \Rightarrow \quad \sigma = \frac{\left\| \mathbf{D}\mathbf{B}^\top(\mathbf{m} - \mathbf{x}_{\mathbf{H}}^*) \right\|}{\sqrt{d - 0.5}}. \tag{3}$$

We can now return to the EGO-CMA description. EGO is stopped and CMA-ES is started at $\mathbf{m}^{(0)} = \mathbf{x}^{best}$. To obtain $\sigma^{(0)}$ and $\mathbf{C}^{(0)}$ from the above quadratic considerations, we take the second order Taylor expansion of the kriging mean (an approximation to the objective function) at point $\mathbf{x}^{best}$:

$$f(\mathbf{x}) \approx f_{\mathbf{H}}(\mathbf{x}) = m(\mathbf{x}^{best}) + \nabla m(\mathbf{x}^{best})^\top(\mathbf{x} - \mathbf{x}^{best}) + \frac{1}{2}(\mathbf{x} - \mathbf{x}^{best})\mathbf{H}(\mathbf{x} - \mathbf{x}^{best}).$$

The initial covariance of CMA-ES is set equal to the inverse of the Hessian of the kriging mean at $\mathbf{x}^{best}$,

$$\mathbf{C}^{(0)} = \mathbf{H}^{-1}. \tag{4}$$

Cases when $\mathbf{H}$ is not strictly positive definite, among which the non invertibility case, are discussed later. Minimization of $f_{\mathbf{H}}$ gives $\mathbf{x}_{\mathbf{H}}^*$, an approximation to the optimum, by which we can complete Eq. (3) and calculate $\sigma^{(0)}$:

$$\mathbf{x}_{\mathbf{H}}^* - \mathbf{x}^{best} = -\mathbf{H}^{-1}(\mathbf{x}^{best})\nabla m(\mathbf{x}^{best})$$

$$\Rightarrow \sigma^{(0)} = \frac{\left\| \mathbf{D}\mathbf{B}^\top \mathbf{H}^{-1}(\mathbf{x}^{best})\nabla m(\mathbf{x}^{best}) \right\|}{\sqrt{d - 0.5}}. \tag{5}$$

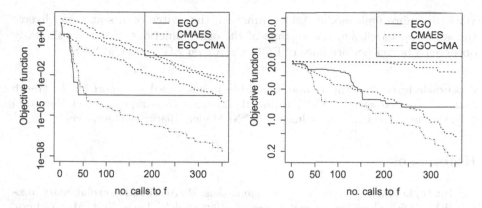

Fig. 2. Median of the best objective function vs. number of calls of EGO, CMA-ES (with three different starting points) and EGO-CMA on the 5 dimensional Sphere (left) and Ackley (right) functions.

We now discuss the cases when the Hessian matrix is not strictly positive definite, i.e., f_H is concave in some directions. f_H is convexified, i.e., the Hessian is forced to be positive definite, by substituting 10^{-6} for the negative eigenvalues in $\mathbf{D}^2$. However, this might increase the condition number of the Hessian matrix that is the ratio of the largest to the smallest eigenvalue, $cond(\mathbf{H}) = \frac{\lambda_{max}}{\lambda_{min}}$. To improve the condition number, we add a positive value, δ, to the main diagonal of the Hessian matrix, $\mathbf{H}_{conv} = \mathbf{B}\mathbf{D}^2_{conv}\mathbf{B}^\top = \mathbf{B}(\mathbf{D}^2 + \delta\mathbf{I})\mathbf{B}^\top$. δ can be calculated by defining an upper bound on the condition number, $\mathcal{CU}$,

$$\frac{\lambda_{max} + \delta}{\lambda_{min} + \delta} \lessapprox \mathcal{CU} \quad \Rightarrow \quad \delta \gtrapprox \frac{\mathcal{CU}\lambda_{min} - \lambda_{max}}{1 - \mathcal{CU}} \tag{6}$$

In our experiments, we set the condition number limit $\mathcal{CU}$ equal to 10^4 and the initial CMA-ES covariance and step size (Eqs. (4) and (5)) are calculated with $\mathbf{H}_{conv}$ and $\mathbf{D}_{conv}$. Finally, the step size is bounded through

$$\frac{0.3 \cdot 10^{-8}}{\sqrt{d}} \times \|\mathbf{D}_{conv}\mathbf{B}^\top(\mathbf{u} - \mathbf{l})\| \leq \sigma^{(0)} \leq \frac{0.3}{\sqrt{d}} \times \|\mathbf{D}_{conv}\mathbf{B}^\top(\mathbf{u} - \mathbf{l})\| . \tag{7}$$

3   Simulation Results

The performance of EGO-CMA is tested by repeating each run of EGO-CMA 5 times on each function, then the results are compared to EGO and CMA-ES. For the sake of brevity, we just illustrate this comparison on the Sphere and Ackley functions in 5 dimensions, see Fig. 2. It is seen that EGO shows a rough yet early location of the global minimum, which allows EGO-CMA to further increase the accuracy. The accuracy of EGO-CMA is about 10^{-8} for the Sphere function with a gain of two orders of magnitude over CMA-ES. The switch from EGO to CMA-ES in EGO-CMA can clearly be seen on the Sphere function before 100 function evaluations as the EGO-CMA curve first follows EGO and then is parallel to CMA.

With the more multimodal Ackley function, the switch occurs at more diverse times of the search. On the average of the other functions tested, we similarly observed a better performance of EGO-CMA over EGO and CMA-ES.

Acknowledgments. The authors would like to acknowledge support by the French national research agency (ANR) within the Modèles Numériques project NumBBO (analysis, improvement and evaluation of "NUMerical BlackBox Optimizers").

References

1. Buche, D., Schraudolph, N.N., Koumoutsakos, P.: Accelerating evolutionary algorithms with Gaussian process fitness function models. Trans. Syst. Man Cybern. **35**, 183–194 (2004)
2. Hansen, N., Auger, A., Ros, R., Finck, S., Pošík, P.: Comparing results of 31 algorithms from the black-box optimization benchmarking BBOB-2009. In: GECCO, pp. 1689–1696 (2010)
3. Hansen, N., Ostermeier, A.: Completely derandomized self-adaptation in evolution strategies. Evol. Comput. **9**(2), 159–195 (2001)
4. Hutter, F., Hoos, H.H., Leyton-Brown, K.: Sequential model-based optimization for general algorithm configuration. In: Coello, C.A.C. (ed.) LION 5. LNCS, vol. 6683, pp. 507–523. Springer, Heidelberg (2011)
5. Idoumghar, L., Melkemi, M., Schott, R.: A novel hybrid evolutionary algorithm for multi-modal function optimization and engineering applications. In: 13th IASTED Conference, vol. 683, pp. 87–93 (2009)
6. Jin, Y.: Surrogate-assisted evolutionary computation: recent advances and future challenges. Swarm Evol. Comput. **1**, 61–70 (2011)
7. Jones, D.R., Schonlau, M., Welch, W.J.: Efficient global optimization of expensive black-box functions. J. Global Optim. **13**(4), 455–492 (1998)
8. Kern, S., Hansen, N., Koumoutsakos, P.: Local meta-models for optimization using evolution strategies. In: Runarsson, T.P., Beyer, H.-G., Burke, E.K., Merelo-Guervós, J.J., Whitley, L.D., Yao, X. (eds.) PPSN IX. LNCS, vol. 4193, pp. 939–948. Springer, Heidelberg (2006)
9. Kruisselbrink, J., Emmerich, M., Deutz, A., Baeck, T.: A robust optimization approach using kriging metamodels for robustness approximation in the CMA-ES. In: Congress on Evolutionary Computation, pp. 1–8 (2010)
10. Loshchilov, I., Schoenauer, M., Sebag, M.: Intensive surrogate model exploitation in self-adaptive surrogate-assisted CMA-ES (saACM-ES). In: GECCO, pp. 439–446 (2013)
11. Rudolph, G.: On correlated mutations in evolution strategies. In: PPSN, pp. 107–116. North-Holland, Amsterdam (1992)
12. Ulmer, H., Streichert, F., Zell, A.: Evolution Strategies Assisted by Gaussian Processes with Improved Pre-Selection Criterion. In: Congress on Evolutionary Computation, pp. 692–699 (2003)

$MO - Mine_{clust}$: A Framework for Multi-objective Clustering

Benjamin Fisset[1]([⊠]), Clarisse Dhaenens[1,2], and Laetitia Jourdan[1,2]

[1] DOLPHIN Project-team, Inria Lille - Nord Europe,
59650 Villeneuve d'Ascq Cedex, France
benjamin.fisset@inria.fr
[2] CRIStAL, Université Lille 1, UMR CNRS 9189,
59650 Villeneuve d'Ascq Cedex, France
{clarisse.dhaenens,laetitia.jourdan}@lifl.fr

Abstract. This article presents $MO - Mine_{clust}$ a first package of the platform in development $MO - Mine$. This platform aims at providing optimization algorithms, and in particular multi-objective approaches, to deal with classical datamining tasks (Classification, association rules...). This package $MO - Mine_{clust}$ is dedicated to clustering. Indeed, it is well-known that clustering may be seen as a multi-objective optimization problem as the goal is both to minimize distances between data belonging to a same cluster, while maximizing distances between data belonging to different clusters. In this paper we present the framework as well as experimental results, to attest the benefit of using multi-objective approaches for clustering.

1 Introduction

Clustering is a very common and popular datamining technique. In a context where data are described by a set of variables, clustering algorithms provide a partition of the dataset, while grouping similar data into clusters. Thus, elements in one cluster are similar among them and different from elements of the other clusters.

This problem may be seen as a combinatorial optimization problem as soon as a criterion able to evaluate the quality of a given clustering can be found. In the literature, many such criteria have been proposed and multi-objective models have been adopted. In this context, genetic and other evolutionary algorithms have been widely used to obtain good solutions regarding the chosen quality measure.

The aim of this paper is to present $MO - Mine_{clust}$, a framework dedicated to multi-objective clustering. This framework is part of a more global one, $MO - Mine$ which will provide to non specialists of datamining or optimization, the ability to execute performant multi-objective algorithms to analyse their data. This framework must be able to be used on any kind of data and has to be generic. The genericity of this framework allows to adopt different models, taking into account several combinations of optimization criteria.

© Springer International Publishing Switzerland 2015
C. Dhaenens et al. (Eds.): LION 9 2015, LNCS 8994, pp. 293–305, 2015.
DOI: 10.1007/978-3-319-19084-6_30

To present the framework and its performance, the rest of the article is presented as follows. In Sect. 2 multi-objective optimization and multi-objective clustering are presented. We focus in particular in the components of existing multi-objective algorithms for clustering. Section 3 presents $MO - Mine_{clust}$, its approach and its implementation. In Sect. 4, results on some classical benchmarks are presented and discussed. The last section gives conclusions and perspectives.

2   Multi-objective Optimisation for Clustering

A clustering solution, that will assign each element to a given cluster, is considered good when elements of each cluster are very similar among them (low intra-cluster variance) and very different from the elements of the other clusters (high inter-cluster variance). This problem is by nature a multi-objective combinatorial optimisation one.

2.1   Multi-objective Combinatorial Optimization

A problem of multi-objective combinatorial optimization can be defined as a problem where a set of $n \geq 2$ objective functions have to be optimized (minimized or maximized) in a finite set of feasible solutions (*decision space Ω*).

$$\begin{cases} \text{optimize} & F(x) = (f_1(x), f_2(x), \dots, f_n(x)) \\ \text{subject to } x \in \Omega \end{cases} \tag{1}$$

Unlike to mono-objective problems, the solution of a multi-objective problem is not unique. It is composed of a set of non-dominated solutions called Pareto solutions. These solutions present the best compromises between the objectives.

2.2   Multi-objective Clustering (MOC)

Given the two natural objectives of clustering, minimizing intra-cluster variance and maximizing inter-cluster variance, several compromise solutions may be obtained. Figure 1 presents an example, where instances are described by two attributes x and y. Their projection in the decision space allows to visualize distances between instances. Several clustering solutions are proposed, in which criteria are either optimized independently or simultaneously [4]. In the multi-objective context, the final solution of a multi-objective clustering is a collection of clustering solutions with different trade-offs between objectives represented in a Pareto set, based on the Pareto dominance (see Fig. 2).

In the literature, many works on multi-objective clustering exist, and in particular, evolutionary approaches have been widely used. For a complete recent survey on Multi-Objective Evolutionary Algorithms (MOEAs) for clustering, the reader may refer to [15]. As indicated in this survey, MOEAs for clustering defer from the underlying MOEA used (PESA-II, NSGA-II, IBEA...), the chromosome representation, the objectives functions used or the evolutionary operators

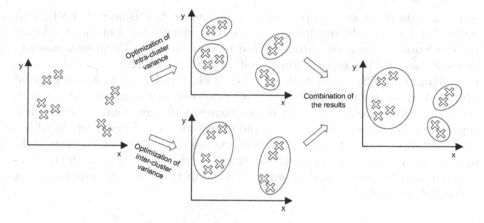

Fig. 1. Optimizing both objectives simultaneously [4].

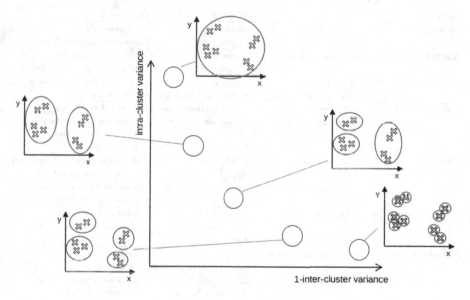

Fig. 2. Multi-objective clustering Pareto set of solutions [4].

implemented. Many combinations have been proposed, some of them may be adapted to specific data [12].

For example, one of the first Multi-Objective Clustering (MOC) approach was VIENNA introduced by Handl and Knowles based on PESA-II incorporating specialized mutation and initialization procedures [9]. The algorithm employs two following internal measures to estimate clustering quality: variance and connectivity. Such clustering quality measures have also been used in many other MOEAs. In particular, MOCK [8] uses overall deviation, that measures the compactness of clusters as a first objective and connectivity that considers whether adjacent data items are grouped into a same cluster. Let us note that VIENNA algorithm, for example, requires to know a-priori the number k of clusters. In the present work, we focus on automatic k-detection, as the platform may adapt to any data set. Table 1 reports some multi-objective clustering approaches while describing their differences; is the number of clusters k fixed in advance? what are the representations (locus-based adjacency, eisen plot, chromosome representation...), objective functions (variance, connectivity, deviation, complete link, separation, global completeness...) and operators used? Regarding the objective functions which determine the clustering model of the problem, most of the approaches are using two types of measures in order to estimate both the compactness and the separation of clusters. Some of these objective measures are presented hereafter.

Table 1. Summary table of some multi-objective clustering methods

Algorithms	k ?	Representation	Objective functions	Operators
VIENNA[9]	Yes	Vorono Cells	-Variance -Connectivity	-No crossover -Directed mutation based NN
MOCK[7,8]	No	Locus-based adjacency	-Deviation -Connectivity	-Uniform crossover -Nearest neighbors mutation
SiMM-TS[1,7]	No	-Eisen plot -Cluster profile plot	-Variance -min dist(2 centroids)	Crossover and mutation (unknown)
MOGA-BF[7,14]	No	Chromosome representation	-Global compactness -Separation	-Crowded binary tournament selection -Single point crossover -Random mutation
MOSSC[18]	No	Chromosome representation	-SSXB $-J'_{fwsc}$	-Binary tournament - crowded comparison operator based -Simulated binary crossover -Polynomial mutation
MOEA/D-Net[6]	No	Locus-based adjacency	-Negative Ratio Association -Ratio Cut	-2-points crossover -Neighbor-based mutation
MOMoDEFC[16]	No	Vector representation	-XB -Global cluster variance	-MoDE crossover -ModiMutation

MOVGA[13]	No	Chromosome
		representation
			-Global compactness
			-Separation based
			  centroid clusters
				-Crowded binary
				  tournament selection
				-crossover point
				-random mutation

2.3   Clustering Quality Measures

As indicated in the survey [15], usually cluster validity indices are used as objective functions. Most of multiobjective clustering algorithms use two validity indices to be simultaneously optimized in order to evaluate two complementary aspects: compactness and separation of clusters. Hence, given C a clustering solution (corresponding to a partition of the dataset), several measures may be computed.

Compactness Measures. Many measures have been proposed to evaluate the compactness. *Variance* and *Deviation* are the most used in the literature. They measure the proximity of data belonging to a same cluster. Therefore, they use the distance between each element and the center of the cluster it belongs to.

Variance. The variance [1,7,9] is computed by $Var(C) = \sum_{C_k \in C} \sum_{i \in C_k} d(i, c_k)^2$, where C is a set of clusters, c_k is the cluster centroid of C_k and $d(\ldots)$ is a distance function to be defined. The variance has to be minimized.

Deviation. The deviation objective function [7,8] is very similar to the variance. It is computed by $Dev(C) = \sum_{C_k \in C} \sum_{i \in C_k} d(i, c_k)$ and has also to be minimized.

Separation. Separation measures evaluate how different the clusters are. The connectivity measure is the most used in the literature.

Connectivity. The connectivity [7,8,17] evaluates the degree to which neighboring data points have been placed in the same cluster. It is computed as follows. Its value lies in the interval $[0, 1]$ and has to be maximized:

$$Conn(C) = \frac{1}{N} \sum_{i=1}^{N} \left(\frac{\sum_{j=1}^{h} \omega_{i,nn_{i(j)}}}{h} \right), \quad \text{where } \omega_{a,b} = \begin{cases} 1 \text{ if } \exists C_k | a, b \in C_k \\ 0 \text{ otherwise.} \end{cases}$$

where $nn_{i(j)}$ is the jth nearest neighbor of data i, h is the number of neighbors used to compute the connectivity. N is a number of data.

Separation Based Centroid Clusters. This measure allows to compute the global sum between clusters centroid [7,14]. It is computed by $SumD(C) = \sum_{C_k \in C, C_l \in C, l \neq k} d(c_k, c_l)$ where c_k and c_l are the cluster centroids of clusters C_k and C_l respectively.

Hence several objective functions exist and choosing a combination of them will define the optimization model. As we will see later in the article, some models may be more efficient than other for some datasets. Therefore $MO - Mine_{clust}$ gives the possibility, for a given dataset, to automatically choose the best one.

3 $MO - Mine_{clust}$

Providing a platform dedicated to non specialists is the goal of $MO - Mine$. Therefore many components have been implemented in order to be able to adapt to the data to analyse. Indeed, as it will be shown in the experiments, a same MOEA, for example, is not always the best to use, according to the dataset. Then, regarding the numerous models (and in particular combinations of objective functions) able to deal with the multi-objective clustering, the objective of the proposed platform is to identify the best combination of

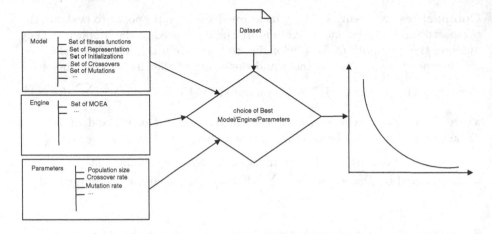

Fig. 3. Presentation of the MO-MINE clustering process part.

model/engine/parameters to a particular dataset in order to offer, to a non specialist, the ability of discovering the best clustering for his/her dataset. Such an approach requires the implementation of several components that can be combined. Therefore, the general approach adopted is described in Fig. 3. This figure shows that given the set of elements necessary to describe the model, the set of available engines and the assoicated parameters, $MO - Mine_{clust}$ will propose, for a given dataset, the best configuration model/engine/parameters. Let us note that this approach is not specific to clustering but can be applied to any optimization problem.

3.1   Components

In this first part of $MO - Mine_{clust}$, the focus is set on the opportunity to provide several MOEAs and a set of objective functions (to propose several multi-objective models).

Available multi-objective models are a combination of the clustering measures presented previously. Ideally, a measure for compactness should be combined with a measure of separation. Currently, the following measures are implemented: Variance, Deviation, Connectivity, Separation based centroid clusters. Other measures may be added easily.

Already available MOEAs in $MO - Mine_{clust}$ are: NSGA II [3], EasyEA, IBEA [19] and PESA II [2].

Concerning the representation and associated operators, in this first study, those presented in MOCK [8] will be used:

- Representation: locus-based adjacency representation,
- Initialization: 1/2 of the population is initialized by the Minimum spanning trees (MSTs), these solutions performing well for compactness, the other 1/2 of the population is initialized using the k-means algorithm, these solutions performing well for separation,

- Mutation operator: Nearest neighbor mutation,
- Crossover operator: Uniform crossover.

In the litterature, several representations exist for clustering problems. In their recent complete work, Garcia-Piquer et al. [5] present an analysis on the efficiency of several representations using PESA-II algorithm. In our work, the focus is given on the different MOEAs (that are called engine), and on the association MOEAs / Model. Therefore we first propose a single representation: the locus based adjacency representation, as it is the one used in MOCK, one of the best Multi-Objective Evolutionary Algorithm for clustering.

3.2 Implementation

All the developments have been realized under ParadisEO[1] and its extension ParadisEO-MOEO [10]. ParadisEO is a C++ white-box object-oriented framework dedicated to the reusable design of metaheuristics. In order to choose the pair of *model/engine*, which has the best performance for a data set, the package IRACE [11] will be used. This package implements an iterated racing procedure, which is an extension of the Iterated F-race procedure.

$MO - Mine_{clust}$ is available at http://mo-mine.gforge.inria.fr/doku.php.

4   Experiments and Discussion

In order to attest the performance of the platform, experiments and comparisons with the literature are proposed.

4.1   Data Sets

In this study, two types of experimental data from the literature are used: hand-crafted two-dimensional data sets (2D) and generated dataset (GD)[2]. Tables 2 and 3 present details of data sets; the number of clusters (k), the number of data (N) and the dimensions which corresponds to the number of attributes characterizing a data.

4.2   Protocol

In [7], Handl and Knowles compare the choice of clustering criteria in multi-objective data clustering thanks to several algorithms. MOCK showed the best performance for the majority of the data sets. Due to these reasons, we decide to compare our approach to MOCK algorithm in the following. Therefore, we will use the same protocol than in [8] to compare the obtained results. In order to be fair in the methods comparison, we use a same number of generations equals to 1000. To compare clustering approaches, it is common to use datasets

---

[1] http://www.paradiseo.gforge.inria.fr.
[2] http://personalpages.manchester.ac.uk/mbs/Julia.Handl/mock.html.

Table 2. Datasets 2D

Data sets	k	N	Dimensions
Square1	4	1000	2
Square4	4	1000	2
Sizes5	4	1000	2
Long1	2	1000	2
Spiral	2	1000	2

Table 3. Datasets GD

Data sets	k	N	Dimensions
2d-4c	4	1572	2
2d-20c	20	1000	2
2d-40c	40	1000	2
10d-4c	4	1289	10
10d-20c	20	1013	10
10d-40c	40	1937	10

where partitions are known a-priori and to evaluate the quality of the found partitions obtained after running of the tested methods in comparison to the original partitions.

The process is detailed in Fig. 4.

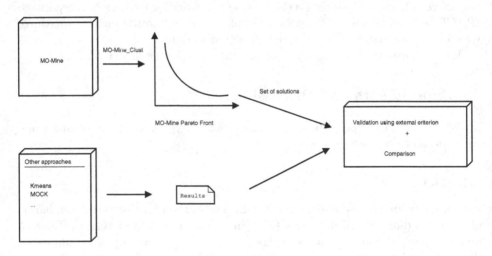

Fig. 4. Comparison process.

4.3 Cluster Evaluation

There exist several performance indices for cluster evaluation. They measure the correspondance between two partitions G and H, corresponding respectively to the clustering solution obtained and to a theoretical solution. Based on these two partitions, it is possible to compute the contingency table that displays the frequency distribution of the variables as follows.

$$a = |\{i, j | C_{G(i)} = C_{G(j)} \wedge C_{H(i)} = C_{H(j)}\}|$$
$$b = |\{i, j | C_{G(i)} = C_{G(j)} \wedge C_{H(i)} \neq C_{H(j)}\}|$$
$$c = |\{i, j | C_{G(i)} \neq C_{G(j)} \wedge C_{H(i)} = C_{H(j)}\}|$$
$$d = |\{i, j | C_{G(i)} \neq C_{G(j)} \wedge C_{H(i)} \neq C_{H(j)}\}|$$

The values a, b, c and d are computed for each possible pair of data elements i and j, and their respective cluster assignments $C_{G(i)}$, $C_{G(j)}$, $C_{H(i)}$ and $C_{H(j)}$. Using these four values several performance indices have been proposed. We will use in this work, the Adjusted Rand index (ARI) as it is one of the most successful cluster validation indices and is the one used in [8]. ARI is the version adjusted for the chance of grouping of elements of the Rand index that is a measure of the similarity between two data clusterings.

The Adjusted Rand Index [4] has to be maximized and is computed as:

$$ARI(G, H) = \frac{\binom{n}{2}(a + d) - [(a + b)(a + c) + (c + d)(b + d)]}{\binom{n}{2}^2 - [(a + b)(a + c) + (c + d)(b + d)]}$$

ARI ranges between -1 and 1. The values of ARI close to 1 indicate an almost perfect concordance between the two compared partitions, whereas the values close to -1 indicate a complete discordance between them.

As previously mentionned, the cluster evaluation is done after the obtention of the results of each clustering methods.

4.1 IRACE Results

As mentioned before, IRACE [11] has been used to obtain the best combination of parameters for each dataset using Adjusted Rand Index of the best solution of the Pareto front as an external criterion to optimize. In these preliminary experiments, only a few ranges of parameters have been proposed to IRACE. The parameters tested, the ranges of values tested, as well as the best parameters obtained are presented in Table 4 for each dataset. We can observe that several combinations of fitness functions are selected: Variance and Separation, Deviation and Separation, Deviation and Connectivity, Variance and Connectivity. It shows that it is not always the same pair of objectives that is selected as the best combination, according to the dataset studied. We can also observe that the best algorithm is not always the same, however, for these datasets, only two of them are selected by IRACE: IBEA and PESA II. These experiments show, that PESA II, used in MOCK, is indeed well adpated for clustering. It also demonstrates the importance to be able to use different MOEAs and to let the framework choose the most adapted one whereas in other works often one single MOEA is tested [5].

Table 4. Components setting obtained by Irace.

Data sets	Neighborhood size	Population size	Objective functions	MOEA
Square1	30 [10;100]	12 [10;20]	Deviation + Connectivity	IBEA
Square4	65 [10;100]	16 [10;20]	Deviation + Connectivity	IBEA
Sizes5	54 [10;100]	19 [10;20]	Deviation + Connectivity	IBEA
Long1	28 [10;100]	15 [10;20]	Deviation + Connectivity	PESA-II
Spiral	42 [10;100]	13 [10;20]	Deviation + Connectivity	PESA-II
2d-4c	31 [10;100]	17 [10;20]	Variance + Connectivity	PESA-II
2d-20c	43 [10;70]	100 [10;100]	Variance + Connectivity	IBEA
2d-40c	63 [10;70]	100 [10;100]	Variance + Connectivity	IBEA
10d-4c	31 [10;100]	20 [10;20]	Deviation + Separation	PESA-II
10d-20c	38 [10;50]	100 [10;100]	Deviation + Separation	IBEA
10d-40c	15 [10;50]	100 [10;100]	Deviation + Separation	PESA-II

4.5   Experimental Results

Table 5 presents results obtained by the platform $MO - Mine_{clust}$. Results obtained by the well-known Kmeans algorithm, as well as those obtained by the state-of-the-art MOCK algorithm based on PESA II are presented. Those results are directly extracted from the original article [8]. The number of clusters computed as well as the average quality (in terms of Adjusted Rand Index (ARI)) are presented. In the table we present the average value of the Adjusted Rand Index value and its standard deviation (Std.) for $MO - Mine_{clust}$ over 10 executions. Let us note that the value of the adjusted rand index is an external criterion used to select the best solution among the Pareto front generated by each algorithm but it is not used within the algorithms as an optimization criterion.

In this table, we can observe that for the majority of the datasets, our approach improves the average Adjusted Rand Index (AV. ARI). The average relative percentage deviation on the other datasets is less than 0.1 %. With a Friedman test, we observe that the algorithms are different with a p-value of 0.001. Concerning the comparison between MOCK and $MO - Mine_{clust}$ the difference is also statistically significative and $MO - Mine_{clust}$ performs better in term of ARI.

Table 5 reports clustering with the best ARI. We can however observe that the number of clusters obtained is sometimes larger than the expected number (square 1, size5, 2d-20c, 10d-20c). In each case, the algorithm selected was IBEA. Let us remark, that for these datasets, PESA-II manages to find good solutions with a correct number of clusters, but their ARI evaluation is not as good as the one proposed by IBEA. This leads to two comments and hypotheses; first IBEA seems to better optimize on one extreme part of the Pareto front and then the best ARI solution leads on the part that favours connectivity. Secondly, we

Table 5. Performance of $MO - Mine_{clust}$.

Data sets	MOCK [8]		Kmeans		$MO - Mine_{clust}$		
Name	k	Av. ARI	k	Av. ARI	k	Av. ARI	Std
Square1	4.22	0.9622	4	0.9651	19.6	**0.9901**	0.013
Square4	4.32	0.7729	4	0.8048	4.4	**0.8196**	0.0225
Sizes5	4.2	0.976	3.92	0.9557	37.8	**0.9838**	0.005
Long1	2	**0.9998**	4.98	0.3562	2	**0.9998**	0.0001
Spiral	2	1	5.12	0.5502	2	1	0
2d-4c	4.12	**0.9893**	3.99	0.9143	4.2	0.988	0.0002
2d-20c	19.94	0.9454	33.79	0.8633	19.2	**0.9832**	0.009
2d-40c	42.14	0.8654	42.36	0.692	19.4	**0.9835**	0.034
10d-4c	4.07	0.9962	3.99	0.9704	4.2	**0.9975**	0.001
10d-20c	20.26	**0.9981**	21.45	0.9820	20.2	0.9979	0.004
10d-40c	42.84	**0.9896**	43.48	0.9678	19.2	0.9859	0.01

can wonder on the capability of the ARI measure to really detect more interesting partitions as sometimes it may prefer solutions with a too large number of clusters. These remarks strengthen the interest of using a generic framework for multi-objective clustering. In particular, it is really easy to modify a selection criterion in the parameter file or to include new additional components and to offer it to the users as the framework will select the best combination.

5 Conclusion

In this article, we have presented a multi-objective framework for clustering data, $MO - Mine_{clust}$. The framework searches for the best association of model/engine/parameter for a dataset without specifiing the number of clusters. $MO - Mine_{clust}$ shows very interesting behavior and shows that the model and the engine have a great importance in the performance of the method and can depend on datasets. There are several directions for future works. First we would like to extend the platform to more models and to different representations. We also want to test our method on real datasets and in particular to datasets from biology where the number of dimensions and observations are large. We would also study how to automatically choose a solution of the pareto front on unseen data. The same kind of methodology could be used to solve other datamining problems and it will be interesting to see if results could also be improved.

Acknowledgements. This work has been realized with the support of the french project ANR-13-TECS-0009.

References

1. Bandyopadhyay, S., Mukhopadhyay, A., Maulik, U.: An improved algorithm for clustering gene expression data. Bioinformatics **23**(21), 2859–2865 (2007)
2. Corne, D., Jerram, N.R., Knowles, J., Oates, M.J.: Pesa-II: region-based selection in evolutionary multiobjective optimization. In: Proceedings of the Genetic and Evolutionary Computation Conference (GECCO 2001), pp. 283–290. Morgan Kaufmann Publishers (2001)
3. Deb, K., Pratap, A., Agarwal, S., Meyarivan, T.: A fast and elitist multiobjective genetic algorithm: NSGA-II. IEEE Trans. Evol. Comput. **6**(2), 182–197 (2002). Cited by(since 1996)7480
4. Piquer, Á.G.: Facing-up challenges of multiobjective clustering based on evolutionary algorithms: representations, scalability and retrieval solutions. Ph.D. thesis, Universitat Ramon Llull (2012)
5. Garcia-Piquer, A., Fornells, A., Bacardit, J., Orriols-Puig, A., Golobardes, E.: Large-scale experimental evaluation of cluster representations for multiobjective evolutionary clustering, pp. 36–53 (2014)
6. Gong, M., Ma, L., Zhang, Q., Jiao, L.: Community detection in networks by using multiobjective evolutionary algorithm with decomposition. Physica A Stat. Mech. Appl. **391**(15), 4050–4060 (2012)
7. Handl, J., Knowles, J.: Clustering criteria in multiobjective data clustering. In: Coello, C.A.C., Cutello, V., Deb, K., Forrest, S., Nicosia, G., Pavone, M. (eds.) PPSN 2012, Part II. LNCS, vol. 7492, pp. 32–41. Springer, Heidelberg (2012)
8. Handl, J., Knowles, J.D.: An evolutionary approach to multiobjective clustering. IEEE Trans. Evol. Comput. **11**(1), 56–76 (2007)
9. Handl, J., Knowles, J.D.: Evolutionary multiobjective clustering. In: Yao, X., et al. (eds.) PPSN 2004. LNCS, vol. 3242, pp. 1081–1091. Springer, Heidelberg (2004)
10. Liefooghe, A., Jourdan, L., Talbi, E.-G.: A software framework based on a conceptual unified model for evolutionary multiobjective optimization: paradiseo-moeo. Eur. J. Oper. Res. **209**(2), 104–112 (2011)
11. López-Ibánez, M., Dubois-Lacoste, J., Stützle, T., Birattari, M.: The Irace package: iterated racing for automatic algorithm configuration. Technical report TR/IRIDIA/2011-004, IRIDIA, Université Libre de Bruxelles, Brussels, Belgium, January 2011
12. Maulik, U., Bandyopadhyay, S., Mukhopadhyay, A.: Multiobjective Genetic Algorithms for Clustering: Applications in Data Mining and Bioinformatics. Springer, Berlin (2011). ISBN 978-3-642-16614-3
13. Mukhopadhyay, A., Maulik, U.: A multiobjective approach to MR brain image segmentation. Appl. Soft Comput. **11**(1), 872–880 (2011)
14. Mukhopadhyay, A., Maulik, U., Bandyopadhyay, S.: Multiobjective genetic algorithm-based fuzzy clustering of categorical attributes. IEEE Trans. Evol. Comput. **13**(5), 991–1005 (2009)
15. Mukhopadhyay, A., Maulik, U., Bandyopadhyay, S., Coello Coello, A.: Survey of multiobjective evolutionary algorithms for data mining: part II. IEEE Trans. Evol. Comput. **18**(1), 20–35 (2014)
16. Saha, I., Maulik, U., Plewczynski, D.: A new multi-objective technique for differential fuzzy clustering: the impact of soft computing for the progress of artificial intelligence. Appl. Soft Comput. **11**(2), 2765–2776 (2011)
17. Cao, H., Zheng, Y., Jia, L.: Multi-objective gene expression programming for clustering. Inf. Technol. Control **41**(3), 283–294 (2012)

18. Zhu, L., Cao, L., Yang, J.: Multiobjective evolutionary algorithm-based soft sub-space clustering. In: IEEE Congress on Evolutionary Computation, pp. 1–8. IEEE (2012)
19. Zitzler, E., Künzli, S.: Indicator-based selection in multiobjective search. In: Yao, X., et al. (eds.) PPSN 2004. LNCS, vol. 3242, pp. 832–842. Springer, Heidelberg (2004)

A Software Interface for Supporting the Application of Data Science to Optimisation

Andrew J. Parkes[⊠], Ender Özcan, and Daniel Karapetyan

ASAP Research Group School of Computer Science, University of Nottingham,
Nottingham, UK
{ajp,exo,dxk}@cs.nott.ac.uk

Abstract. Many real world problems can be solved effectively by meta-heuristics in combination with neighbourhood search. However, implementing neighbourhood search for a particular problem domain can be time consuming and so it is important to get the most value from it. Hyper-heuristics aim to get such value by using a specific API such as 'HyFlex' to cleanly separate the search control structure from the details of the domain. Here, we discuss various longer-term additions to the HyFlex interface that will allow much richer information exchange, and so enhance learning via data science techniques, but without losing domain independence of the search control.

Keywords: Combinatorial optimisation · Metaheuristics · Data science · Machine learning

1 Introduction

Over the last few decades many highly-effective metaheuristic search methods, working on numerous target problem domains, have been developed. They are generally based on neighbourhood improvement search in which a solution is iteratively changed by using moves taken from one or more neighbourhoods. The generation and acceptance/rejection of the moves is generally controlled by a metaheuristic. The neighbourhoods are often quite sophisticated and involve a fairly deep insight into the domain. However, all-too-often the metaheuristics are relatively simple, rather static, and do not exploit the specifics of the interactions between the neighbourhood search operators. Hyper-heuristics are a technique, and a software architecture, that separates the control (the meta-heuristic) from the details of the domain and the neighbourhoods [2]. A key aim allowing learning and statistical techniques, 'data science', to be applied to optimisation without them having to be re-implemented separately for every problem domain; essentially giving a plug-and-play version of sophisticated adaptive metaheuristics. The goal is to lift the control from the domain level up to the higher hyper-heuristic level so that data science methods have access to the details of the search process but in a problem domain independent manner.

© Springer International Publishing Switzerland 2015
C. Dhaenens et al. (Eds.): LION 9 2015, LNCS 8994, pp. 306–311, 2015.
DOI: 10.1007/978-3-319-19084-6_31

In a sense, this is refactoring of standard algorithms leading to better 'separation of concerns'; search control agents should not know about the domain details.

We discuss the support of this goal by using the 'HyFlex' (*Hyper-heuristics Flexible framework*) interface[1] [4] to separate the hyper-heuristic control from the details of the domain. In the initial limited interface, the hyper-heuristic simply selects neighbourhood moves (the domain-level heuristics) and in return all it learns about the current solution(s) is the objective value. Although such an interface is narrow, one should note that this is sufficient for some well-known meta-heuristics; e.g. standard simulated annealing can be implemented as a simple hyper-heuristic. If data science techniques for optimisation are to become both easy-to-use and still effective, then a drive should be to extend the interfaces (APIs) towards a clean separation but supporting a rich control and information flow. (There are a few existing examples, e.g. [5], that consider a limited broadening of the interface, allowing increased information flow.) The point of this paper, (which is necessarily brief, 'positional', and with only a few key references), is to strengthen and promote the general point that a significantly richer information flow is still consistent with a clean separation between the control and the domain.

The interest in frameworks enabling implementation of general purpose algorithms is growing; for example, Ryser-Welch and Miller [7] provided an overview of some of those frameworks, including Snappy, SATzilla, ParHyFlex, Hyperion and HyFlex. We focus on *selection hyper-heuristics* which mix and control a pre-defined set of low level heuristics during the search process [2]. Corresponding to these, an initial version of an interface, HyFlex v1.0 was implemented using Java and used in the first Cross-domain Heuristic Search Competition; CHeSC 2011 [4]. HyFlex connects the high level control layer managing a set of low level heuristics via a *domain barrier* but does not allow any problem specific information flow from the domain to the control level. Problem domain implementation details are hidden from the users so that they could focus on the design of the higher level method that will mix and control the low level heuristics and their settings; giving a for researchers, as well as practitioners, to develop new cross-domain solution methods and solve their problems with reduced effort.

The implementation of a metaheuristic is a special case which is supported by HyFlex. The only restriction is that the metaheuristic has to use the operators provided for a problem domain, or the problem domain implementation needs to be extended to include new operators. HyFlex can already be used as a benchmark to evaluate the performance of metaheuristic/hyper-heuristic methods. HyFlex also allows data science techniques and metaheuristics to be employed at the hyper-heuristic level to build, tune or refine hyper-heuristics via analysis (data collection) and execution modes of operation.

The interface was extended to HyFlex v1.1 [1] to enable treating the problem instances collectively as a batch, and was used in the second Cross-domain Heuristic Search Competition; CHeSC 2014. The extension supports balancing of computational effort between instance; if some instances are much "easier"

---

[1] http://www.hyflex.org/.

than others then it seems reasonable that they should be allocated less computational time. More importantly, it also allows inter-instance learning: If some of the instances are from the same domain then it makes sense that the hyper-heuristics should be able to learn from the earlier instances in order to perform better on the latter ones. The implementations of HyFlex also provide implementations of multiple problem domains allowing ideas for search control to be tried and tested with much reduced time and effort. Each problem domain includes implementation of a set of low level heuristics (operators), categorised as 'local search' (guarantees a non-worsening solution), 'mutation' (might be worsening), 'ruin-recreate', and 'crossover'. As well as selecting the heuristic, the hyper-heuristic may aslo need to control the heuristics via parameters. For example, local search can be controlled via a 'Depth of Search' parameter, and mutation/ruin-recreate by an 'Intensity of Mutation' parameters.

HyFlex v1.1 also considers the recent developments in the CPUs by supporting multi-core mode of operation, and also allows solution exchange via external memory. In particular, it allows multiple instances of the same solver to be working on the same problem instance, and to share solutions between the instances via a central pool of solutions. Naturally, this means that the system should aim to learn which solutions are most useful, and so it would be helpful for it to have more information about them. (This is part of the motivation for the 'solution features' discussed in the next section.)

2   Future Extensions to HyFlex

Here we discuss future extensions to HyFlex, including support for better annotations, instance/solution features, distance metrics and multi-objective optimisation. In many situations, metaheuristics are run as time contract algorithms, i.e., they terminate after a given time limit. HyFlex v1.0 has full support for this type of operation, returning the final solution and its quality. In certain situations, time limit can be irrelevant or relaxed and running an algorithm on and off, even running different algorithms at any phase might be preferable. This would require a hyper-heuristic (HH) reading from and writing into a file. The next HyFlex version will support saving of a (set of) solution(s) into file(s) and initialising a (set of) solution(s) from a (set of) file(s). Additionally, we will investigate ways of supporting delta/incremental evaluation, enabling fast computation of the objective values (fitness/cost) of a given solution.

In order to give more context, in Fig. 1 we give a more refined picture of the kind of structure that often (but not necessarily always) occurs within the hyper-heuristic. Specifically, it can (often) be split into two portions, "reactive" and "reflective". The reactive or 'dispatch' portion of the HH is directly responsible for calling the low level operators; typically, it make such decisions based on some control parameters. The parameters used by the dispatch side are controlled by the reflective portion that 'monitors' the sequence of actions (heuristics selected, etc.) and their effects (changes in the objectives, etc.). It uses data science techniques, aiming to dynamically set the control parameters to better values.

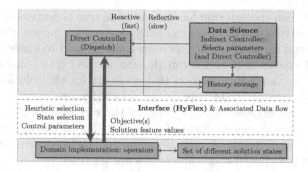

Fig. 1. Proposed general architecture of HyFlex 2.

As an example, the hyper-heuristic might be a form of adaptive reactive simulated annealing, and the parameter could be the temperature. The reflective portion of the HH could then try to observe the search progress in order to decide cooling rates, and make reheats. In this view, a metaheuristic may often be rather static and so could be considered as a hyper-heuristic that lacks the reflective data-science component. The reactive portion bears the responsibility for the 'selection of low-level heuristics' whereas it is the combination of both reactive and reflective that might be said to be closer to doing a 'search the space of meta-heuristics'.

Instance Features: Each problem instance in a given domain carries a set of features reflecting its specific characteristics. There are existing optimisation approaches, such as, algorithm portfolios, that make use of the instance features for choosing the best approach and/or best setting of an algorithm to solve a given instance. Our aim is that HyFlex should support the use of machine learning techniques to relate such characteristics to the choice of hyper-heuristic components or their parameter settings. HyFlex v1.0 does not support instance features; however, the basics for this are already in HyFlex v1.1: In CHeSC 2014, *size* of an instance was provided as an instance feature which can be used for better balancing of computational effort across the instances, i.e. allowing allocation of computational time for the 'smaller' instances. Other instance features could be graph density, number of constraints or planning horizon. Importantly, no domain-specific semantic information will be provided; the hyper-heuristic level will treat the set of feature values features as abstract vectors describing the instance, but will need to reflect and discover for itself how these relate to the search control.

Solution Features: Within domain-specific methods it is quite likely that some features of the current solution would be exploited to guide the search. It seems reasonable that in many cases these could be exposed to the hyper-heuristic as (for example) a vector of values. The meaning of the values would not be known to the hyper-heuristic, however, it could still extract information about the patterns that occur and use that in order to guide the search; e.g. spotting correlations between the solution features and the true objective(s). If the objective is uninformative (due to plateaux) or expensive to compute, then some solution features could be used as cheaper surrogates or used to guide the search. A more advanced extension of HyFlex might also all the hyper-heuristic to control the individual heuristics in a fashion that accounts for the solution features — e.g. the hyper-heuristic could supply the selected move operators with some bounds or preferences/goals on the values of the mix of multiple objective and solution features. One specific form of this would be for the operators to internally be optimising a weighted sum, but allowing the hyper-heuristic to control the weights. If the exposed solution features are regarded as a chromosome then one can potentially link with the mixed black-white box concepts [6], and so the hyper-heuristics become closer to the realm of evolutionary computation — and lead towards combining evolutionary/genetic algorithms with other metaheuristics. For example, solution features might have the potential to permit search control that captures the essence of EDA (estimation of distribution) algorithms.

Distances: An natural and straightforward extension to HyFlex is to allow the domain level to provide some measure of the difference between different solutions. This is algorithmically useful, for example, to support methods to maintain diversity within populations. This extension has been considered previously [5] for the simplest case of a single distance metric. However, there is no reason not to also permit multiple distance metrics. An annotation system can also say what kinds of properties they satisfy (such as triangle inequality). Note that this does not break the domain barrier as the actual nature of the solutions and the precise meaning of the metric itself still remains hidden. The task of data science would then be to extract useful information so as to control the search.

Multi-objectivity: Another natural and obvious extension to HyFlex is that it should allow the hyper-heuristic access to multiple objectives rather than a single one (e.g. see [3]). There are already studies on approaches that mix and control multi-objective evolutionary algorithms, which is currently not possible with HyFlex. The support for multiple objectives should then enable implementation of evolutionary search methods within the HyFlex and hyper-heuristic context. The primary difference would be that the details of the mutations (or perturbations) are implemented in the domain level, and not visible to the hyper-heuristic. Usually, techniques for diversity are done at the phenotype level (i.e. the objective values), but equivalents of genotypic diversity could be done by also using the distances discussed earlier. Since the hyper-heuristic will have access to the distance metrics, multiple objectives and multiple solutions, then it can measure the quality of the Pareto front, and control the search accordingly.

Annotations: Currently, HyFlex annotates low level heuristics with labels 'mutation', 'local search', 'ruin-recreate' and 'crossover', but an extended typology and annotation system should allow improved implementation of techniques such as iterated local search or memetic algorithms. For example, crossover operators could be annotated by whether they act as 'local search'; whether or not they never generate worsening solutions. Similarly, a ruin and recreate operator could act as a mutational operator in a given domain and as a local search operator in another domain. Furthermore, annotations can be extended to cover instance/solutions features, objectives and distance metrics.

3   Conclusion

The crucial conclusion is that HyFlex domain barrier can be modified to permit a much richer search control and information flow, but without losing the essential advantage of the designer of a hyper-heuristic still not needing to become an expert in the specific domain, but instead be able to apply and exploit data science techniques. An obvious task is to continue with the work in [1] in order to implement this and provide appropriate implemented domain solvers. The advantage, and major challenge, for those studying the application of data science methods to optimisation, is then to find techniques to exploit the rich streams of data that will result during runs of the solvers. We believe that the popular metaheuristics of today barely scratch the surface of what is possible in such a system. We intend to continue with the extensions to HyFlex, including initially support for initialisation from a solution file, saving of a solution, better annotations, instance and solution features, distance metrics, multi-objectivity. In order to reach a wider audience/users, training and teaching material will also be provided. Finally, we remark that using learning at the hyper-heuristic level does not exclude also learning at the domain level; though, expect this adds the challenge of the use of learning to control systems whose behaviour is itself changing due to their own internal learning.

References

1. Asta, S., Özcan, E., Parkes, A.J.: Batched mode hyper-heuristics. In: Nicosia, G., Pardalos, P. (eds.) LION 7. LNCS, vol. 7997, pp. 404–409. Springer, Heidelberg (2013)
2. Burke, E.K., Gendreau, M., Hyde, M., Kendall, G., Ochoa, G., Özcan, E., Qu, R.: Hyper-heuristics: a survey of the state of the art. J. Oper. Res. Soc. **64**(12), 1695–1724 (2013)
3. Maashi, M., Özcan, E., Kendall, G.: A multi-objective hyper-heuristic based on choice function. Expert Syst. Appl. **41**(9), 4475–4493 (2014)
4. Ochoa, G., Hyde, M., Curtois, T., Vazquez-Rodriguez, J.A., Walker, J., Gendreau, M., Kendall, G., McCollum, B., Parkes, A.J., Petrovic, S., Burke, E.K.: HyFlex: a benchmark framework for cross-domain heuristic search. In: Hao, J.-K., Middendorf, M. (eds.) EvoCOP 2012. LNCS, vol. 7245, pp. 136–147. Springer, Heidelberg (2012)
5. Ochoa, G., Walker, J., Hyde, M., Curtois, T.: Adaptive evolutionary algorithms and extensions to the hyflex hyper-heuristic framework. In: Coello, C.A.C., Cutello, V., Deb, K., Forrest, S., Nicosia, G., Pavone, M. (eds.) PPSN 2012, Part II. LNCS, vol. 7492, pp. 418–427. Springer, Heidelberg (2012)
6. Parkes, A.J.: Combined blackbox and algebraic architecture (CBRA). In: Proceedings of the 8th International Conference on the Practice and Theory of Automated Timetabling (PATAT 2010), pp. 535–538 (2010)
7. Ryser-Welch, P., Miller, J.F.: A review of hyper-heuristic frameworks. In: Proceedings of the Evo20 Workshop, AISB 2014 (2014)

Author Index

Printed in the United States
By Bookmasters